Superhalogens and Superalkalis

Superhalogens and Superalkalis is a comprehensive volume designed as the go-to resource on the exciting and evolving topics of these special classes of atomic clusters and the acid salt that results from their interactions. The book details how these substances possess not only unusual structures but also unique properties that can be exploited for various applications.

Superhalogens' strong oxidizing capacity, resulting from their high-electron affinity, leads to their applications in the design of superacids, organic superconductors, and ionic liquids. The low ionization energy of superalkalis enables them to act as strong reducing agents, making them useful in the design of superbases and alkalides. Illustrated throughout, this timely book provides an overview of the research and development on these and other aspects of superhalogens and superalkalis.

Key features:

- Offers a basic introduction of superatoms that is accessible for readers to understand.
- Includes extensive study questions after each chapter.
- Provides a systematic presentation of the existing literature on this increasingly trending topic.
- Presents the latest developments in the field, offering readers state-of-art knowledge.

This book is a key reference guide for graduate students, postdocs, upper-level undergraduate students, academic professionals, and researchers who are interested in this fascinating topic.

Series Name: Atoms, Molecules, and Clusters: Structure, Reactivity, and Dynamics

Series Editor: Pratim Kumar Chattaraj

About the Series:

While atoms and molecules constitute the fundamental building blocks of matter, atomic and molecular clusters lie somewhere between actual atoms and molecules and extended solids. Helping to elucidate our understanding of this unique area with its abundance of valuable applications, this series includes volumes that investigate the structure, property, reactivity, and dynamics of atoms, molecules, and clusters. The scope of the series encompasses all things related to atoms, molecules, and clusters including both experimental and theoretical aspects. The major emphasis of the series is to analyze these aspects under two broad categories: approaches and applications.

The approaches category includes different levels of quantum mechanical theory with various computational tools augmented by available interpretive methods, as well as state-of-the-art experimental techniques for unraveling the characteristics of these systems including ultrafast dynamics strategies. Various simulation and quantitative structure-activity relationship (QSAR) protocols will also be included in the area of approaches.

The applications category includes topics like membranes, proteins, enzymes, drugs, biological systems, atmospheric and interstellar chemistry, solutions, zeolites, catalysis, aromatic systems, materials, and weakly bonded systems. Various devices exploiting electrical, mechanical, optical, electronic, thermal, piezoelectric, and magnetic properties of those systems also come under this purview.

***Quantum Trajectories*, 1st Edition**
Edited By Pratim Kumar Chattaraj

***Aromaticity and Metal Clusters*, 1st Edition**
Edited By Pratim Kumar Chattaraj

Superhalogens and Superalkalis
Bonding, Reactivity, Dynamics and Applications
Edited By Pratim Kumar Chattaraj and Ambrish Kumar Srivastava

Superhalogens and Superalkalis

Bonding, Reactivity, Dynamics and Applications

Edited by
Pratim Kumar Chattaraj and
Ambrish Kumar Srivastava

CRC Press is an imprint of the
Taylor & Francis Group, an **informa** business

Designed cover image: © Shutterstock

First edition published 2024
by CRC Press
6000 Broken Sound Parkway NW, Suite 300, Boca Raton, FL 33487–2742

and by CRC Press
4 Park Square, Milton Park, Abingdon, Oxon, OX14 4RN

CRC Press is an imprint of Taylor & Francis Group, LLC

ISBN: 978-1-032-46623-1 (hbk)
ISBN: 978-1-032-47013-9 (pbk)
ISBN: 978-1-003-38420-5 (ebk)

DOI: 10.1201/9781003384205

Typeset in Times
by Apex CoVantage, LLC

Contents

Contributors

Alexander I. Boldyrev
Department of Chemistry and Biochemistry, Utah State University
USA

Pratim Kumar Chattaraj
Department of Chemistry, Birla Institute of Technology
Ranchi, India

Jyotirmoy Deb
Department of Chemistry, Birla Institute of Technology
Jharkhand, India

Ankit Kargeti
Department of Applied Sciences, BML Munjal University
India

Abhishek Kumar
Department of Geology
Babasaheb Bhimrao Ambedkar University,
India

Milan Milovanović
Faculty of Physical Chemistry
University of Belgrade

Neeraj Misra
Department of Geology, Babasaheb Bhimrao Ambedkar University
India

Anton S. Pozdeev
Department of Chemistry and Biochemistry, Utah State University
USA

Tabish Rasheed
Department of Applied Sciences, BML Munjal University
India

Ranajit Saha
Institute for Chemical Reaction Design & Discovery (ICReDD)
Japan

Utpal Sarkar
Department of Physics, Assam University
India

Shamoon Ahmad Siddiqui
Department of Physics, Faculty of Science, Integral University
India

Ambrish Kumar Srivastava
Department of Physics, Deen Dayal Upadhyaya Gorakhpur University
India

Harshita Srivastava
Department of Physics, Deen Dayal Upadhyaya Gorakhpur University
India

Ruby Srivastava
Bioinformatics, CCMB-CSIR, Hyderabad
India

Wei-Ming Sun
The Department of Basic Chemistry, School of Pharmacy, Fujian Medical University
P. R. China

Jitendra Kumar Tripathi
Department of Physics, Deen Dayal Upadhyaya Gorakhpur University
India

Suzana Velickovic
University of Belgrade, Belgrade
Serbia

Filip Veljkovic
"VINCA" Institute of Nuclear Sciences – National Institute of the Republic of Serbia, University of Belgrade, Belgrade
Serbia

About the Editors

Dr. Pratim Kumar Chattaraj is a Distinguished Visiting Professor at the Birla Institute of Technology Mesra. Formerly he was an Institute Chair Professor at Indian Institute of Technology (IIT) Kharagpur. He received his PhD from IIT Bombay. He was a research associate at the University of North Carolina, Chapel Hill, USA, and FAU, Erlangen-Nürnberg, Germany. He has been actively engaged in research in the areas of density functional theory, ab-initio calculations, nonlinear dynamics, aromaticity in metal clusters, hydrogen storage, noble gas compounds, machine learning, confinement, fluxionality, chemical reactivity and quantum trajectories. He was honored with: University Gold and Bardhaman Sammilani medals; INSA (Young Scientist) medal; CRSI Bronze and Silver medals; IAAM medal; Associate, Indian Academy of Sciences; BM Birla Science Prize; BC Deb Memorial award; Institute Chair Professorship Award (IIT Kharagpur); Acharya P. C. Ray Memorial Award (Indian Chemical Society); Professor Sadhan Basu Memorial Lecture Award of INSA; Professor R.P. Mitra Memorial Lecture Award, Delhi University; Professor S.K. Siddhanta Memorial Lecture Award, Burdwan University; Professor Prafulla Chandra Ray Memorial Lecture award (NASI). He is a Fellow of The World Academy of Sciences (TWAS), Italy, Royal Society of Chemistry, UK, Indian National Science Academy; Indian Academy of Sciences; National Academy of Sciences, India; West Bengal Academy of Science and Technology; and FWO, Belgium. He is a Sir J.C. Bose National Fellow. He was Convener, Center for Theoretical Studies, Convener, Kharagpur Local Chapter, INSA, Council member of CRSI, Member, Joint Science Education Panel of Science Academies of India, Dean (Faculty) and Head of the Chemistry Department of IIT Kharagpur. He was a Distinguished Visiting Professor of IIT Bombay. He was in the Editorial Board of a number of journals published by the American Chemical Society, Elsevier, Frontiers Group, etc. Several of his papers have become hot/most accessed/most cited/cover/Editors' choice articles.

Dr. Ambrish Kumar Srivastava is an Assistant Professor at the Department of Physics in Deen Dayal Upadhyaya Gorakhpur University, Gorakhpur, India. He was a Junior Research Fellow (with All India Rank 18) and Senior Research Fellow of the Council of Scientific and Industrial Research (CSIR), India. He earned his PhD on the topic entitled "Computational Studies on Biologically Active Molecules and Small Clusters: DFT and TDDFT Approaches" from the University of Lucknow, India, and subsequently, worked as a National Postdoctoral Fellow of the Science & Engineering Research Board

(SERB) at D.D.U. Gorakhpur University. He has published over 120 research papers in various journals of international repute with an h-index of 24 and a citation index of 1750. In addition, he has authored/edited four books and two more books are under preparation for Taylor & Francis/CRC Press. He is an active reviewer for various leading journals and reviewed more than 40 research papers so far. He is an Associate Editor of Frontiers in Physics for the section Physical Chemistry and Chemical Physics and also serves on the Editorial Board of a few journals. He is a member of various scientific societies and organizations including the American Chemical Society, Royal Society of Chemistry, Indian Chemical Society, Materials Research Society of India, etc. He has recently received the prestigious NASI-Young Scientist Platinum Jubilee Award – 2022 from the National Academy of Sciences, India. His broad research interests include Superatomic Clusters, Computational Materials Science and Biophysics.

1 Aromatic Superatoms

Utpal Sarkar, Jyotirmoy Deb and Pratim Kumar Chattaraj

1.1 INTRODUCTION

Aromaticity [1–3] phenomenon is exhibited by cyclic, planar, conjugated molecules such as benzene, having (4n + 2) delocalized electrons and is used to describe the molecules' and coordination compounds' [4,5] unusual stability in spectroscopic and magnetic properties [6,7] and reactivity pattern of the planar organic molecules [8,9]. Although the concept of aromaticity has been well known in the research community for many decades, it has not been able to provide us with a precise definition for determining the aromaticity of a molecule due to the lack of measuring parameters both experimentally and theoretically [10]. The aromaticity is closely related to the high symmetry of organic molecules, as their structural stability strongly correlates to closed-shell structure [11,12]. Earlier, the concept of aromaticity was restricted to planar organic molecules but now this concept is extended to various inorganic compounds including three-dimensional compounds. After that, this concept has been widely used to explore all-metal species [13–18] such as cluster anions of Al, Ga, Hg, In, Si and Sn apart from polyacene analogues of inorganic ring compounds through spectroscopic measurements or theoretical calculations. Significant efforts have been made to set up a criterion to determine the aromaticity in metal clusters [16,19–23] and to indirectly correlate the aromaticity by experimentally measuring the chemical shifts in nuclear magnetic resonance [24]. Many challenges remain in understanding the aromaticity/antiaromaticity of metal systems and their complexes by the existing concepts. The spacing between the energy levels of metal clusters is very low as a result of which the σ-π separation is difficult that establishes a Hückel-type molecular orbital theory like benzene and cyclobutadiene. Aromatic clusters are considered possible building blocks for various new compounds such as Al_4^{4-}, Al_6^{2-}, $Al_4TiAl_4^{2-}$, *etc.* [16,25–27]. Experimental study confirms that some aromatic species such as $C_5H_5^-$, $C_7H_7^{3-}$ and Si_5^{6-} [28,29] have been found as key structural units for crystalline solids. Apart from that the cubic geometry of different systems has been explored to investigate the stability by correlating the electron delocalization using the concept of cubic aromaticity. Some examples where cubic aromaticity has been implemented are Zn_8 [30], Mn_8 [31], $Co_{13}O_8$ [32] and CBe_8H_{12} [33] clusters.

It has been a well-known fact that clusters with specific sizes and compositions showing similar behaviour as that of atoms are termed as superatoms [34–41]. These unusual species can retain their identity even when they are connected with an extended nanostructure and offering a new direction for serving as a

DOI: 10.1201/9781003384205-1

building block for new nanomaterials with highly tunable properties promising a wide variety of potential applications [42–44]. Therefore, searching for novel superatomic building blocks become an appealing interest among worldwide researchers. Superalkalis are the famous subset of superatoms having lower ionization potentials (IPs) compared to alkali metal atoms [45]. Superalkalis exhibit excellent reducibility and for that reason, they are useful in synthesizing unusual charge-transfer salts, constructing cluster-assembled salts, where the steric hindrance restricts the use of alkali metal atoms, designing effective hydrogen storage materials because of their strong hydrogen trapping ability [46] and nonlinear optical materials due to their remarkable hyperpolarizabilities [47,48]. The presence of intriguing features in superalkali clusters has gained significant attention and enormous efforts have been made to predict and experimentally fabricate new superalkalis. Moreover, superalkali cations remain intact in their structural and electronic integrity when they are encapsulated within other systems, such as $(BLi_6)^+(X)^-$ ($X = F$, LiF_2, BeF_3 and BF_4) [47], $(Li_3)^+(SH)^-$ ($SH = LiF_2$, BeF_3 and BF_4) [49], *etc.*

Conceptual density functional theory (CDFT) has been widely utilised [50–56] in analysing the concept of aromaticity and antiaromaticity in many superatoms. Some well-known concepts such as electronegativity [57–61], electrophilicity [62–66], hardness [58,67–71], chemical potential [58], *etc.* have been utilised in investigating the stability, reactivity, chemical bonding and interactions in superatoms. CDFT provides a basis for all these concepts and associated principles such as the hard-soft acids and bases (HSAB) principle [72–75], electronegativity equalization principle (EEP) [76–78], the maximum hardness principle (MHP) [79–81], the minimum polarizability principle (MPP) [82–84], the minimum electrophilicity principle (MEP) [85–88], the minimum magnetizability principle (MMP) [89] and many others. In this chapter, we have provided a comprehensive discussion on the aromatic and antiaromatic behaviour of various superatoms in light of CDFT.

1.2 THEORETICAL BACKGROUND

In the context of CDFT, the global reactivity descriptors such as electronegativity (χ), hardness (η), electrophilicity (ω) and chemical potential (μ) can be defined as:

$$\chi = -\left(\frac{\partial E}{\partial N}\right)_{v(\vec{r})} = -\mu \tag{1.1}$$

$$\eta = \left(\frac{\partial^2 E}{\partial N^2}\right)_{v(\vec{r})} = \left(\frac{\partial \mu}{\partial N}\right)_{v(\vec{r})} \tag{1.2}$$

$$\omega = \frac{\mu^2}{2\eta} \tag{1.3}$$

where E is the total energy, N is the total number of electrons and $v(\vec{r})$ is the external potential of the system.

Site selectivity in a molecule can be analysed using local reactivity descriptors such as the Fukui function (f^{α}) and the philicity (ω^{α}). They are of three different types, namely nucleophilic (f^{+}/ω^{+}), electrophilic (f^{-}/ω^{-}) and radical (f^{0}/ω^{0}) attacks. Using finite-difference and frozen core approximations, it can be defined as:

$$f^{+}(\vec{r})=\left(\frac{\partial\rho}{\partial N}\right)^{+}_{v(\vec{r})}\cong\rho_{N+1}(\vec{r})-\rho_{N}(\vec{r})\approx\rho_{LUMO}(\vec{r})\text{ for nucleophilic attack}\quad(1.4a)$$

$$f^{-}(\vec{r})=\left(\frac{\partial\rho}{\partial N}\right)^{-}_{v(\vec{r})}\cong\rho_{N}(\vec{r})-\rho_{N-1}(\vec{r})\approx\rho_{HOMO}(\vec{r})\text{ for electrophilic attack}\quad(1.4b)$$

$$f^{0}(\vec{r})=\left(\frac{\partial\rho}{\partial N}\right)^{0}_{v(\vec{r})}\cong\frac{1}{2}\left(\rho_{N+1}(\vec{r})-\rho_{N-1}(\vec{r})\right)\approx\frac{1}{2}\left(\rho_{HOMO}(\vec{r})+\rho_{LUMO}(\vec{r})\right)\quad(1.4c)$$

for radical attack

and $\omega^{\alpha}(\vec{r})=\omega f^{\alpha}(\vec{r})$; $\alpha=+1,-1$ *and* 0 stands for nucleophilic, electrophilic and radical attacks, respectively.

Here, ρ_{N+1} represents electron densities of N+1, ρ_{N} represents electron densities of N, ρ_{N-1} represents electron densities of N-1 electronic systems and ρ_{LUMO}, ρ_{HOMO} are the electron densities of the lowest unoccupied molecular orbital (LUMO) and highest occupied molecular orbital (HOMO), respectively.

Also, if f^{+}, f^{0} and f- have higher magnitudes at any reaction site, the change in chemical potential is also significantly higher.

Using the finite-difference approximation method, the working equations of global reactivity descriptors can be expressed as:

$$\chi=-\mu=\frac{I+A}{2}\quad(1.5)$$

$$\eta=I-A\quad(1.6)$$

where I and A represent the ionization potential (I) and electron affinity (A) of the system, respectively, and can be calculated using the following relations

$$I=E(N-1)-E(N)\quad(1.7)$$

$$A=E(N)-E(N+1)\quad(1.8)$$

where $E(N-1)$, $E(N)$ and $E(N+1)$ stand for the electronic energy of $N-1$, N and $N+1$ electron systems, respectively.

With the help of Koopmans' theorem I and A can be further expressed as

$$I = -E_{HOMO} \tag{1.9}$$

$$A = -E_{LUMO} \tag{1.10}$$

where E_{HOMO} and E_{LUMO} are the frontier molecular orbital energies.

1.3 AROMATIC SUPERATOMS

We start by analysing different parameters of two sets of doped Al clusters; one is formed by replacing an Al atom with a C atom (Al_nC^-) (Figure 1.1) and the other by substituting an Al atom with an O atom (Al_nO^-) (Figure 1.2), where $n = 5-8$. The stability and reactivity of the clusters are checked by studying the electronic structure principles including the chemical reactivity parameters. The possible aromatic behaviour of these clusters is also investigated via nucleus-independent chemical shift (NICS), which has been calculated in the ring centre. All the clusters have minimum energy structure as confirmed by the absence of any negative frequencies. Also, to be noted that geometrically, for $n = 7$, *i.e.*, for X(Al_7C^-), C is endohedrally placed, whereas, for Y(Al_7O^-), O is exohedrally placed on the respective clusters. Bond lengths of both clusters are in close agreement with the previously reported results [37]. The calculated χ values of X and Y have been compared with that of their ionic counterparts and the result indicates that $X^+(Y^+)$ has a much higher tendency to accommodate additional electrons in comparison to their neutral analogues. Moreover, due to having negative χ values of X- and Y- it is very difficult to take further electrons into the systems. Their stability is further confirmed via the maximum hardness principle and minimum electrophilicity principle.

The variation of chemical hardness and electrophilicity with the number of Al atoms present in the Al_nM^- ($M = C, O$ *and* $n = 5-8$) clusters is presented in Figure 1.3. The above analysis confirms that both Al_7C- and Al_7O- systems are found to be stable in comparison to the other clusters of their own sets as well as when compared to each other. Both these clusters are also strongly aromatic in nature due to having large negative NICS (0) values. To know about the potent sites of electrophilic or nucleophilic attack on the clusters, their philicities and NPA charges at different atomic centres of the two most stable clusters, *viz.*, Al_7C- and Al_7O^-, are calculated. It has been noticed that in Al_7C- system, all the Al atoms are preferable sites for being attacked by some anion or hard nucleophile, while the C atom can be attacked by a cation or any hard electrophile. On the other hand, in Al_7O- cluster, some of the Al atoms along with the O atom are potent sites for getting attacked by a cation or a hard electrophile, although some of the remaining Al atoms are good sites for anionic or nucleophilic attack. This result is further supported by molecular frontier orbitals plots (Figure 1.1 and Figure 1.2). If a comparison is drawn between Al_7O- and Al_7C^-, it is observed that the C-containing Al cluster with a comparatively larger electrophilicity value is naturally considered to be more electrophilic than its O-containing counterpart.

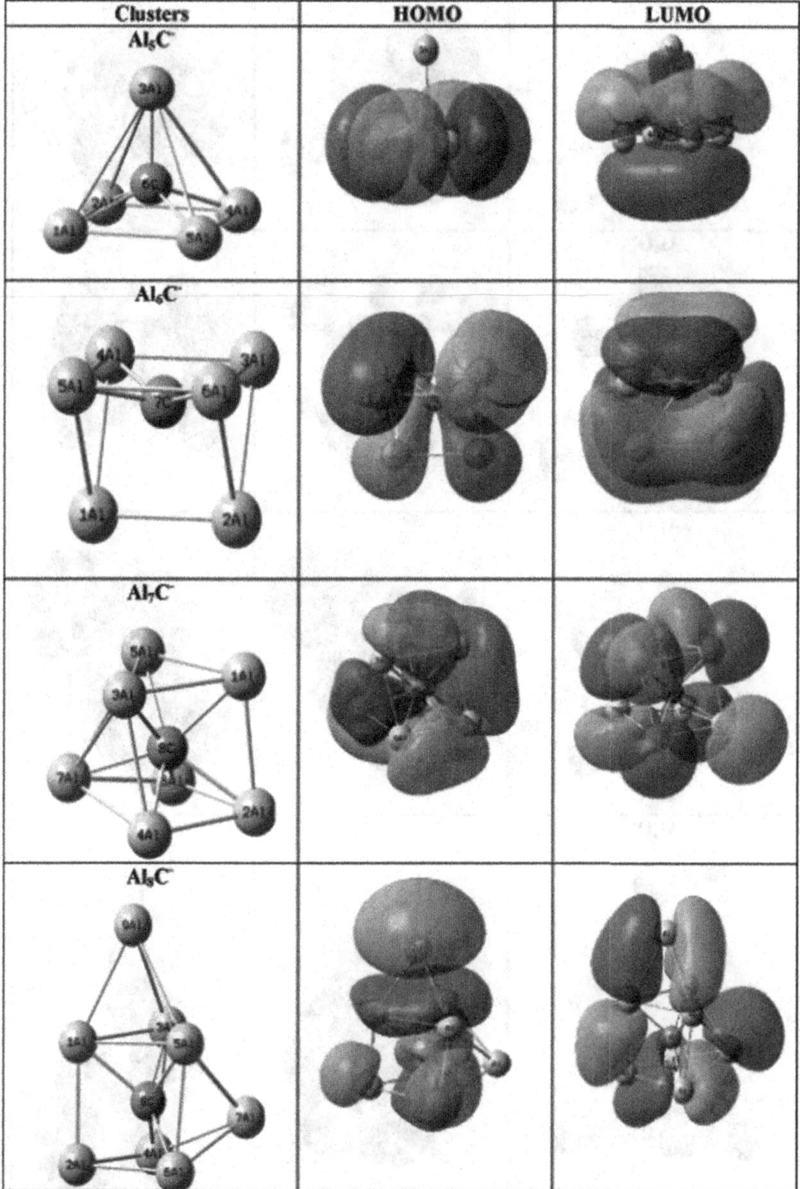

FIGURE 1.1 Optimised geometries and frontier molecular orbitals of Al_nC- clusters calculated using B3LYP/6–311+G** level of theory [Reproduced from Ref. 90 with permission of the American Chemical Society].

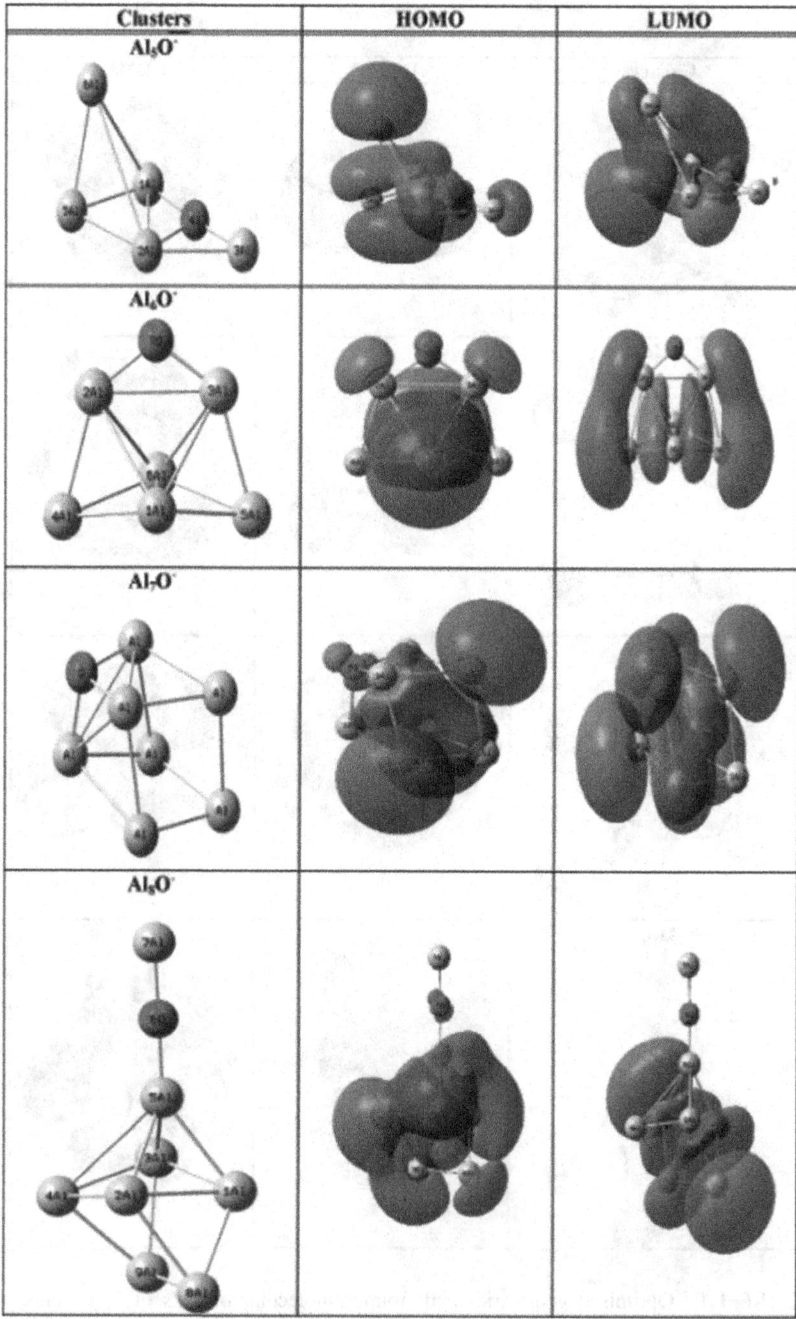

FIGURE 1.2 Optimised geometries and frontier molecular orbitals of Al_nO- clusters calculated using B3LYP/6–311+G** level of theory [Reproduced from Ref. 90 with permission of the American Chemical Society].

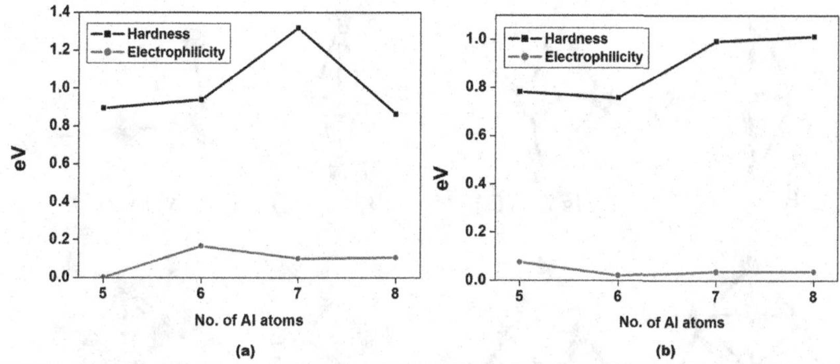

FIGURE 1.3 Variation of chemical hardness and electrophilicity with number of Al atoms in (a) Al_nC- clusters and (b) Al_nO- clusters [Reproduced from Ref. 90 with permission of the American Chemical Society].

Also, from the discussion as we can see that C and O atomic centres are not suitable for attack by soft electrophiles, although they are dominant for being attacked by hard electrophiles. However, overall, the atomic centres of Al_7C- are found to be more reactive than that of the Al_7O- cluster [90].

Next, another class of aromatic superatom cations, MLi_{n+1}^{+} have been studied. Here, M is the central aromatic core and n represents the anionic charge of M. This study deals with a series of both inorganic (Be_3^{2-}, B_3^{-}, Mg_3^{2-}, Al_3^{-}, Al_4^{2-}, Al_6^{2-}, Si_4^{4-}, Si_5^{6-} *etc.*) and organic ($C_4H_4^{2-}$, $C_5H_5^{-}$, $C_7H_7^{3-}$, $C_8H_8^{2-}$) anions for M. On systematic analysis of the polynuclear MLi_{n+1}^{+} cations, it has been observed that by combining suitable aromatic anions with that alkali metal cations, aromatic superatoms can be obtained. The optimised structures of MLi_{n+1}^{+} cations have been depicted in Figure 1.4. Figure 1.4 shows that $Be_3Li_3^+$ exhibits a C_{2v} symmetry and leads to the trigonal bipyramid structure of Be_3Li_2 with a third Li ligand connected to one side of the Be_3 ring while $Mg_3Li_3^+$ structure shows lower C_s symmetry as here the third Li ligand is situated on the face of the Mg_3Li_2 segment. The discrepancy in their structures can be well understood by the different electronic structure principles. In $Be_3Li_3^+$ cluster, each Li atom has an NBO charge of +0.8 e and reduces the repulsive interaction between the three Li atoms by occupying a far position from each other. Interestingly, the charge on each Li atom in the $Mg_3Li_3^+$ cluster is significantly smaller compared to $Be_3Li_3^+$ cluster. Moreover, the incorporation of Li_3^+ cation results in the shortening of Be-Be and Mg-Mg bonds in comparison to their isolated counterparts. Further study shows that all the considered aromatic anions, except N_4^{2-} and $C_4H_4^{2-}$ keep their identities intact inside the MLi_{n+1}^{+} cations. All these cations except for $N_4Li_3^{+}$, $C_4H_4Li_3^{+}$ are showing aromatic behaviour. The proposed MLi_{n+1}^{+} cations in this work, have lower vertical electron affinities and they represent a different class of superalkali (also known as pseudoalkali) cation. These systems also possess large thermodynamic stability when

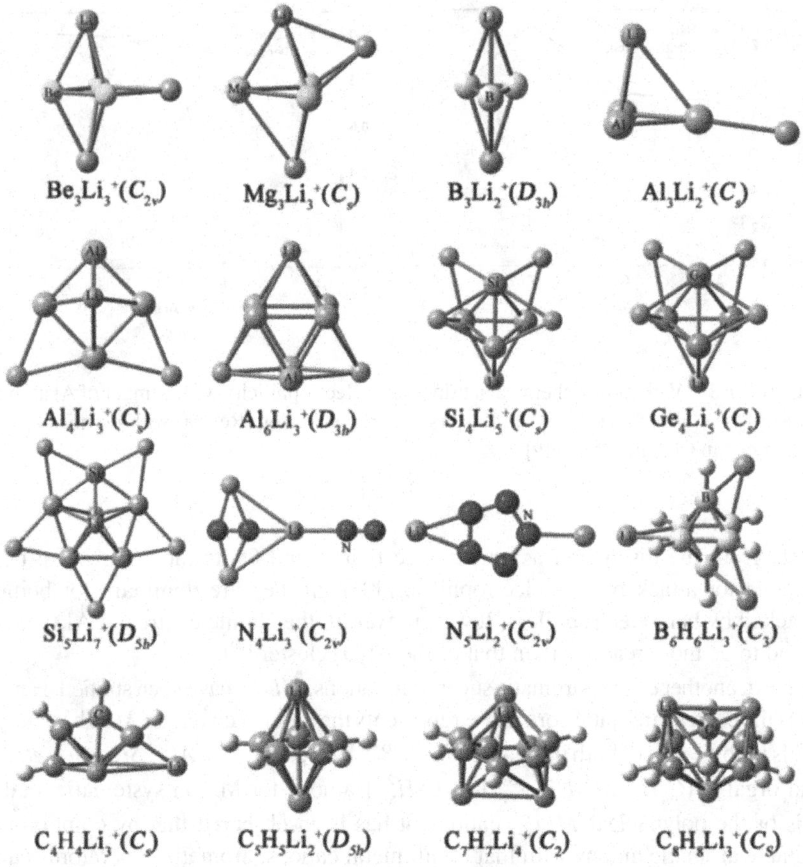

FIGURE 1.4 Optimised geometries of MLi_{n+1}^+ cations calculated using B3LYP/6–311++G(3df,3pd) level of theory [Reproduced from Ref. 91 with permission of the American Chemical Society].

get dissociated. They show larger HOMO-LUMO gaps in the range of 1.94–7.95 eV suggesting their enhanced chemical stability. Further, these cations also have a tendency to combine with that of (super)halogen anions; thereby, forming stable superatom compound systems [91].

The electronic structure of gold-doped superatoms, $[M@Au_8L_8]^q$, where M = Pd, Pt, Ag, *etc.*; L = PPh_3; and q = 0, +1, +2 and the magnetic response of their valence electrons have been studied. The magnetic responses (current susceptibility tensor and magnetically induced current density) are calculated by using the gauge including magnetically induced currents (GIMIC), under the framework of density functional theory. It has been found that both diatropic and paratropic magnetically induced currents are sensitive to the atomic structure of the studied clusters. These currents are even sensitive to the number of superatomic electrons as well as the chemical nature of the dopant metal used. To note here that the strongest aromaticity is found for Pd and Pt doped $[M@Au_8L_8]^0$ systems, among cubic-like structures. Interestingly, this study shows that doping by Pd and Pt atoms in cubic structure, an increment in the aromaticity is observed as compared to that of pure Au case that is valid not only near the central atom but also covering the whole metallic core; the aromatic trend follows, Pd > Pt > Au [92]. The magnetically induced currents in two more gold-based hydrogen-containing nanoclusters with stabilise ligand, *viz.*, $\left[H\,Au_9\left(PPh_3\right)_8\right]^{2+}$ and $\left[PtH\,Au_8\left(PPh_3\right)_8\right]^{+}$ have been analysed by Lopez-Estrada et al. Interestingly, both clusters possess "superatomic" electrons, which are acquired by the presence of Au atoms. It is observed that doping by the Pt atom in gold makes an increment in the diatropic behaviour of the superatomic electrons when subjected to an external magnetic field, as seen from the analysis of magnetically induced currents (MICs) in the metal cores. Thus, it produces an enhancement in the aromaticity of $\left[PtHAu_8\left(PPh_3\right)_8\right]^{+}$ cluster. A reason may be attributed to a stronger shielding of a proton present in the metal core of the cluster in comparison to $\left[HAu_9\left(PPh_3\right)_8\right]^{2+}$ cluster, which causes a significant up-field proton chemical shift and is fortunately in agreement with the experimental proton NMR data as found for these two systems. However, the overall effect produced from the shielding currents is found to be paratropic in one plane of the hydrogen present in $\left[HAu_9\left(PPh_3\right)_8\right]^{2+}$ cluster, which induces de-shielding effect along with the down-field chemical shift of the proton [93].

It is established from many studies on different systems that aromaticity plays a leading role in explaining the stabilities of the systems including all-metal compounds, various clusters and solid states. For an intermetalloid cluster $[Pd_3Sn_8Bi_6]^{4-}$, patterns based on spherical aromaticity have been observed, which is consistent with that of Hirsch $2(n+1)^2$ rule. This is also found to be responsible for the internal triangle network of Pd_3, connecting the external Sn-Bi cage. Notably, the clusters which show spherical aromaticity are also known as "superatoms" and hence are used in nanotechnology and material science due to their persistency and scalability as building blocks in these fields [94].

1.4 CONCLUDING REMARKS

In a nutshell, we have presented here a brief discussion on aromaticity in superatoms using the CDFT techniques. Initially, the structural stability, reactivity and aromaticity of Al_7C- and Al_7O- clusters have been investigated. The stability of these clusters is corroborated by the maximum hardness principle and the minimum electrophilicity principle. NICS analysis further confirms their strong aromatic behaviour. Next, the thermodynamic stability, energy gaps, electron affinities and aromaticity have been disclosed for aromatic superatom cations (MLi^+_{n+1}). Moreover, the GIMIC investigation has been performed on ligand-stabilised gold-based superatoms to infer about their aromatic characteristics. Lastly, the concept of spherical aromaticity in experimentally synthesized $[Pd_3Sn_8Bi_6]^{4-}$ is explored.

ACKNOWLEDGMENTS

P.K.C. would like to thank Professor Srivastava for inviting him to write this book chapter. He also thanks DST, New Delhi for the J.C. Bose Fellowship.

Conflict of interests

The authors declare no conflict of interest.

REFERENCES

1. Burkhardt, A. M.; Loomis, R. A.; Shingledecker, C. N.; Lee, K. L. K.; Remijan, A. J.; McCarthy, M. C.; McGuire, B. A. Ubiquitous Aromatic Carbon Chemistry at the Earliest Stages of Star Formation. *Nat. Astron.*, 2021, 5, 181–187.
2. Schleyer, P. V. R. Introduction: Aromaticity. *Chem. Rev.*, 2001, 101, 1115–1118.
3. Garratt, P. J. *Aromaticity*; Wiley: New York, 1986.
4. Chen, D.; Hua, Y.; Xia, H. Metallaaromatic Chemistry: History and Development. *Chem. Rev.*, 2020, 120, 12994–13086.
5. Zhu, J. Open Questions on Aromaticity in Organometallics. *Commun. Chem.*, 2020, 3, 161.
6. Haags, A.; Reichmann, A.; Fan, Q.; Egger, L.; Kirschner, H.; Naumann, T.; Werner, S.; Vollgraff, T.; Sundermeyer, J.; Eschmann, L.; et al. Kekulene: On-Surface Synthesis, Orbital Structure, and Aromatic Stabilization. *ACS Nano*, 2020, 14, 15766–15775.
7. Gomes, J. A. N. F.; Mallion, R. B. Aromaticity and Ring Currents. *Chem. Rev.*, 2001, 101, 1349–1384.
8. Roglans, A.; Pla-Quintana, A.; Solà, M. Mechanistic Studies of Transition-Metal-Catalyzed [2 + 2 + 2] Cycloaddition Reactions. *Chem. Rev.*, 2020, 121, 1894–1979.
9. Zhu, J. Rational Design of a Carbon-Boron Frustrated Lewis Pair for Metal-Free Dinitrogen Activation. *Chem. Asian J.*, 2019, 14, 1413–1417.
10. Solà, M. Why Aromaticity Is a Suspicious Concept? Why? *Front. Chem.*, 2017, 5, 22.
11. Hirsch, A., Chen, Z. H., Jiao, H. Spherical Aromaticity in IH Symmetrical Fullerenes: The 2(N + 1)2 Rule. *Angew. Chem.*, 2000, 39, 3915–3917.
12. Solà, M. Connecting and Combining Rules of Aromaticity: Towards a Unified Theory of Aromaticity. *Wiley Interdiscip. Rev. Comput. Mol. Sci.*, 2019, 9, e1404.
13. Li, X.; Kuznetsov, A. E.; Zhang, H.-F.; Boldyrev, A. I.; Wang, L.-S. Observation of All-Metal Aromatic Molecules. *Science*, 2001, 291, 859.
14. Kuznetsov, A. E.; Zhai, H. J.; Wang, L.-S.; Boldyrev, A. I. Peculiar Antiaromatic Inorganic Molecules of Tetrapnictogen in $Na^+Pn_4^-$ (Pn = P, As, Sb) and Important Consequences for Hydrocarbons. *Inorg. Chem. Commun.*, 2002, 41, 6062.
15. Zhan, C.-G.; Zheng, F.; Dixon, D. A. Electron Affinities of Al_n-Clusters and Multiple-Fold Aromaticity of the Square Al_4^{2-} Structure. *J. Am. Chem. Soc.*, 2002, 124, 14795.
16. Boldyrev, A. I.; Wang, L. S. All-Metal Aromaticity and Antiaromaticity. *Chem. Rev.*, 2005, 105, 3716.
17. Chattaraj, P. K.; Roy, D. R.; Elango, M.; Subramanian, V. Stability and Reactivity of All-Metal Aromatic and Antiaromatic Systems in Light of the Principles of Maximum Hardness and Minimum Polarizability. *J. Phys. Chem. A*, 2005, 109, 9590.
18. Chattaraj, P. K.; Roy, D. R. Aromaticity in Polyacene Analogues of Inorganic Ring Compounds. *J. Phys. Chem. A*, 2007, 111, 4684.

19. Pham, H. T.; Nguyen, M. T. Aromaticity of Some Metal Clusters: A Different View from Magnetic Ring Current. *J. Phys. Chem. A*, 2018, 122, 1378–1391.
20. Boldyrev, A. I.; Wang, L.-S. Beyond Organic Chemistry: Aromaticity in Atomic Clusters. *Phys. Chem. Chem. Phys.*, 2016, 18, 11589–11605.
21. Badri, Z.; Pathak, S.; Fliegl, H.; Rashidi-Ranjbar, P.; Bast, R.; Marek, R.; Foroutan-Nejad, C.; Ruud, K. All-Metal Aromaticity: Revisiting the Ring Current Model Among Transition Metal Clusters. *J. Chem. Theory Comput.*, 2013, 9, 4789–4796.
22. Chattaraj, P. K. *Aromaticity and Metal Clusters*; CRC Press: Boca Raton, FL, 2010.
23. Paul, D.; Sarkar, U. Conceptual Density Functional Theory and All Metal Aromaticity. In *Atomic Clusters with Unusual Structure, Bonding and Reactivity*; Elsevier: Amsterdam, The Netherlands, 2023, 87–98.
24. Mitchell, R. H. Measuring Aromaticity by NMR. *Chem. Rev.*, 2001, 101, 1301–1316.
25. Kuznetsov, A. E.; Birch, K.; Boldyrev, A. I.; Li, X.; Zhai, H.; Wang, L.-S. All-Metal Antiaromatic Molecule: Rectangular Al_4^{4-} in the Li_3Al^{4-} Anion. *Science*, 2003, 300, 622.
26. Kuznetsov, A. E.; Boldyrev, A. I.; Zhai, H.-J.; Li, X.; Wang, L.-S. Al_6^{2-} – Fusion of Two Aromatic Al3- Units: A Combined Photoelectron Spectroscopy and ab Initio Study of $M^+[Al_6^{2-}]$ (M = Li, Na, K, Cu, and Au). *J. Am. Chem. Soc.*, 2002, 124, 11791.
27. Mercero, J. M.; Ugalde, J. M. Sandwich-Like Complexes Based on "All-Metal" (Al_4^{2-}) Aromatic Compounds. *J. Am. Chem. Soc.*, 2004, 126, 3380.
28. Tamm, M. Synthesis and Reactivity of Functionalized Cycloheptatrienyl – Cyclopentadienyl Sandwich Complexes. *Chem. Commun.*, 2008, 3089.
29. Kuhn, A.; Sreeraj, P.; Pöttgen, R.; Wiemhöfer, H.-D.; Wilkening, M.; Heitjans, P. Angew. Li NMR Spectroscopy on Crystalline $Li_{12}Si^7$: Experimental Evidence for the Aromaticity of the Planar Cyclopentadienyl-Analogous Si_5^{6-} Rings. *Chem., Int. Ed.*, 2011, 50,12099.
30. Cui, P.; Hu, H.-S.; Zhao, B.; Miller, J. T.; Cheng, P.; Li, J. A Multicentre-Bonded $[Zn^I]_8$ Cluster with Cubic Aromaticity. *Nat. Commun.*, 2015, 6, 6331.
31. Hu, H.; Hu, H.; Zhao, B.; Cui, P.; Cheng, P.; Li, J. Metal – Organic Frameworks (MOFs) of a Cubic Metal Cluster with Multicentered Mn^I-Mn^I Bonds. *Angew. Chem.*, 2015, 54, 11681–11685.
32. Geng, L.; Weng, M.; Xu, C.-Q.; Zhang, H.; Cui, C.; Wu, H.; Chen, X.; Hu, M.; Lin, H.; Sun, Z.-D.; et al. $Co_{13}O_8$–Metalloxocubes: A New Class of Perovskite-Like Neutral Clusters with Cubic Aromaticity. *Natl. Sci. Rev.*, 2020, 8, nwaa201.
33. Guo, J.-C.; Feng, L.-Y.; Dong, C.; Zhai, H.-J. A Designer 32-Electron Superatomic CBe_8H_{12} Cluster: Core–Shell Geometry, Octacoordinate Carbon, and Cubic Aromaticity. *New J. Chem.*, 2020, 44, 7286–7292.
34. Khanna, S. N.; Jena, P. Assembling Crystals from Clusters. *Phys. Rev. Lett.*, 1992, 69, 1664.
35. Khanna, S. N.; Jena, P. Atomic Clusters: Building Blocks for a Class of Solids. *Phys. Rev. B*, 1995, 51, 13705.
36. Bergeron, D. E.; Castleman, A. W.; Morisato, T.; Khanna, S. N. Formation of $Al_{13}I^-$: Evidence for the Superhalogen Character of Al_{13}. *Science*, 2004, 304, 84.
37. Bergeron, D. E.; Roach, P. J.; Castleman, A. W.; Jones, N. O.; Khanna, S. N. Al Cluster Superatoms as Halogens in Polyhalides and as Alkaline Earths in Iodide Salts. *Science*, 2005, 307, 231.
38. Reveles, J. U.; Khanna, S. N.; Roach, P. J.; Castleman, A. W. Multiple Valence Superatoms. *Proc. Natl. Acad. Sci. U. S. A.*, 2006, 103, 18405.
39. Reveles, J. U.; Clayborne, P. A.; Reber, A. C.; Khanna, S. N.; Pradhan, K.; Sen, P.; Pederson, M. R. Designer Magnetic Superatoms. *Nat. Chem.*, 2009, 1, 310.

40. Zhang, X.; Wang, Y.; Wang, H.; Lim, A.; Gantefoer, G.; Bowen, J. K. H.; Reveles, J. U.; Khanna, S. N. On the Existence of Designer Magnetic Superatoms. *J. Am. Chem. Soc.*, 2013, 135, 4856.
41. Jena, P. Beyond the Periodic Table of Elements: The Role of Superatoms. *J. Phys. Chem. Lett.*, 2013, 4, 1432.
42. Claridge, S. A.; Castleman, A. W.; Khanna, S. N.; Murray, C. B.; Sen, A.; Weiss, P. S. Cluster-Assembled Materials. *ACS Nano*, 2009, 3, 244.
43. Qian, M.; Reber, A. C.; Ugrinov, A.; Chaki, N. K.; Mandal, S.; Saavedra, H. M.; Khanna, S. N.; Sen, A.; Weiss, P. S. Cluster-Assembled Materials: Toward Nanomaterials with Precise Control Over Properties. *ACS Nano*, 2010, 4, 235.
44. Mandal, S.; Reber, A. C.; Qian, M.; Weiss, P. S.; Khanna, S. N.; Sen, A. Controlling the Band Gap Energy of Cluster-Assembled Materials. *Acc. Chem. Res.*, 2013, 46, 2385–2395.
45. Lias, S. G.; Bartmess, J. E.; Liebman, J. F.; Homes, J. L.; Levin, R. D.; Mallard, W. G. Gas-Phase Ion and Neutral Thermochemistry. *J. Phys. Chem. Ref. Data Suppl.*, 1988, 17, 1285.
46. Pan, S.; Merino, G.; Chattaraj, P. K. The Hydrogen Trapping Potential of Some Li-Doped Star-Like Clusters and Super-Alkali Systems. *Phys. Chem. Chem. Phys.*, 2012, 14, 10345.
47. Li, Y.; Wu, D.; Li, Z. R. Compounds of Superatom Clusters: Preferred Structures and Significant Nonlinear Optical Properties of the BLi_6-X (X = F, LiF_2, BeF_3, BF_4) Motifs. *Inorg. Chem.*, 2008, 47, 9773.
48. Yang, H.; Li, Y.; Wu, D.; Li, Z.-R. Structural Properties and Nonlinear Optical Responses of Superatom Compounds BF_4-M (M = Li, FLi_2, OLi_3, NLi_4). *Int. J. Quantum Chem.*, 2012, 112, 770.
49. Wang, F. F.; Li, Z. R.; Wu, D.; Sun, X. Y.; Chen, W.; Li, Y.; Sun, C. C. Novel Superalkali Superhalogen Compounds $(Li_3)^+(SH)^-$ (SH=LiF_2, BeF_3, and BF_4) with Aromaticity: New Electrides and Alkalides. *ChemPhysChem*, 2006, 7, 1136.
50. Parr, R. G.; Weitao, Y. *Density-Functional Theory of Atoms and Molecules*; Oxford University Press: New York, 1995, 5–15.
51. Geerlings, P.; De Proft, F.; Langenaeker, W. Conceptual Density Functional Theory. *Chem. Rev.*, 2003, 103, 1793–1874.
52. Geerlings, P.; Chamorro, E.; Chattaraj, P. K.; De Proft, F.; Gázquez, J. L.; Liu, S.; Morell, C.; Toro-Labbé, A.; Vela, A.; Ayers, P. Conceptual Density Functional Theory: Status, Prospects, Issues. *Theor. Chem. Acc.*, 2020, 139, 36.
53. Sarkar, U.; Chattaraj, P. K. Reactivity Dynamics. *J. Phys. Chem. A*, 2021, 125, 2051–2060.
54. Chakraborty, D.; Chattaraj, P. K. Conceptual Density Functional Theory Based Electronic Structure Principles. *Chem. Sci.*, 2021, 12, 6264–6279.
55. Sarkar, U.; Chattaraj, P. K. Conceptual DFT Based Electronic Structure Principles in a Dynamical Context. *J. Ind. Chem. Soc.*, 2021, 98, 100098.
56. Sarkar, U.; Giri, S.; Chattaraj, P. K. Dirichlet Boundary Conditions and Effect of Confinement on Chemical Reactivity. *J. Phys. Chem. A.*, 2009, 113, 10759–10766.
57. Parr, R. G.; Donnelly, R. A.; Levy, M.; Palke, W. E. Electronegativity: The Density Functional Viewpoint. *J. Chem. Phys.*, 1978, 68, 3801–3807.
58. Parr, R. G.; Pearson, R. G. Absolute Hardness: Companion Parameter to Absolute Electronegativity. *J. Am. Chem. Soc.*, 1983, 105, 7512–7516.
59. Deb, J.; Paul, D.; Sarkar, U.; Ayers, P. W. Characterizing the Sensitivity of Bonds to the Curvature of Carbon Nanotubes. *J. Mol. Model.*, 2018, 24, 249.
60. Sarkar, U.; Padmanabhan, J.; Parthasarathi, R.; Subramanian, V.; Chattaraj, P. K. Toxicity Analysis of Polychlorinated Dibenzofurans Through Global and Local Electrophilicities. *J. Mol. Struct. Theochem.*, 2006, 758, 119–125.

61. Singh, N. B.; Sarkar, U. A Density Functional Study of Chemical, Magnetic and Thermodynamic Properties of Small Palladium Clusters. *Mol. Simul.*, 2014, 40, 1255–1264.
62. Parr, R. G.; Szentpály, L. V.; Liu, S. Electrophilicity Index. *J. Am. Chem. Soc.*, 1999, 121, 1922–1924.
63. Chattaraj, P. K.; Sarkar, U.; Roy, D. R.; Elango, M.; Parthasarathi, R.; Subramanian, V. Is Electrophilicity a Kinetic or a Thermodynamic Concept? *Indian J. Chem. Sect. A*, 2006, 45, 1099–1112.
64. Chattaraj, P. K.; Sarkar, U.; Roy, D. R. Electrophilicity Index. *Chem. Rev.*, 2006, 106, 2065–2091.
65. Chattaraj, P. K.; Sarkar, U.; Parthasarathi, R.; Subramanian, V. DFT Study of Some Aliphatic Amines Using Generalized Philicity Concept. *Int. J. Quantum Chem.*, 2005, 101, 690–702.
66. Chattaraj, P. K.; Maiti, B.; Sarkar, U. Philicity: A Unified Treatment of Chemical Reactivity and Selectivity. *J. Phys. Chem. A*, 2003, 107, 4973–4975.
67. Pearson, R. G. *Chemical Hardness*; Wiley: New York, 1997.
68. Miranda-Quintana, R. A.; Franco-Pérez, M.; Gázquez, J. L.; Ayers, P. W.; Vela, A. Chemical Hardness: Temperature Dependent Definitions and Reactivity Principles. *J. Chem. Phys.*, 2018, 149, 124110.
69. Paul, D.; Sarkar, U. Designing of PC 31 BM-Based Acceptors for Dye-Sensitized Solar Cell. *J. Phys. Org. Chem.*, 2022. https://doi.org/10.1002/poc.4419.
70. Ayers, P. W. The Physical Basis of the Hard/Soft Acid/Base Principle. *Faraday Discuss*, 2007, 135, 161–190.
71. Sarkar, U.; Roy, D. R.; Chattaraj, P. K.; Parthasarathi, R.; Padmanabhan, J.; Subramanian, V. A Conceptual DFT Approach Towards Analysing Toxicity. *J. Chem. Sci.*, 2005, 117, 599–612.
72. Pearson, R. G. Recent Advances in the Concept of Hard and Soft Acids and Bases. *J. Chem. Educ.*, 1987, 64, 561.
73. Chattaraj, P. K.; Lee, H.; Parr, R. G. HSAB Principle. *J. Am. Chem. Soc.*, 1991, 113, 1855–1856.
74. Chattaraj, P. K.; Maiti, B. HSAB Principle Applied to the Time Evolution of Chemical Reactions. *J. Am. Chem. Soc.*, 2003, 125, 2705–2710.
75. Chattaraj, P. K.; Sarkar, U. Ground- and Excited-States Reactivity Dynamics of Hydrogen and Helium Atoms. *Int. J. Quantum Chem.*, 2003, 91, 633–650.
76. Sanderson, R. T. Partial Charges on Atoms in Organic Compounds. *Science*, 1955, 121, 207–208.
77. Sanderson, R. T. *Chemical Bonds and Bond Energy*; Academic Press: New York, 1976.
78. Sanderson, R. T. *Polar Covalence*; Academic Press: New York, 1983.
79. Parr, R. G.; Chattaraj, P. K. Principle of Maximum Hardness. *J. Am. Chem. Soc.*, 1991, 113, 1854–1855.
80. Franco-Pérez, M.; Gázquez, J. L.; Ayers, P. W.; Vela, A. Thermodynamic Hardness and the Maximum Hardness Principle. *J. Chem. Phys.*, 2017, 147, 074113.
81. Khatua, M.; Sarkar, U.; Chattaraj, P. K. Reactivity Dynamics of a Confined Molecule in Presence of an External Magnetic Field. *Int. J. Quantum Chem.*, 2014, 115, 144–157.
82. Chattaraj, P. K.; Sengupta, S. Popular Electronic Structure Principles in a Dynamical Context. *J. Phys. Chem.*, 1996, 100, 16126–16130.
83. Ghanty, T. K.; Ghosh, S. K. A Density Functional Approach to Hardness, Polarizability, and Valency of Molecules in Chemical Reactions. *J. Phys. Chem.*, 1996, 100, 12295–12298.

84. Saha, S. K.; Bhattacharya, B.; Sarkar, U.; Shankar Rao, D. S.; Paul, M. K. Unsymmetrical Achiral Four Ring Hockey Stick Shaped Mesogens Based on 1,3,4-Oxadiazole: Photophysical, Mesogenic and DFT Studies. *J. Mol. Liq.*, 2017, 241, 881–896.
85. Chamorro, E.; Chattaraj, P. K.; Fuentealba, P. Variation of the Electrophilicity Index Along the Reaction Path. *J. Phys. Chem. A.*, 2003, 107, 7068–7072.
86. Chattaraj, P. K.; Gutiérrez-Oliva, S.; Jaque, P.; Toro-Labbé, A. Towards Understanding the Molecular Internal Rotations and Vibrations and Chemical Reactions Through the Profiles of Reactivity and Selectivity Indices: An ab Initio SCF and DFT Study. *Mol. Phys.*, 2003, 101, 2841–2853.
87. Singh, N. B.; Sarkar, U. Structure, Vibrational, and Optical Properties of Platinum Cluster: A Density Functional Theory Approach. *J. Mol. Model.*, 2014, 20, 2537.
88. Miranda-Quintana, R. A.; Chattaraj, P. K.; Ayers, P. W. Finite Temperature Grand Canonical Ensemble Study of the Minimum Electrophilicity Principle. *J. Chem. Phys.*, 2017, 147, 124103.
89. Tanwar, A.; Roy, D. R.; Pal, S.; Chattaraj, P. K. Minimum Magnetizability Principle. *J. Chem. Phys.*, 2006, 125, 056101.
90. Chattaraj, P. K.; Giri, S. Stability, Reactivity, and Aromaticity of Compounds of a Multivalent Superatom. *J. Phys. Chem. A*, 2007, 111, 11116–11121.
91. Sun, W.-M.; Li, Y.; Wu, D.; Li, Z.-R. Designing Aromatic Superatoms. *J. Phys. Chem. C*, 2013, 117, 24618–24624.
92. López-Estrada, O.; Selenius, E.; Zuniga-Gutierrez, B.; Malola, S.; Häkkinen, H. Cubic Aromaticity in Ligand-Stabilized Doped Au Superatoms. *J. Chem. Phys.*, 2021, 154, 204303.
93. López-Estrada, O.; Zuniga-Gutierrez, B.; Selenius, E.; Malola, S.; Häkkinen, H. Magnetically Induced Currents and Aromaticity in Ligand-Stabilized Au and AuPt Superatoms. *Nat. Commun.*, 2021, 12, 2477.
94. Fedik, N.; Kulichenko, M.; Boldyrev, A. I. Two Names of Stability: Spherical Aromatic or Superatomic Intermetalloid Cluster $[Pd_3Sn_8Bi_6]^{4-}$. *Chem. Phys.*, 2019, 522, 134–137.

2 Doubly Charged Superhalogen Systems

Anton S. Pozdeev and Alexander I. Boldyrev

2.1 CONCEPT

The concept of superhalogens was first proposed by Boldyrev and Gutsev in 1981 [1]; since those times, such systems have been intensively studied both theoretically and experimentally. Molecular systems are classified as superhalogens if their excess electron binding energy goes beyond that of halogens; more specifically, the highest value of electron affinity (EA) of 3.61 eV for the chlorine atom [2]. The high value of electron affinity of a neutral specie or high detachment energy of the corresponding anionic system can characterize superhalogens systems. Vertical detachment energy (VDE) is the energy difference between the neutral system in the equilibrium geometry of a negatively charged system and the negatively charged system in its equilibrium geometry. Adiabatic detachment energy (ADE) is the energy difference between the neutral system in its equilibrium geometry and the negatively charged system in its equilibrium geometry. The difference between the ADE and VDE is called an adiabatic correction.

Some molecules like WF_6, UF_6 and C_6H_6 have minor adiabatic correction (within 0.1–0.6 eV), which can be associated with removing the extra electron from nonbonding MO [3]. However, large adiabatic correction within the range of 1.5–3.0 eV may be investigated for systems where the removal of an extra electron occurs from an antibonding MO. For example, for PF_5 and AsF_5, adiabatic correction ≈ 2.8 eV, with the extra electron located in the antibonding orbital. Experiments showed that their ionic structures are square pyramidal. In contrast, the neutral molecules are trigonal bipyramidal, *i.e.*, the detachment of an extra electron from the HOMO orbital of an anion results in significant geometry rearrangement [4].

Anomalous characteristics of anionic systems, such as high VDE (ADE), should be associated with some essential properties. According to ref [4], several factors should be considered. Firstly, the nature of atoms making up the molecule makes sense. A higher EA of atoms should correspond to a higher EA of the molecule. Moreover, in the case of a significant difference in electronegativity of ligands and a central atom, greater charge separation is observed. Significant Coulomb interaction between the positively charged central atom and negatively charged ligands additionally stabilizes the system; fluorine and oxygen atoms are the most suitable simple ligands. Secondly, the typical order of IP in molecular orbitals for any system is IP (bonding MO) > IP (nonbonding MO) > IP (antibonding MO).

DOI: 10.1201/9781003384205-2

Thus, it requires the absence (or negligible value) of destabilizing antibonding ligand–central atom interactions in the HOMO orbital of an anion. Thirdly, there should be a large number of ligands for the effective delocalization of an extra negative charge. The smaller the fraction of an extra δe added to an atom, the easier for an atom to retain it. And in the limiting case of an unlimited number of ligands, the δe approaches zero, which corresponds the IP of an anion tending to the IP of a neutral system that is usually high.

Usually, these three factors are adequate to predict the superhalogen properties of a molecular system. In the original work [1], general formulae MF_{k+1} and $MO_{(k+1)/2}$ were suggested (M is a main-group atom, k is the maximal valence). It includes common species as BO_2^-, NO_3^-, BF_4^-, PO_{3-}, *etc.*

2.1.1 Theoretical and Experimental Study of Singly Charged Superhalogens

Both theoretical and experimental methods can effectively investigate superhalogen systems. Experimental methods include photoelectron spectroscopy [5–7], surface ionization [8,9], measurement of the energy of crystal lattice [10], charge-transfer reactions [11,12], *etc.* Today, a large amount of experimental data has been accumulated [13–23]. Moreover, the accuracy of modern theoretical methods often leads to a consensus between experimental and theoretical results.

We do not intend to demonstrate all superhalogen systems that were studied by this time, but provide a short summary to illustrate the variety of superhalogen systems. The original article by Boldyrev and Gutsev in 1981 [1] studied anions of the formulae MF_{k+1} and $MO_{(k+1)/2}$. Specifically, the molecular structures included: NO_3 (EA: 3.7 eV), MgF_3 (3.2 eV), SiF_5 (6.4 eV), PF_6 (6.8 eV) and many others. The data were obtained at SCF DVM-Xα level and double-ζ quality basis set. Later, first experimental photoelectron spectra for a series of MK_2^- (M = Li, Na; K = Cl, Br, I) were recorded in 1999 [24]. Measured VDEs were 5.92±0.04 eV ($LiCl_2^-$), 5.86±0.06 eV ($NaCl_2^-$), 5.42±0.03 eV ($LiBr_2^-$), *etc.* For the system of BeX_3^- (X = F, Cl, Br), VDE of BeF_{3-} exceeds 8.256 eV at OVGF/6–311++G(3df) level [25]; MX_3^- (M = B, Al; X = F, Cl, Br) with VDE of AlF_4^- exceeds 9.256 eV [26]. VDE of SiF_5^- is 9.32 eV [27] and VDE of PF_6^- is 9.43 eV [28]. For ReF_n (n = 0,7) system EA of ReF_7 exceeds 4.82 eV [29]. More sophisticated systems include species where superhalogens are used as ligands, such as $Gd(AlF_4)_4^-$ (VDE: 9.66 eV) [30], $Gd(BF_4)_4^-$ (10.22 eV) [31] and $Fe(BO_2)_4^-$ (6.73 eV) [32]. Apart from superhalogen systems with one central atom, polynuclear systems also have been investigated, exhibiting significant detachment energy values. Such systems include $Al_2F_7^-$ (11.16 eV) [33], $Fe_3(CN)_7^-$ (6.02 eV) [34], $Sb_3F_{16}^-$ (13.27 eV) [35] and many others. Sometimes, exotic species like $H_9F_{10}^-$ (12.97 eV), $H_{12}F_{13}^-$ (13.87 eV) [36] and $B_5O_6^-$ (5.91 eV) [37] also may possess specific superhalogen properties.

Molecular systems exhibiting extremal properties are often an object of intensive theoretical and experimental study because they allow us to suggest and develop new ideas and considerably extend the range of technologies. Today superhalogens as an example of such systems with extremal properties, holding interest beyond mere academic inquiry. They have found different applications in various fields. They are used for the synthesis of organic superconductors [38,39]; for instance, it was shown in ref. [40] that the concept of superhalogens is very useful in designing potential candidates for organic superconductors on the example of Bechgaard salts known as the first organic superconductors with formula $(TMTSF)_2PF_6$ (TMTSF = tetramethyl tetraselenafulvalene) and Fabre salts with formula $(BETS)_2GaCl_4$ (BETS = bisethylenedithio tetraselenafulvalene). Superhalogens are widely used for oxidizing substances with high ionization potentials like O_2 and noble gases [41–45]. Also, they contribute as a component of common ionic liquids [46,47], such common anions like BF_4^-, PF_6^- and $AlCl_4^-$ are applied as counterions to large asymmetric organic cations forming more stable complexes compared to their halogen counterparts [48]. Superhalogens are building blocks of superacids [49–53]. On the example of BX_3, where X is superhalogen ligands $C_2BNO(CN)_3$ and $C_2BNS(CN)_3$, a new approach to the design of strong Lewis acids was introduced in ref. [54].

Thus, since 1981 various superhalogen systems have been studied, theoretically interpreted, experimentally investigated and even found real applications in different fields of science and industry. Inspired and motivated by this success, we decided to go beyond and extend the area of superhalogen systems.

2.1.2 Doubly Charged Superhalogen Systems

This study is a kind of extension of the definition of superhalogens because it additionally includes di-anions in classical superhalogen systems, which consist only of singly charged anions. The examples above demonstrate that VDE and ADE of properly designed singly charged superhalogens systems can reach several times higher values than the reference value of 3.62 eV. However, our knowledge of doubly charged superhalogens is still limited. In this work, we aimed to focus on some approaches and ideas effectively used for singly charged superhalogens and apply them to the design of doubly charged superhalogens. We performed the theoretical investigation of MF_6^{2-} (M = Si, Ge, Sn, Pb) series, TeF_8^{2-}, AlF_5^{2-}–$Al_2F_8^{2-}$ series, SiF_6^{2-}–$Si_2F_{10}^{2-}$–$Si_3F_{14}^{2-}$ series, $B_{12}H_{12}^{2-}$ and $B_{12}F_{12}^{2-}$.

For some anions, we carried out a global minimum search using the Coalescence Kick (CK) algorithm written by Averkiev [55,56]. For SiF_6^{2-}, SiF_6^{1-}, AlF_5^{2-} and AlF_5^{1-}, we generated 1,000 random structures; for TeF_8^{2-} and TeF_8^{1-} 1500 structures; for $Si_2F_{10}^{2-}$, $Si_2F_{10}^{1-}$, $Al_2F_8^{2-}$ and $Al_2F_8^{1-}$2,000 structures, for $Si_3F_{14}^{2-}$ and $Si_3F_{14}^{1-}$4,500 structures to find the global minimum (GM) structures. All randomly generated structures were optimized using unrestricted Kohn–Sham formalism with MN-15 density functional and LANL2DZ basis set. Our motivation for the functional was based on its good applicability to various nontrivial electronic structure problems [57]. We chose the LANL2DZ basis set as a balance of speed and accuracy [58] and we applied the Hay–Wadt pseudopotential to Ge, Pb, Sn and Te atoms to account for scalar relativity correction [59]. We used Gaussian 16 for the preoptimization of randomly generated structures in the CK algorithm [60].

In the next step, we reoptimized all non-dissociated low-lying isomers in the 10–15 kcal/mol energy window obtained after the CK search. We chose a more accurate def2-TZVPPD basis set [61,62] with Stuttgart-Dresden effective core potential for Pb, Sn and Te atoms [63] and density functional TPSSh [64] chosen for its good energy ordering of isomers. All finally optimized structures were checked by Hessian nuclear calculation at the TPSSh/def2-TZVPPD level of theory. Using obtained local minima geometries, single point energies were recalculated at the DLPNO-CCSD(T) (quasi-restricted orbitals formalism) level of theory with the aug-cc-pVTZ-DK [65,66] basis set (obtained results are consistent with those employing SK-MCDHF-RSC pseudopotential on Pb, Sn and Te atoms [67,68]). Since we involved large molecular systems in this work ($B_{12}F_{12}^{2-}$, $Si_3F_{14}^{2-}$ and others), we chose DLPNO-CCSD(T) method supporting a much larger molecular system than classical CCSD(T) with almost no loss in quality [69]. Douglas-Kroll-Hess 2nd (DKH) order scalar relativistic approximation [70,71] was performed at this step for a more accurate energy comparison of species with heavy elements (lead and tin) and light elements (silicon and aluminum). Some data of the chosen basis set were downloaded from the BSE database [72–74]. The re-optimization step and all subsequent single-point calculations were performed with the ORCA 5.03 software [75,76]. Thus, the final GM structures were chosen as the lowest energy-lying structures at the DLPNO-CCSD(T)/aug-cc-pVTZ-DK level. For $B_{12}H_{12}^{2-}$ and $B_{12}F_{12}^{2-}$ the initial geometries were taken from [77].

VDE values for each di-anion were calculated by two approaches. The first, an indirect approach, measured VDE as the energy difference between the singly charged anion in the equilibrium geometry of a doubly charged anion and the doubly charged anion in its equilibrium geometry at DLPNO-CCSD(T)/aug-cc-pVTZ-DK (including DKH correction) level. Energy calculation of both doubly charged and singly charged anions were carried out with QRO formalism. The alternative direct approach employed the outer valence Green function method (OVGF) [78,79] with aug-cc-pVTZ basis set for each di-anions in the MF_6^{2-} (M = Si, Ge, Sn, Pb) series, TeF_8^{2-} and 6–311++G(3df) basis set for all other di-anions. The choice of aug-cc-pVTZ basis set was determined by the necessity of comparison of each MF_6^{2-} di-anion at the same level. For other doubly charged anions, we chose the 6–311++G(3df) basis set because it demonstrated the most accurate result for the OVGF method [25]. ADE was calculated as an energy difference between the singly charged anion in its equilibrium geometry and the doubly charged anion in its equilibrium geometry at DLPNO-CCSD(T)/aug-cc-pVTZ-DK (including DKH correction) level.

2.2 DOUBLY CHARGED SUPERHALOGENS

In the following sections, we share our results of a primary search of doubly charged superhalogens. Where possible, we reference data known in the literature. The main object of this work is to demonstrate some general approaches that can be used to design doubly charged super halogen systems. For all presented systems, we chose fluorine as a ligand because this suitable choice allows us to compare different systems with each other more adequately. This ligand is the most effective in stabilizing molecular systems toward electron detachment among a list of simple ligands.

2.2.1 MF_6^{2-} Series

Based on the steps described in section 1.3, the lowest-lying non-dissociated structures found for SiF_6^{2-} and SiF_6^{1-} are considered the global minima (GM) geometries (both structures are presented in Figure 2.1). Further, the GM geometries of SiF_6^{2-} and SiF_6^{1-} were employed as the initial structures for optimizing other anions in the MF_6^{2-} and MF_6^{1-} (M = Si, Ge, Si, Pb) series. Doubly charged structures are found to be octahedral complexes (O_h), whereas the singly charged anions exhibit C_{2v} point group symmetry as reported in ref. [80]. Corresponding O_h symmetry of the singly charged anions displays one imaginary frequency, which indicates they are not a local minimum on the PES, and the singly charged anions undergo Jahn-Teller Distortion. The intensive search of the lowest-lying geometry of singly charged anions is a key step in the calculation of ADE and classification of selected doubly charged anions as superhalogens. For the MF_6^{2-} series, we found bond lengths of Si-F is 1.72 Å (the experimental bond length of SiF_6^{2-} in $FeSiF_6{*}6H_2O$ salt is 1.702 Å according to [81]), for Ge-F it is 1.83 Å, for Sn-F it is 2.00 Å, for Pb-F it is 2.11 Å, *i.e.*, we observe a gradual increase of bond lengths.

The first VDEs of MF_6^{2-} (M = Si, Sn, Ge, Pb) anions calculated at DLPNO-CCSD(T)/aug-cc-pVTZ-DK level of theory and OVGF/aug-cc-PVTZ are collected in Table 2.1, ADEs are calculated at DLPNO -CCSD(T)/aug-cc-pVTZ-DK level of theory. According to ref. [25], DLPNO-CCSD(T) level tends to underestimate the VDE values, whereas the OVGF level tends to overestimate the values.

Both theoretical methods show the increase of VDE from silicon to tin, but for lead it is approximately at the same level as germanium. Based on VDE

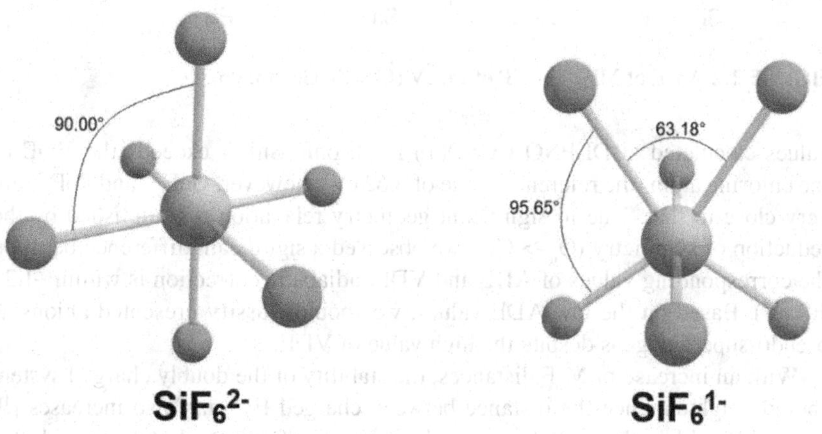

FIGURE 2.1 Optimised GM geometries of SiF_6^{2-} and SiF_6^{1-}.

TABLE 2.1
The vertical detachment energy (VDE) in eV of MF_6^{2-} (M = Si, Sn, Ge, Pb) series calculated at DLPNO-CCSD(T)/aug-cc-pVTZ-DK and OVGF/aug-cc-pVTZ levels of theory, adiabatic detachment energy (ADE) in eV calculated at DLPNO-CCSD(T)/aug-cc-pVTZ-DK level.

Anion	DLPNO-CCSD(T)/aug-cc-pVTZ-DK		OVGF/aug-cc-pVTZ
	ADE, eV	VDE, eV	VDE, eV
SiF_6^{2-}	1.61	3.19	3.57
GeF_6^{2-}	1.89	3.60	4.09
SnF_6^{2-}	2.37	4.01	4.19
PbF_6^{2-}	2.37	3.55	3.71

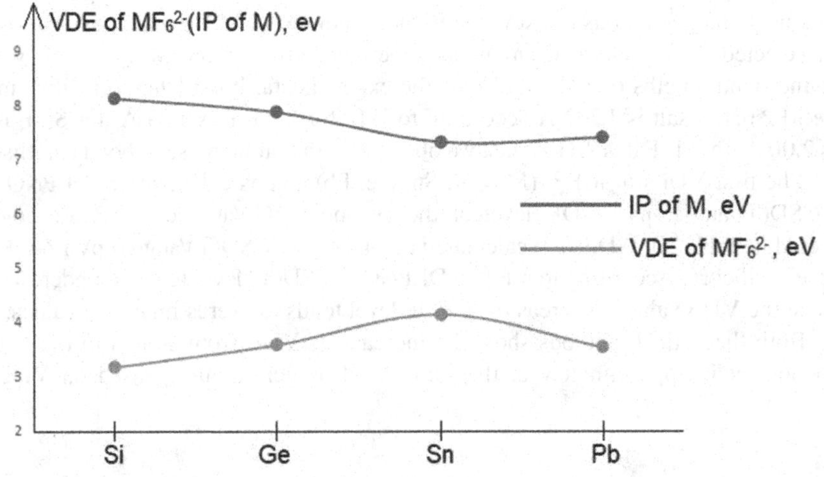

FIGURE 2.2 VDE of MF_6^{2-} and IP of M, eV (M = Si, Ge, Sn, Pb).

values calculated at DLPNO-CCSD(T) level, only SnF_6^{2-} exceeds the VDE of the chlorine atom (the reference value of 3.62 eV); however, GeF_6^{2-} and PbF_6^{2-} are very close to that. Due to significant geometry relaxation accomplished by the reduction of symmetry (O_h -> C_{2v}), we observed a significant difference between the corresponding values of ADE and VDE (adiabatic correction is within ~1.2–1.8 eV). Based on the low ADE values, we should classify presented anions as pseudo-superhalogens despite the high value of VDE.

With an increase of M-F distances, the stability of the doubly charged system should heighten since the distance between charged F atoms also increases [4]. Beyond bond lengths, the change in the property of a central atom should also be considered. In Figure 2.2, we presented the dependence of VDEs values on IP

TABLE 2.2
The NBO charges of central atoms and the F atoms of the MF_6^{2-} (M = Si, Sn, Ge, Pb) series.

Anion	NBO charge of a central atom	NBO charge of F
SiF_6^{2-}	2.373	−0.729
GeF_6^{2-}	2.533	−0.756
SnF_6^{2-}	2.657	−0.776
PbF_6^{2-}	2.495	−0.749

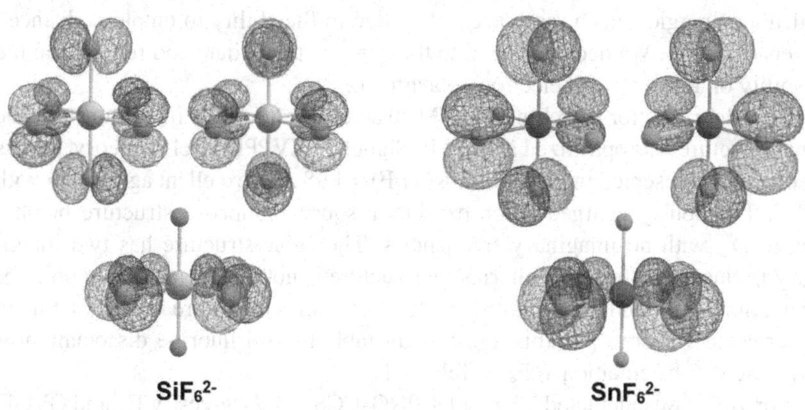

FIGURE 2.3 The trice degenerate HOMO orbitals of SiF_6^{2-} and SnF_6^{2-}.

(data taken from [82–85]) of a central atom. According to the graph, the lower the IP value of M corresponds to a higher VDE value of MF_6^{2-}.

Additionally, we calculated NBO charges on each atom in MF_6^{2-} anions. Sn has the highest charge among presented elements as a sequence of the lowest IP value. Thus, in the MF_6^{2-} series, SnF_6^{2-} di-anion is the most stable toward electron detachment, and fluorine ligands have the highest extra negative charges. These results agree with the concept of superhalogen systems, the extra negative charge is predominantly distributed on the ligands rather than a central atom [4].

In Figure 2.3, we presented the trice-degenerated HOMO molecular orbitals of SiF_6^{2-} and SnF_6^{2-} (GeF_6^{2-} and PbF_6^{2-} have the same patterns). The figure illustrates that all HOMO orbitals are exclusively composed of atomic orbitals of fluorine atoms, with no contribution of the central atom. It indicates the nonbonding character of these HOMO orbitals relating to the central atom–fluorine interactions. According to the general rules in ref. [4], such characters of the HOMO orbital can be associated with high values of VDE.

We did not estimate the thermodynamic stability of the presented doubly charged anion as they have been experimentally identified (SiF_6^{2-} in the form of $[Cu_2(C_6H_4N_3(C_3H_5))_2(H_2O)_2SiF_6]\cdot 2H_2O$ [86], GeF_6^{2-} in the form of Na_2GeF_6 [87], SnF_6^{2-} in the form of Cs_2GeF_6 [88] and PbF_6^{2-} in the form of K_2PbF_6 [89]).

2.2.2 TeF_8^{2-}

The delocalization of an extra charge among a large number of ligands is a possible key ingredient for achieving high detachment energy values. Therefore, the next step from MF_6^{2-} series could be an increase in the number of fluorine atoms. In 1992, TeF_8- was proposed by Boldyrev and Simons as a potential doubly charged system with high detachment energy, calculating VDE above 5 eV at the SCF/DZ level of theory [90]. Modern computational capabilities allow us to calculate energies much more accurately due to the ability to employ advanced levels of theory. We decided to refine the computational data and recalculate the stability of TeF_8^{2-} toward electron detachment.

The algorithm for searching the GM structure is described in section 1.3. The final structure was optimized at the TPSSh/def2-TZVPPD level of theory, and its geometry is presented in Figure 2.4. R(Ti-F) = 1.98 Å in excellent agreement with [90,91]. The doubly charged anion exhibits a square antiprism structure belonging to D_{4d} with no imaginary frequencies. The cubic structure has two imaginary frequencies; thus, this alternative structure is not a local minimum on PES. Our efforts to identify a non-dissociated TeF_{8-} structure were unsuccessful. In accordance with ref. [92], this anion is unstable toward fluorine dissociation as expressed in the equation $TeF_8^- = TeF_7^- + F$.

For TeF_8^{2-} we calculated VDE at DLPNO-CCSD(T)/aug-cc-PVTZ and OVGF/cc-PVTZ-PP levels of theory; the data are presented in Table 2.3. Given that TeF_8^{1-} is not a stable structure, and we calculate ADE relatively to singly charged anions relaxed within one adiabatic surface, we omitted the calculations of ADE for TeF_8^{2-}.

We can observe a slight increase of VDE compared with SnF_6^{2-} (4.01 eV at DLPNO-CCSD(T) level) as the best representative of the MF_6^{2-} series. This case illustrates that it is essential considering several parameters for the design

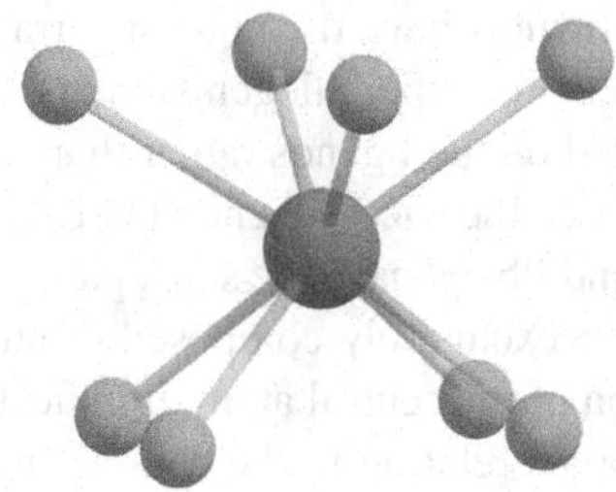

FIGURE 2.4 Optimised GM geometry of TeF_8^{2-}.

of superhalogen structures. Specifically, tellurium has high IP of 9.00 eV [93]; for comparison, the IP of Sn is 7.34 eV. It can be vividly illustrated with NBO charges, we found Te atom has +3.214e, F atoms have -0.652e (for MF_6^{2-} series charges of F atoms are within (-0.729e, -0.776e), Table 2.2). The bond length Te-F is very close to Sn-F, so this parameter can be neglected here. Figure 2.5 demonstrates the HOMO orbital of TeF_8^{2-}. This is a pure nonbonding orbital with respect to tellurium-fluorine interaction and is entirely localized on fluorine atoms.

TABLE 2.3
The vertical detachment energy (VDE) in eV of TeF_8^{2-} calculated at DLPNO-CCSD(T)/aug-cc-pVTZ-DK and OVGF/aug-cc-pVTZ levels.

Anion	CCSD(T)/aug-cc-pVTZ-DK	OVGF/aug-cc-pVTZ
TeF_8^{2-}	4.11	4.22

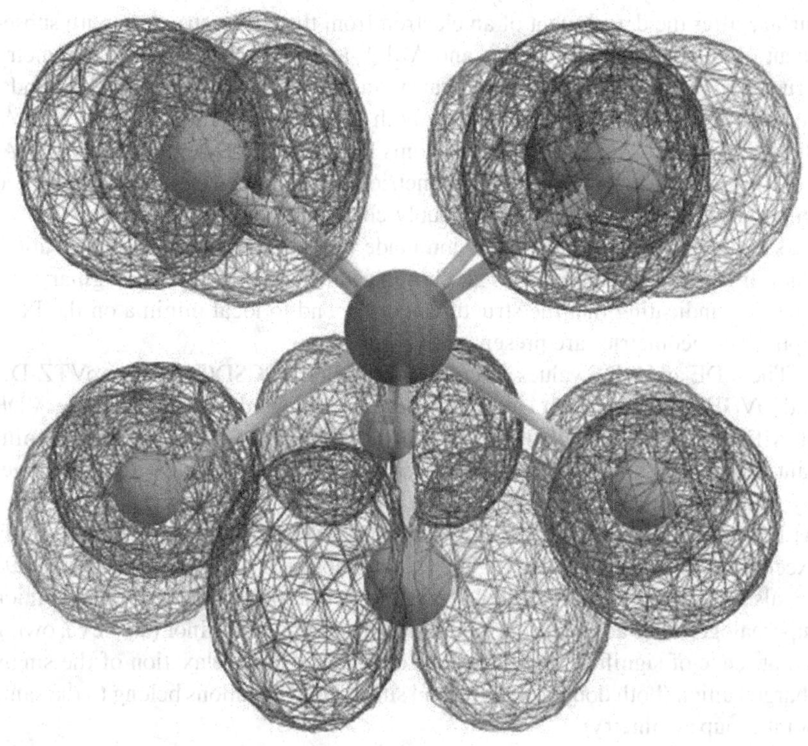

FIGURE 2.5 The singly degenerate HOMO orbitals of TeF_8^{2-}.

Comparison of TeF_8^{2-} with the data obtained for MF_6^{2-} series indicates that merely increasing number of fluorine atoms bonded with one central atom is insufficient to stabilize the doubly charged anion toward electron detachment. As mentioned in section 1.3, we aimed to use the general approach applied for stabilizing singly charged anions. Let us proceed further and employ another decisive modification: increasing both the number of central atoms and fluorine atoms, *i.e.*, design of polynuclear superhalogens.

2.2.3 AlF_5^{2-}-$Al_2F_8^{2-}$ Series

We proposed to introduce the idea of increasing the VDE value by increasing the number of central atoms as it was successfully applied to singly charged anions [33,34,94]. Initially, we theoretically investigated the anions AlF_5^{2-} and $Al_2F_8^{2-}$; both anions have been found experimentally as a counterion in salts $SrAlF_5$ [95] and (Mg, Ca)Al_2F_8*$2H_2O$ [96].

We applied the steps described in section 1.3 to find the GM structures of singly and doubly charged anions. Final geometries were obtained at the TPSSh/def2-TZVPPD level of theory. AlF_5^{2-} is a trigonal bipyramidal structure of D_{3h} symmetry, where R(Al-F_{eq}) = 1.78 Å and R(Al-F_{ax}) = 1.82 Å. For AlF_5^-, we found a lower symmetry C_{2v} structure, which can be obtained within the same adiabatic surface after the detachment of an electron from the AlF_5^{2-} structure with subsequent optimization. Both $Al_2F_8^{2-}$ and $Al_2F_8^{1-}$ are the C_{2h} point group symmetry structures. For $Al_2F_8^{2-}$ we found several symmetrically nonequivalent Al-F bonds, *i.e.*, for two fluorine atoms connecting both aluminum atoms R(Al-F_1) = 1.80 Å, R(Al-F_2) = 1.89 Å; for other fluorine atoms R(Al-F_3) = 1.68 Å, R(Al-F_4) = 1.73 Å, R(Al-F_5) = 1.76 Å. For both stoichiometries, the bonds Al-F of singly charged anion are shorter than that of the doubly charged anion average by 0.1–0.15 Å. This is due to the electronic repulsion made by the extra electrons' distribution. Nuclear Hessian calculations for all chosen geometries provide no imaginary frequencies, indicating that the structures correspond to local minima on the PES. Optimized geometries are presented in Figure 2.6.

The VDE and ADE values calculated at DLPNO-CCSD(T)/aug-cc-pVTZ-DK and OVGF/6–311++G(3df) level of theory are presented in Table 2.4. The VDE of AlF_5^{2-} is 2.70 eV, ADE is 1.04 eV at DLPNO-CCSD(T) level. The significant difference (1.66 eV) between the two values is associated with the geometry transformation accomplished by the reduction of symmetry (D_{3h} -> C_{2h}). For $Al_2F_8^{2-}$ the calculated VDE is 5.41 eV at the DLPNO-CCSD(T) level, significantly exceeding the threshold value of 3.62 eV for chlorine. Moreover, the ADE of 4.94 eV also overreaches the threshold value, which allows us to call this di-anion superhalogen. The adiabatic correction for this di-anion is minor (0.47 eV), owing the absence of significant geometric changes during the relaxation of the singly charged anion (both doubly charged and singly charged anions belong to the same point group symmetry).

Thus, the threshold value has been achieved due to the increased number of central atoms and ligands stabilizing the di-anion toward electron detachment.

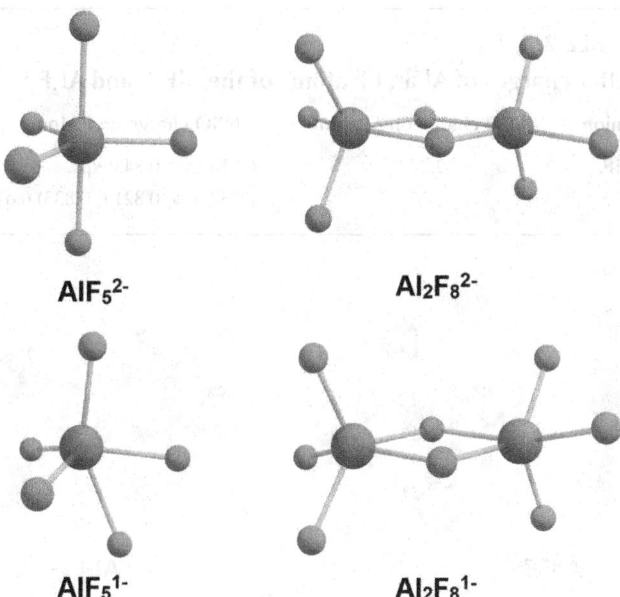

FIGURE 2.6 Optimised geometries of AlF_5^{2-}, AlF_5^{1-}, $Al_2F_8^{2-}$ and $Al_2F_8^{1-}$.

TABLE 2.4
The vertical detachment energy (VDE) in eV of AlF_5^{2-} and $Al_2F_8^{2-}$ calculated at DLPNO-CCSD(T)/aug-cc-pVTZ-DK and OVGF/6–311++G(3df) levels of theory, adiabatic detachment energy (ADE) in eV calculated at DLPNO-CCSD(T)/aug-cc-pVTZ-DK level.

Anion	DLPNO-CCSD(T)/aug-cc-pVTZ-DK		OVGF/6–311++G(3df)
	ADE, eV	VDE, eV	VDE, eV
AlF_5^{2-}	1.04	2.70	3.24
$Al_2F_8^{1-}$	4.94	5.41	6.00

AlF_5^{2-} has a too low value of VDE(ADE) to be classified as a superhalogen; however, $Al_2F_8^{2-}$ completely satisfies the criterion.

The calculated NBO charges for both doubly charged anions are presented in Table 2.5. According to the data, the nature of charge distribution in both double-charged anions is the same, *i.e.*, Al atoms have ~+2.250e, and each F atom in the $Al_2F_8^{2-}$ di-anion has a bit higher charge (-0.782e, -0.833e), than in the AlF_5^{2-} di-anion (-0.858e, -0.843e). As we expected, the extra charge localized among more F atoms. Moreover, NBO charges on fluorine atoms in both aluminum structures are much lower than in the MF_6^{2-} series (-0.73e, -0.78e) and especially

TABLE 2.5
NBO charges of Al and F atoms of the AlF_5^{2-} and $Al_2F_8^{1-}$.

Anion	NBO charge on Al atom	NBO charge on F atoms
AlF_5^{2-}	2.247	–0.858(ax)/–0.843(eq)
$Al_2F_8^{2-}$	2.258	–0.782(ax)/–0.821 (–0.833) (eq)

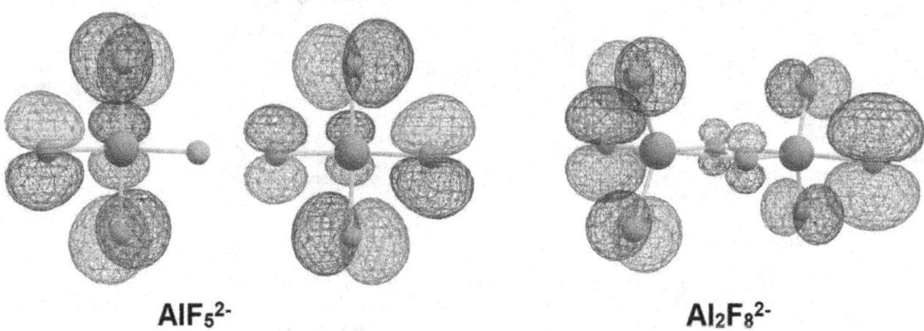

FIGURE 2.7 The doubly degenerated HOMO orbitals of AlF_5^{2-} and single degenerated HOMO orbital of $Al_2F_8^{2-}$.

TeF_8^{2-}(-0.65e). It is in total agreement with the IP of Al (5.99 eV [97]), which is much lower than that of Sn (7.34 eV) and Te (9.00 eV), indicating that this aluminum fluoride system is much more ionic than the systems previously discussed.

It is noteworthy that in the sequence AlF_5^{2-}-SnF_6^{2-}-TeF_8^{2-}, each successive dianion has a higher VDE than the previous, but this difference is substantial from Al to Sn (1.31 eV) and tiny from Sn to Te (0.10 eV), even despite the aluminium fluoride has one F atom less than the tin fluoride, whereas it has two F atoms less than the tellurium fluoride. It demonstrates that adding more ligands can enhance VDE, though the effect is not linear.

Additionally, we presented the visualized HOMO orbitals in Figure 2.7. As in the previous cases, they are nonbonding toward the interaction of a central atom–fluorine atoms; with no contribution from a central atom.

Thus, the concept of increasing VDE (ADE) value by increasing the number of both fluorine atoms and central atoms can be successfully applied not only to a singly charged anions but also to di-anions. We decided to investigate the series of SiF_6^{2-}-$Si_2F_{10}^{2-}$-$Si_3F_{14}^{2-}$ as another demonstration of the effectiveness of the design of polynuclear superhalogen systems.

2.2.4 SiF_6^{2-}-$Si_2F_{10}^{2-}$-$Si_3F_{14}^{2-}$ Series

GM non-dissociated geometries for every anion and dianions of each stoichiometry were obtained via the CK GM search algorithm, as described in section 1.3.

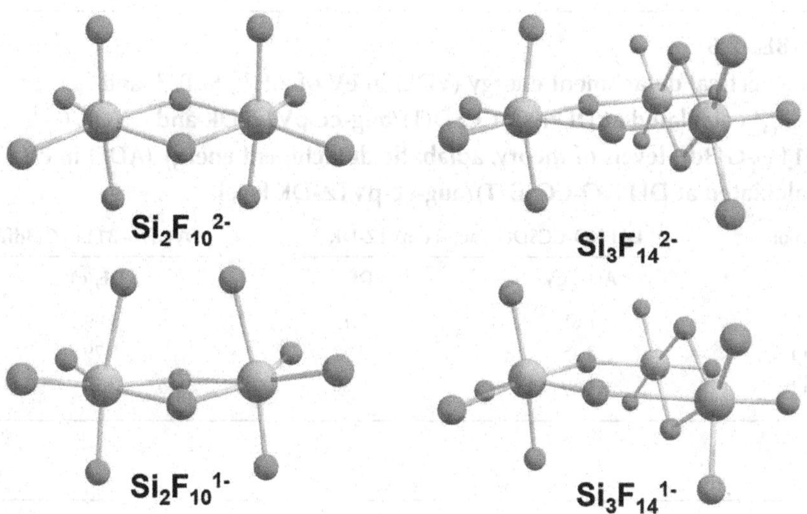

FIGURE 2.8 Optimised GM geometries of $Si_2F_{10}^{2-}$, $Si_2F_{10}^{1-}$, $Si_3F_{14}^{2-}$ and $Si_3F_{14}^{1-}$.

The optimized doubly charged and singly charged anions of Si_2F_{10} and Si_3F_{14} stoichiometries are presented in Figure 2.8, SiF_6 stoichiometry is presented in Figure 2.1.

SiF_6^{2-} is described in section 2.1. $Si_2F_{10}^{2-}$ is a D_{2h} structure, R(Si-F) = 1.87 Å for Si-F-Si, R(Si-F) = 1.66 Å for other fluorine atoms, *i.e.*, some Si-F bonds are a bit shorter, other bonds are a bit longer than those of SiF_6^{2-} (1.72 Å). $Si_2F_{10-}^{1}$ has reduced symmetry of $C_{2v,}$ which is associated with an increase of two Si-F bond lengths from 1.66 Å up to 1.83 Å and slight decrease of other bond lengths within the range of 0.01–0.03 Å. $Si_3F_{14}^{2-}$ is a bit more sophisticated case; it is a C_{2v} structure R(Si-F) = 1.84 Å for Si-F-Si within Si_2F_8 fragment, R(Si-F) = 1.76 Å for Si-F-Si between two fragments. Within the SiF_6 fragment, R(Si-F) = 1.94 Å; within the Si_2F_8 fragment, R(Si-F) is in the range of 1.63–1.65 Å. Singly charged anion is a C_1 structure; after electron detachment, mostly only SiF_6 fragment undergoes relaxation.

The VDE and ADE values calculated at DLPNO-CCSD(T)/aug-cc-pVTZ-DK and OVGF/6–311++G(3df) level of theory are presented in Table 2.6. We can observe a graduate increase of both VDE and ADE as the number of central atoms increases. For $Si_2F_{10}^{2-}$ the VDE value of 5.62 eV overreaches the reference value of 3.62 eV but its ADE of 3.45 eV is just a bit lower. During the detachment of an electron, the reduction of symmetry D_{2h} -> C_{2v} occurs, which leads to a large adiabatic correction of 2.17 eV. For $Si_3F_{14}^{2-}$ both VDE (6.80 eV) and ADE (4.60 eV) greatly overreach the reference value. This di-anion has high detachment energy, so even a large adiabatic correction of 2.20 eV allows it to overcome the reference value. As in the aluminum series, the approach of polynuclear superhalogen is successful.

TABLE 2.6
The vertical detachment energy (VDE) in eV of SiF_6^{2-}, $Si_2F_{10}^{2-}$ and $Si_3F_{14}^{2-}$calculated at DLPNO-CCSD(T)/aug-cc-pVTZ-DK and OVGF/6–311++G(3df) levels of theory, adiabatic detachment energy (ADE) in eV calculated at DLPNO-CCSD(T)/aug-cc-pVTZ-DK level.

Anion	DLPNO-CCSD(T)/aug-cc-pVTZ-DK		OVGF/6–311++G(3df)
	ADE, eV	VDE, eV	VDE, eV
SiF_6^{2-}	1.61	3.19	3.49
$Si_2F_{10}^{2-}$	3.45	5.62	5.79
$Si_3F_{14}^{2-}$	4.60	6.80	7.10

TABLE 2.7
NBO charges of Si and F atoms of SiF_6^{2-}, $Si_2F_{10}^{2-}$ and $Si_3F_{14}^{2-}$

Anion	NBO charge on Si atom	NBO charge on F atoms
SiF_6^{2-}	2.373	–0.729
$Si_2F_{10}^{2-}$	2.603	–0.700 (Si-**F**-Si), -0.738 (eq)/-0.729 (ax)
$Si_3F_{14}^{2-}$	2.614 (2Si)/2.649(1Si)	–0.653, -0.676 (Si-**F**-Si), -0.730, -0.719 (eq)/-0.720 (ax)

Let us focus on the pairwise comparison. For AlF_5^{2-} and SiF_6^{2-} pair, the silicon fluoride has a higher VDE (ADE) by ~0.55 eV (their adiabatic corrections are both about 1.55 eV). IP of Si (8.15 eV) is much higher than that of Al (5.99 eV); thus, here, we observe two opposite tendencies by shifting from AlF_5^{2-} to SiF_6^{2-}. The first tendency is the increase of the number of F atoms, resulting in more effective delocalization of extra negative charge. The second one is the increase of IP of the central atom with a consequence decrease in the effectiveness of delocalization of the extra charges. The first tendency prevails. In the second pair of $Al_2F_8^{2-}$ and $Si_2F_{10}^{2-}$ silicon fluoride again has higher VDE at DLPNO-CCSD(T) level, but due to minor adiabatic correction for $Al_2F_8^{2-}$ its ADE is much higher (~1.5 eV). In this work, we aimed ADE to overreach the reference value of 3.62 eV to classify a di-anion as a superhalogen. By this reason, in the silicon series, we go further and analyse $Si_3F_{14}^{2-}$.

NBO charges presented in Table 2.7 demonstrate a slight increase of NBO charge on Si atoms as the number of atoms increases. Naturally, the NBO charges on F atoms decrease. This evaluation of the series may be interpreted in terms of more significant charge separation, *i.e.*, there is stronger stabilization of the extra charges for $Si_3F_{14}^{2-}$ than for SiF_6^{2-}.

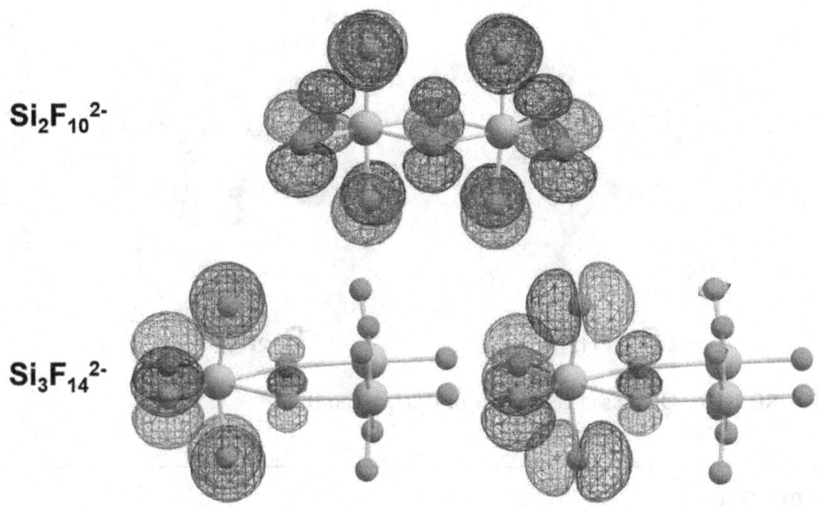

FIGURE 2.9 The singly degenerate HOMO orbitals of $Si_2F_{10}^{2-}$ and doubly degenerate HOMO orbital of $Si_3F_{14}^{2-}$.

As in all previous cases, we visualized the HOMO orbitals for each di-anion (Figure 2.9). SiF_6^{2-} is described in section 2.1. For $Si_2F_{10}^{2-}$ the singly degenerate HOMO orbital is localized on F atoms without any influence of silicon atoms. For $Si_3F_{14}^{2-}$ the doubly denigrated HOMO orbitals are also localized on F atoms, but this localization is almost totally on F atoms of the SiF_6 fragment, not on the Si_2F_{10} fragment. This agrees with observed geometry relaxation during the detachment of an electron; geometry rearrangements were observed mainly on SiF_6 fragment.

Thus, on the example of this series the approach of design of polynuclear doubly charged anions also occurred to be effective. We observe a gradual increase in detachment energy as the number of atoms increase.

2.2.5 $B_{12}H_{12}^{2-}$ AND $B_{12}F_{12}^{2-}$

Finally, we investigated $B_{12}H_{12}^{2-}$ and $B_{12}F_{12}^{2-}$ systems. Various boron hydrides are well-investigated theoretically [69,98–100] and experimentally [101–103]. Due to a kind of unique properties of boron, it is a good model of nonclassical bonding pattern (perhaps, diborane should be considered as the most famous example of compounds with 3c-2e bonds) that produces a wide variety of B_xH_y structures. We chose the B_{12} cluster, *i.e.*, highly stable icosahedron (I_h) $B_{12}H_{12}^{2-}$ and its derivatives $B_{12}F_{12}^{2-}$ as a model of poly-atomic structures with potential high stability toward the electron detachment.

In [104], Longuet-Higgins and Roberts predicted doubly charged anion $B_{12}H_{12}^{2-}$ to be stable based on molecular orbital symmetry analysis. The closed electron

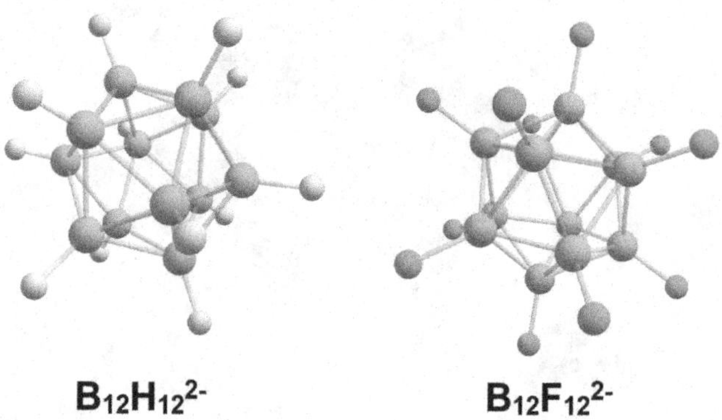

FIGURE 2.10 Optimised geometries of $B_{12}H_{12}^{2-}$ and $B_{12}F_{12}^{2-}$.

TABLE 2.8
The vertical detachment energy (VDE) in eV of $B_{12}H_{12}^{2-}$ and $B_{12}F_{12}^{2-}$ calculated at DLPNO-CCSD(T)/aug-cc-pVTZ-DK and OVGF/6–311++G(3df) levels of theory, adiabatic detachment energy (ADE) in eV calculated at DLPNO-CCSD(T)/aug-cc-pVTZ-DK level.

Anion	CCSD(T)/aug-cc-pVTZ-DK	OVGF/6–311++G(3df)
$B_{12}H_{12}^{2-}$	1.32	1.42
$B_{12}F_{12}^{2-}$	2.17	2.34

shell of $B_{12}H_{12}$ can be formed only in case of exactly 26 electrons are available to fill the MO, including quadruple degenerate HOMO orbitals; it is possible only in case of a doubly charged anionic system. Later, Pitochelli and Hawthorne experimentally isolated icosahedral $B_{12}H_{12}^{2-}$ doubly charged anion [[105]]. Today chemical and physical properties of the di-anion are intensively investigated [[106,107]]. The fluorinated derivative $B_{12}F_{12}^{2-}$ was also isolated and investigated [[108,109]]. Thus, based on previous research and the complexity of CK search for such a large system, we took the icosahedral geometry $B_{12}H_{12}^{2-}$ for both $B_{12}H_{12}^{2-}$ and $B_{12}F_{12}^{2-}$ for further re-optimization at TPSSh/def2-TZVP level of theory (section 1.3). The structures are presented in Figure 2.10.

The average R(B-B) = 1.80 Å for both di-anions, whereas R(B-H) = 1.19 Å and R(B-F) = 1.39 Å. For both cases, we found no imaginary frequencies. Detachment of one electron from $B_{12}H_{12}^{2-}$ leads to partially filled quadruple degenerate HOMO orbitals, which should lead to Jahn-Teller distortion. Our attempts to find a stable (with no imaginary frequencies) singly charged structure within the same adiabatic surface were unsuccessful; thus, we limited our task to calculating VDEs. As in the previous sections, we calculated VDEs at DLPNO-CCSD(T)/aug-cc-pVTZ-DK and OVGF/6–311++G(3df) levels of theory; data are presented in Table 2.8.

TABLE 2.9
NBO charges of B, H and F atoms of $B_{12}H_{12}^{2-}$ and $B_{12}F_{12}^{2-}$.

Anion	NBO charge of B atom	NBO charge of H/F atoms
$B_{12}H_{12}^{2-}$	~(-0.140)	~(-0.025)
$B_{12}F_{12}^{2-}$	~(+0.408)	~(-0.575)

VDE values are significantly far from the reference value of 3.62 eV of the chlorine atom for both di-anions. It is understandable in the case of $B_{12}H_{12}^{2-}$ because the concept of superhalogen is based on the suggestion of delocalizing negative charge on electronegative ligands, whereas hydrogen is a very weak non-metal. However, the exchange of H on F leads to a slight increase of VDE within 0.8–0.9 eV. We suggest several explanations for the phenomenon.

Firstly, in the example of MF_6^{2-} (M = Si, Ge, Sn, Pb) series, we found the correlation of VDE values with the IP of a central atom. B has high IP (8.30 eV); for example, it is 8.15 eV for Si. This can be illustrated by the NBO charges of F atoms (Table 2.9). For all presented cases in Sections 2.1–2.4, the NBO charges of F atoms were in the range of (-0.65; -0.86)e, but for $B_{12}F_{12}^{2-}$ it is just (-0.575) e. Boron atoms are involved much more significantly in the delocalization of the extra negative charges, which makes the di-anion much less stable toward electron detachment than other studied species.

Secondly, ideally, superhalogen should have a large number of ligands, but the number of central atoms also matters. It should have a high ratio of N_F(number of F atoms)/N_M(number of central atoms). The ratio is within (4–8) for all studied cases in this work. However, in $B_{12}F_{12}^{2-}$ the ratio is just 1. Moreover, as described above, the small ratio is also accompanied by a relatively high IP of the B atom.

Finally, for all investigated cases in this work, we checked the nature of HOMO; they were always nonbonding toward the interaction of the fluorine atom–a central atom. In the case of $B_{12}F_{12}^{2-}$ these quadruple degenerate HOMO orbitals are more challenging to interpret; however, we can say they are, to a certain extent, partly antibonding toward boron-fluorine interaction (Figure 2.11).

All described reasons do not allow the $B_{12}F_{12}^{2-}$ species to be stable enough toward electron detachment to be classified as a superhalogen. Furthermore, this system is a good illustration of violations of the rules described in section 1.3.

Thus, despite $B_{12}F_{12}^{2-}$ ($B_{12}H_{12}^{2}$) appearing like a good candidate to be classified as a doubly charged superhalogen due to its polynuclear nature, successful experimental detection and detailed experimental studies, theoretical calculations demonstrate this system has too low detachment energy.

2.3 FINAL REMARKS

In this work, we aimed to extend the traditional definition of superhalogen systems by incorporating doubly charged systems. Despite the challenge of

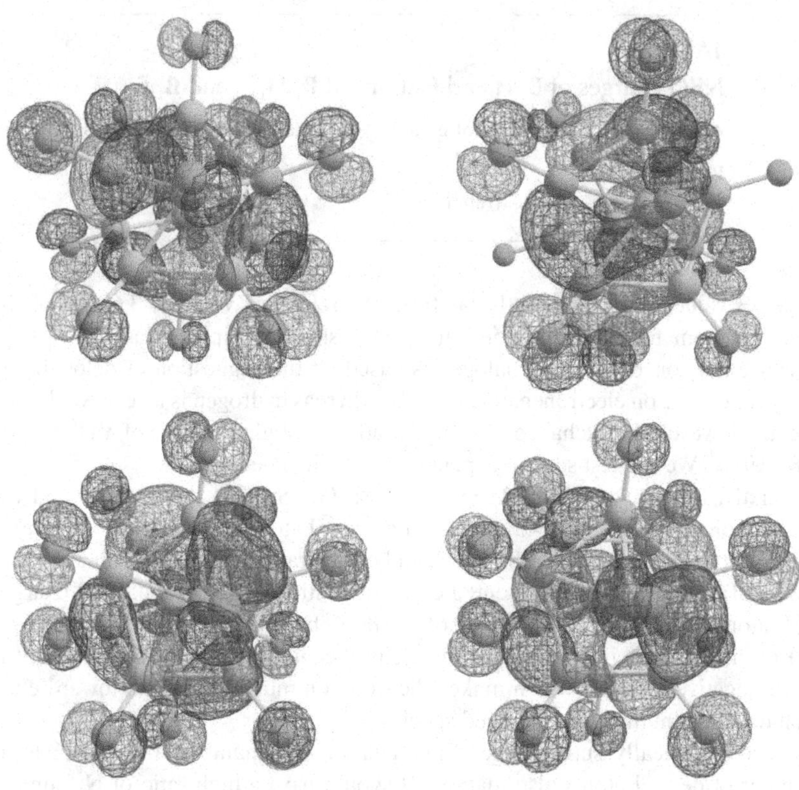

FIGURE 2.11 The quadruple degenerate HOMO orbitals of $B_{12}F_{12}^{2-}$.

stabilizing the di-anion against electron detachment, our study shows that polynuclear species can exhibit sufficient stability. We found that the general approaches used for the design of singly charged superhalogen systems are also effective for constructing doubly charged superhalogen systems. They include (i) the use of ligands with high IP (we used fluorine atoms as the most effective simple ligand), (ii) the nonbonding character of HOMO orbitals and (iii) employing as many ligands as possible. The design of polynuclear species emerged as a highly effective approach for creating doubly charged superhalogens in our research. We hope these results will motivate further investigation in the field of superhalogens, potentially leading to the development of even triply or quadruply charged superhalogens, thus enriching the diversity and application spectrum of superhalogen species.

ACKNOWLEDGMENTS

The support and resources from the Centre for High Performance Computing at the University of Utah are gratefully acknowledged.

REFERENCES

1. Gutsev, G. L.; Boldyrev, A. I. DVM-XQ Calculations on the Iionization Potentials of MX^{-}_{k+a} Complex Anions and the Electron Affinities of MX_{k+1} 'Superhalogens'. *Chem. Phys.*, 1981, 277–283. https://doi.org/10.1016/0301-0104(81)80150-4.
2. Hotop, H. Binding Energies in Atomic Negative Ions. *JPCRD*, 2009, *4* (3), 539. https://doi.org/10.1063/1.555524.
3. Karlsson, L.; Mattsson, L.; Jadrny, R.; Bergmark, T.; Siegbahn, K. Valence Electron Spectra of Benzene and the Hexafluorides of Sulphur, Molybdenum, Tungsten and Uranium. An Application of Multichannel Detector Technique to UV-Valence Electron Spectroscopy. 1976, *14*, 230. https://doi.org/10.1088/0031-8949/14/5/007.
4. Gutsev, G. L.; Boldyrev, A. I. The Electronic Structure of Superhalogens and Superalkalies. *Russ. Chem. Rev.*, 1987, *56* (6), 519–531. https://doi.org/10.1070/RC1987v056n06ABEH003287.
5. Berry, R. S. Small Free Negative Ions. *Chem. Rev.*, 1969, *69* (4), 533–542. https://doi.org/10.1021/cr60260a003.
6. Yang, J.; Wang, X.-B.; Xing, X.-P.; Wang, L.-S. Photoelectron Spectroscopy of Anions at 118.2 nm: Observation of High Electron Binding Energies in Superhalogens MCl_4^- (M=Sc, Y, La). *The Journal of Chemical Physics*, 2008, *128* (20), 201102. https://doi.org/10.1063/1.2938390.
7. Zhao, L.-J.; Xu, H.-G.; Feng, G.; Wang, P.; Xu, X.-L.; Zheng, W.-J. Superhalogen Properties of BS_2^- and BSO^-: Photoelectron Spectroscopy and Theoretical Calculations. *Phys. Chem. Chem. Phys.*, 2016, *18* (8), 6175–6181. https://doi.org/10.1039/C5CP07673K.
8. Sutton, P. P.; Mayer, J. E. A Direct Experimental Determination of Electron Affinities, the Electron Affinity of Iodine. *The Journal of Chemical Physics*, 1935, *3* (1), 20–28. https://doi.org/10.1063/1.1749548.
9. Farragher, A. L.; Page, F. M.; Wheeler, R. C. Electron Affinities of the Oxides of Nitrogen. *Discuss. Faraday Soc.*, 1964, *37*, 203–208. https://doi.org/10.1039/DF9643700203.
10. Milovanović, M. Small Lithium-Chloride Clusters: Superalkalis, Superhalogens, Supersalts and Nanocrystals. *Comput. Chem.*, 2021, *42* (26), 1895–1904. https://doi.org/10.1002/jcc.26722.
11. Berkowitz, J.; Chupka, W. A.; Gutman, D. Electron Affinities of O_2, O_3, NO, NO_2, NO_3 by Endothermic Charge Transfer. *Chem. Phys.*, 1971, *55* (6), 2733–2745. https://doi.org/10.1063/1.1676488.
12. Chupka, W. A.; Berkowitz, J.; Gutman, D. Electron Affinities of Halogen Diatomic Molecules as Determined by Endoergic Charge Transfer. *Chem. Phys.*, 1971, *55* (6), 2724–2733. https://doi.org/10.1063/1.1676487.
13. Sidorov, L. N.; Borshchevsky, A. Y.; Rudny, E. B.; Butsky, V. D. Electron Affinities of Higher Molybdenum Fluorides as Determined by the Effusion Technique. *Chem. Phys.*, 1982, *71* (1), 145–156. https://doi.org/10.1016/0301-0104(82)87014-6.
14. Joseph, J.; Pradhan, K.; Jena, P.; Wang, H.; Zhang, X.; Jae Ko, Y.; Bowen, K. H. Evolution of Superhalogen Properties in $PtCl_n$ Clusters. *J. Chem. Phys.*, 2012, *136* (19), 194305. https://doi.org/10.1063/1.4719089.
15. Ko, Y. J.; Wang, H.; Pradhan, K.; Koirala, P.; Kandalam, A. K.; Bowen, K. H.; Jena, P. Superhalogen Properties of Cu_mCl_n Clusters: Theory and Experiment. *J. Chem. Phys.*, 2011, *135* (24), 244312. https://doi.org/10.1063/1.3671457.
16. Weaver, A.; Metz, R. B.; Bradforth, S. E.; Neumark, D. M. Spectroscopy of the I + HI Transition-State Region by Photodetachment of IHI^-. *J. Phys. Chem.*, 1988, *92* (20), 5558–5560. https://doi.org/10.1021/j100331a004.

17. Chen, E. C. M.; Shuie, L.; Desai D'sa, E.; Batten, C. F.; Wentworth, W. E. The Negative Ion States of Sulfur Hexafluoride. *J. Chem. Phys.*, 1988, *88* (8), 4711–4719. https://doi.org/10.1063/1.454710.
18. Taylor, T. R.; Xu, C.; Neumark, D. M. Photoelectron Spectra of the $C_{2n}H^-$ (N=1–4) and $C_{2n}D^-$ (N=1–3) Anions. *J. Chem. Phys.*, 1998, *108* (24), 10018–10026. https://doi.org/10.1063/1.476462.
19. Chen, J.; Yang, H.; Wang, J.; Cheng, S.-B. Revealing the Effect of the Oriented External Electronic Field on the Superatom-Polymeric Zr_3O_3 Cluster: Superhalogen Modulation and Spectroscopic Characteristics. *Spectrochim. Acta Part A*, 2020, *237*, 118400. https://doi.org/10.1016/j.saa.2020.118400.
20. Lu, S.-J.; Wu, L.-S.; Lin, F. Structural, Bonding, and Superhalogen Properties of $Au_4X_4^{-/0}$ (X = F, Cl, Br, and I) Clusters. *Theor. Chem. Acc.*, 2019, *138* (4), 51. https://doi.org/10.1007/s00214-019-2442-1.
21. Feng, Y.; Xu, H.-G.; Zhang, Z.-G.; Gao, Z.; Zheng, W. Photoelectron Spectroscopy and Density Functional Calculations of Fe_nBO^{2-} Clusters. *J. Chem. Phys.*, 2010, *132* (7), 074308. https://doi.org/10.1063/1.3299290.
22. Ito, S.; Tasaka, Y.; Nakamura, K.; Fujiwara, Y.; Hirata, K.; Koyasu, K.; Tsukuda, T. Electron Affinities of Ligated Icosahedral M_{13} Superatoms Revisited by Gas-Phase Anion Photoelectron Spectroscopy. *J. Phys. Chem. Lett.*, 2022, *13* (22), 5049–5055. https://doi.org/10.1021/acs.jpclett.2c01284.
23. Korobov, M. V.; Kuznetsov, S. V.; Sidorov, L. N.; Shipachev, V. A.; Mit'kin, V. N. Gas-Phase Negative Ions of Platinum Metal Fluorides. II. Electron Affinity of Platinum Metal Hexafluorides. *Int. J. Mass Spectrom. Ion Process.*, 1989, *87* (1), 13–27. https://doi.org/10.1016/0168-1176(89)80002-3.
24. Wang, X.-B.; Ding, C.-F.; Wang, L.-S.; Boldyrev, A. I.; Simons, J. First Experimental Photoelectron Spectra of Superhalogens and Their Theoretical Interpretations. *J. Chem. Phys.*, 1999, *110* (10), 4763–4771. https://doi.org/10.1063/1.478386.
25. Anusiewicz, I.; Skurski, P. An ab Initio Study on BeX_3^- Superhalogen Anions (X = F, Cl, Br). *Chem. Phys. Lett.*, 2002, 426–434. https://doi.org/10.1016/S0009-2614(02)00666-8.
26. Sikorska, C.; Smuczyńska, S.; Skurski, P.; Anusiewicz, I. BX_4^- and AlX_4^- Superhalogen Anions (X = F, Cl, Br): An ab Initio Study. *Inorg. Chem.*, 2008, *47* (16), 7348–7354. https://doi.org/10.1021/ic800863z.
27. Marchaj, M.; Freza, S.; Skurski, P. Why Are SiX $_5$ – and GeX $_5$ – (X = F, Cl) Stable but Not CF_5– and CCl_5– ? *J. Phys. Chem. A*, 2012, *116* (8), 1966–1973. https://doi.org/10.1021/jp300251t.
28. Hay, P. J. The Relative Energies of SF_{6-} and SF_6 as a Function of Geometry. *J. Chern. Phys.*, 1982, *76*, 502–504. https://doi.org/10.1063/1.442696.
29. Srivastava, A. K.; Pandey, S. K.; Misra, N. Superhalogen Properties of ReF_n (n≥ 6) Species. *Chem. Phys. Lett.*, 2015, *624*, 15–18. https://doi.org/10.1016/j.cplett.2015.01.056.
30. Paduani, C. DFT Study of Gadolinium Aluminohydrides and Aluminofluorides. *Chem. Phys.*, 2013, *417*, 1–7. https://doi.org/10.1016/j.chemphys.2013.03.011.
31. Paduani, C.; Jena, P. Super and Hyperhalogen Behavior in MgX_n and GdX_n (X = F, BF_4) Clusters. *J. Nanopart. Res.*, 2012, *14* (9), 1035. https://doi.org/10.1007/s11051-012-1035-x.
32. Koirala, P.; Pradhan, K.; Kandalam, A. K.; Jena, P. Electronic and Magnetic Properties of Manganese and Iron Atoms Decorated with BO_2 Superhalogens. *J. Phys. Chem. A*, 2013, *117* (6), 1310–1318. https://doi.org/10.1021/jp307467j.
33. Sobczyk, M.; Sawicka, A.; Skurski, P. Theoretical Search for Anions Possessing Large Electron Binding Energies. *Eur. J. Inorg. Chem.*, 2003, *2003* (20), 3790–3797. https://doi.org/10.1002/ejic.200300180.

34. Ding, L.-P.; Shao, P.; Lu, C.; Zhang, F.-H.; Liu, Y.; Mu, Q. Prediction of the Iron-Based Polynuclear Magnetic Superhalogens with Pseudohalogen CN as Ligands. *Inorg. Chem.*, 2017, *56* (14), 7928–7935. https://doi.org/10.1021/acs.inorgchem.7b00646.
35. Czapla, M.; Skurski, P. Strength of the Lewis–Brønsted Superacids Containing In, Sn, and Sb and the Electron Binding Energies of Their Corresponding Superhalogen Anions. *J. Phys. Chem. A*, 2015, *119* (51), 12868–12875. https://doi.org/10.1021/acs.jpca.5b10205.
36. Freza, S.; Skurski, P. Enormously Large (Approaching 14 eV!) Electron Binding Energies of $[H_nF_{n+1}]$- (n= 1–5, 7, 9, 12) Anions. *Chem. Phys. Lett.*, 2010, *487* (1–3), 19–23. https://doi.org/10.1016/j.cplett.2010.01.022.
37. Gao, S.-J.; Guo, J.-C.; Zhai, H.-J. Boron Oxide $B_5O_6^-$ Cluster as a Boronyl-Based Inorganic Analog of Phenolate Anion. *Front. Chem.*, 2022, *10*, 868782. https://doi.org/10.3389/fchem.2022.868782.
38. Wudl, F. From Organic Metals to Superconductors: Managing Conduction Electrons in Organic Solids. *Acc. Chem. Res.*, 1984, *17* (6), 227–232. https://doi.org/10.1021/ar00102a005.
39. Williams, J. M.; Beno, M. A.; Wang, H. H.; Leung, P. C. W.; Emge, T. J.; Geiser, U.; Carlson, K. D. Organic Superconductors: Structural Aspects and Design of New Materials. *Acc. Chem. Res.*, 1985, *18* (9), 261–267. https://doi.org/10.1021/ar00117a001.
40. Srivastava, A. K.; Kumar, A.; Tiwari, S. N.; Misra, N. Application of Superhalogens in the Design of Organic Superconductors. *New J. Chem.*, 2017, *41* (24), 14847–14850. https://doi.org/10.1039/C7NJ02868G.
41. Sladkey, F. O.; Bulliner, P. A.; Bartlett, N.; DeBoer, B. G.; Zalkin, A. Xenon Difluoride as a Fluoride Ion Donor and the Crystal Structure of $[Xe_2F_3]^+[AsF_6]$–. *Chem. Commun. (London)*, 1968, 1048–1049. https://doi.org/10.1039/C19680001048.
42. Christe, K. O.; Wilson, W. W. Synthesis and Properties of ClF6BF4. *Inorg. Chem.*, 1983, *22* (13), 1950–1951. https://doi.org/10.1021/ic00155a024.
43. Leary, K.; Zalkin, A.; Bartlett, N. Crystal Structure of $[Xe_2F_{11}]^+[AuF_6]^-$. *J. Chem. Soc., Chem. Commun.*, 1973, 131–132. https://doi.org/10.1039/C39730000131.
44. Bougon, R.; Wilson, W. W.; Christe, K. O. Synthesis and Characterization of Tetrafluoroammonium Hexafluorochromate and Reaction Chemistry of Chromium Pentafluoride. *Inorg. Chem.*, 1985, *24* (14), 2286–2292. https://doi.org/10.1021/ic00208a032.
45. Christe, K. O. Bartlett's Discovery of Noble Gas Fluorides, a Milestone in Chemical History. *Chem. Commun.*, 2013, *49* (41), 4588. https://doi.org/10.1039/c3cc41387j.
46. Zhang, H.; Xu, S.; Guo, T.; Du, D.; Tao, Y.; Zhang, L.; Liu, G.; Chen, X.; Ye, J.; Guo, Z.; et al. Dual Effect of Superhalogen Ionic Liquids Ensures Efficient Carrier Transport for Highly Efficient and Stable Perovskite Solar Cells. *ACS Appl. Mater. Interfaces*, 2022, *14* (25), 28826–28833. https://doi.org/10.1021/acsami.2c04993.
47. Srivastava, A. K. Prediction of Novel Liquid Crystalline Molecule Based on BO_2 Superhalogen. *J. Mol. Liq.*, 2021, *344*, 117968. https://doi.org/10.1016/j.molliq.2021.117968.
48. Srivastava, A. K.; Kumar, A.; Misra, N. Superhalogens as Building Blocks of Ionic Liquids. *J. Phys. Chem. A*, 2021, *125* (10), 2146–2153. https://doi.org/10.1021/acs.jpca.1c00599.
49. Kulsha, A. V.; Sharapa, D. I. Superhalogen and Superacid. *J. Comput. Chem.*, 2019, *40* (26), 2293–2300. https://doi.org/10.1002/jcc.26007.
50. Srivastava, A. K.; Kumar, A.; Misra, N. A Path to Design Stronger Superacids by Using Superhalogens. *J. Fluorine Chem.*, 2017, *197*, 59–62. https://doi.org/10.1016/j.jfluchem.2017.03.001.

51. Srivastava, A. K.; Misra, N. Hydrogenated Superhalogens Behave as Superacids. *Polyhedron*, 2015, *102*, 711–714. https://doi.org/10.1016/j.poly.2015.09.072.
52. Srivastava, A. K.; Kumar, A.; Misra, N. Superhalogens as Building Blocks of a New Series of Superacids. *New J. Chem.*, 2017, *41* (13), 5445–5449. https://doi.org/10.1039/C7NJ00129K.
53. Zhou, F.-Q.; Zhao, R.-F.; Li, J.-F.; Xu, W.-H.; Li, C.-C.; Luo, L.; Li, J.-L.; Yin, B. Constructing Organic Superacids from Superhalogens Is a Rational Route as Verified by DFT Calculations. *Phys. Chem. Chem. Phys.*, 2019, *21* (5), 2804–2815. https://doi.org/10.1039/C8CP07313A.
54. Reddy, G. N.; Parida, R.; Jena, P.; Jana, M.; Giri, S. Superhalogens as Building Blocks of Super Lewis Acids. *ChemPhysChem*, 2019, *20* (12), 1607–1612. https://doi.org/10.1002/cphc.201900267.
55. Saunders, M. Stochastic Search for Isomers on a Quantum Mechanical Surface. *Comput. Chem.*, 2004, *25* (5), 621–626. https://doi.org/10.1002/jcc.10407.
56. Coalescence Modification to Saunders KICK Version. https://kick.science/KICK_Saunders_coalescence.html (accessed on 24 July 2023).
57. Yu, H. S.; He, X.; Li, S. L.; Truhlar, D. G. MN15: A Kohn–Sham Global-Hybrid Exchange – Correlation Density Functional with Broad Accuracy for Multi-Reference and Single-Reference Systems and Noncovalent Interactions. *Chem. Sci.*, 2016, *7* (8), 5032–5051. https://doi.org/10.1039/C6SC00705H.
58. Dunning, T. H.; Hay, P. J. *Modern Theoretical Chemistry*, Ed. H. F. Schaefer III; Plenum: New York, 1977; Vol. 3.
59. Wadt, W. R.; Hay, J. Ab Initio Effective Core Potentials for Molecular Calculations. Potentials for Main Group Elements Na to Bi. *J. Chem. Phys.*, 1985, *82*, 284–298. https://doi.org/10.1063/1.448800.
60. Frisch, M. J.; Trucks, G. W.; Schledel, H. B.; Scuseria, G. E.; Robb, M. A.; Cheeseman, J. R.; Scalmani, G.; Barone, V.; et al. *Gaussian 16*. https://gaussian.com/citation/.
61. Weigend, F.; Ahlrichs, R. Balanced Basis Sets of Split Valence, Triple Zeta Valence and Quadruple Zeta Valence Quality for H to Rn: Design and Assessment of Accuracy. *Phys. Chem. Chem. Phys.*, 2005, *7* (18), 3297. https://doi.org/10.1039/b508541a.
62. Rappoport, D.; Furche, F. Property-Optimized Gaussian Basis Sets for Molecular Response Calculations. *The Journal of Chemical Physics*, 2010, *133* (13), 134105. https://doi.org/10.1063/1.3484283.
63. Martin, J. M. L.; Sundermann, A. Correlation Consistent Valence Basis Sets for Use with the Stuttgart–Dresden–Bonn Relativistic Effective Core Potentials: The Atoms Ga–Kr and In–Xe. *The Journal of Chemical Physics*, 2001, *114* (8), 3408–3420. https://doi.org/10.1063/1.1337864.
64. Tao, T.; Perdew, J. P.; Staroverov, V. N.; Scuseria, G. E. Climbing the Density Functional Ladder: Nonempirical Meta – Generalized Gradient Approximation Designed for Molecules and Solids. *Phys. Rev. Lett*, 2003, *91*, 146401. https://doi.org/10.1103/PhysRevLett.91.146401.
65. Dunning, T. H. Gaussian Basis Sets for Use in Correlated Molecular Calculations. I. The Atoms Boron Through Neon and Hydrogen. *J. Chem. Phys.*, 1989, *90* (2), 1007–1023. https://doi.org/10.1063/1.456153.
66. Woon, D. E.; Dunning, T. H. Gaussian Basis Sets for Use in Correlated Molecular Calculations. III. The Atoms Aluminum through Argon. *J. Chem. Phys.*, 1993, *98* (2), 1358–1371. https://doi.org/10.1063/1.464303.
67. Metz, B.; Stoll, H.; Dolg, M. Small-Core Multiconfiguration-Dirac–Hartree–Fock-Adjusted Pseudopotentials for Post- ***d*** Main Group Elements: Application to PbH and PbO. *J. Chem. Phys.*, 2000, *113* (7), 2563–2569. https://doi.org/10.1063/1.1305880.

68. Peterson, K. A. Systematically Convergent Basis Sets with Relativistic Pseudopotentials. I. Correlation Consistent Basis Sets for the Post- *d* Group 13–15 Elements. *J. Chem. Phys.*, 2003, *119* (21), 11099–11112. https://doi.org/10.1063/1.1622923.
69. Aprà, E.; Warneke, J.; Xantheas, S. S.; Wang, X.-B. A Benchmark Photoelectron Spectroscopic and Theoretical Study of the Electronic Stability of $[B_{12}H_{12}]^{2-}$. *J. Chem. Phys.*, 2019, *150* (16), 164306. https://doi.org/10.1063/1.5089510.
70. Jansen, G.; Hess, B. A. Revision of the Douglas-Kroll Transformation. *Phys. Rev. A*, 1989, *39* (11), 6016–6017. https://doi.org/10.1103/PhysRevA.39.6016.
71. Hess, B. A. Relativistic Electronic-Structure Calculations Employing a Two-Component No-Pair Formalism with External-Field Projection Operators. *Phys. Rev. A*, 1986, *33* (6), 3742–3748. https://doi.org/10.1103/PhysRevA.33.3742.
72. Pritchard, B. P.; Altarawy, D.; Didier, B.; Gibson, T. D.; Windus, T. L. New Basis Set Exchange: An Open, Up-to-Date Resource for the Molecular Sciences Community. *J. Chem. Inf. Model.*, 2019, *59* (11), 4814–4820. https://doi.org/10.1021/acs.jcim.9b00725.
73. Feller, D. The Role of Databases in Support of Computational Chemistry Calculations. *J. Comput. Chem.*, 1996, *17* (13), 1571–1586. https://doi.org/10.1002/(SICI)1096-987X(199610)17:13<1571::AID-JCC9>3.0.CO;2-P.
74. Schuchardt, K. L.; Didier, B. T.; Elsethagen, T.; Sun, L.; Gurumoorthi, V.; Chase, J.; Li, J.; Windus, T. L. Basis Set Exchange: A Community Database for Computational Sciences. *J. Chem. Inf. Model.*, 2007, *47* (3), 1045–1052. https://doi.org/10.1021/ci600510j.
75. Neese, F. The ORCA Program System. *Wiley Interdiscip. Rev. Comput. Mol. Sci*, 2012, *2* (1), 73–78. https://doi.org/10.1002/wcms.81.
76. Neese, F. Software Update: The ORCA Program System – Version 5.0. *WIREs Comput. Mol. Sci.*, 2022, *12* (5). https://doi.org/10.1002/wcms.1606.
77. Bukovsky, E. V.; Peryshkov, D. V.; Wu, H.; Zhou, W.; Tang, W. S.; Jones, W. M.; Stavila, V.; Udovic, T. J.; Strauss, S. H. Comparison of the Coordination of $B_{12}F_{12}^{2-}$, $B_{12}Cl_{12}^{2-}$, and $B_{12}H_{12}^{2-}$ to Na^+ in the Solid State: Crystal Structures and Thermal Behavior of $Na_2(B_{12}F_{12})$, $Na_2(H_2O)_4$ $(B_{12}F_{12})$, $Na_2(B_{12}$ $Cl_{12})$, and $Na_2(H_2O)_6(B_{12}Cl_{12})$. *Inorg. Chem.*, 2017, *56* (8), 4369–4379. https://doi.org/10.1021/acs.inorgchem.6b02920.
78. Ortiz, J. V. Electron Binding Energies of Anionic Alkali Metal Atoms from Partial Fourth Order Electron Propagator Theory Calculations. *The Journal of Chemical Physics*, 1988, *89* (10), 6348–6352. https://doi.org/10.1063/1.455401.
79. Cederbaum, L. S. One-Body Green's Function for Atoms and Molecules: Theory and Application. *J. Phys. B: At. Mol. Phys.*, 1975, *8* (2), 290–303. https://doi.org/10.1088/0022-3700/8/2/018.
80. Gutsev, G. L. Theoretical Investigation on the Existence of the SiF^-_6 Anion. *Chem. Phys. Lett.*, 1991, *184* (4), 305–309. https://doi.org/10.1016/0009-2614(91)85128-J.
81. Hamilton, W. C. Bond Distances and Thermal Motion in Ferrous Fluosilicate Hexahydrate: A Neutron Diffraction Study. *Acta. Cryst.*, 1962, *15* (4), 353–360. https://doi.org/10.1107/S0365110X62000870.
82. Martin, W. C.; Zalubas, R. Energy Levels of Silicon, Si I Through Si XIV. *J. Phys. Chem. Ref. Data*, 1983, *12*, 323. https://doi.org/10.1063/1.555685.
83. Kessler, T.; Bruck, K.; Baktash, C.; Beene, J. R.; Geppert, Ch.; Havener, C. C.; Krause, H. F.; Liu, Y.; Schultz, D. R.; Stracener, D. W.; et al. Three-Step Resonant Photoionization Spectroscopy of Ni and Ge: Ionization Potential and Odd-Parity Rydberg Levels. *JPCRD*, 2007, *40*, 4413. https://doi.org/10.1088/0953-4075/40/23/002.
84. Brown, C. M.; Tilford, S. G.; Giner, M. L. Absorption Spectrum of Sn_1 between 1580 and 2040 Å. *J. Opt. Soc. Am.*, *67*, 1977. https://doi.org/10.1364/JOSA.67.000607.

85. Dembczyński, J.; Stachowska, E.; Wilson, M.; Buch, R.; Ertmer, W. Measurement and Interpretation of the Odd-Parity Levels of Pb I. *Phys. Rev. A*, 1994, 745–754. https://doi.org/10.1103/PhysRevA.49.745.
86. Goreshnik, E. A.; Slyvka, Yu. I.; Mys'kiv, M. G. The First Example of a Direct Cu+– Bond. Synthesis and Crystal Structure of Two Closely Related Copper(I) Hexafluorosilicate π-Complexes with 1-Allylbenzotriazole of $[Cu_2(C_6H_4N_3(C_3H_5))_2(H_2O)$-$2SiF_6]\cdot 2H_2O$ and $[Cu_2(C_6H_4N_3(C_3H_5))_2(CH_3OH)_2(H_2O)_2]SiF_6$ Composition. *Inorg. Chim. Acta*, 2011, *377* (1), 177–180. https://doi.org/10.1016/j.ica.2011.08.008.
87. Xu, Y. K.; Adachi, S. Properties of Na2SiF6:Mn^{4+} and Na_2GeF_6:Mn^{4+} Red Phosphors Synthesized by Wet Chemical Etching. *J. Appl. Phys.*, 2009, *105* (1), 013525. https://doi.org/10.1063/1.3056375.
88. Gabuda, S. P.; Kavun, V. Y.; Kozlova, S. G.; Tverskikh, V. V. Structure of Hexafluorostannate Ion $[SnF_6]^{2-}$: The ^{119}Sn and ^{19}F NMR MAS-Spectroscopic Studies and ab Initio Calculations. *Russ. J. Coord*, 2003, *29* (1). https://doi.org/10.1023/A:1021826513857.
89. Bandemehr, J.; Baumann, D.; Seibald, M.; Kraus, F. Alkali Metal Hexafluorido Plumbates(IV) $A_2[PbF_6]$ (A =Na–Cs) and Luminescence of the Mn $^{4+}$-Substituted Compounds A_2 $[PbF_6]$: Mn (A =Li–Cs) and $Li_2[MF_6]$: Mn (M =Ti, Ge, Sn). *Eur. J. Inorg. Chem.*, 2021, *2021* (37), 3870–3877. https://doi.org/10.1002/ejic.202100577.
90. Boldyrev, A. I.; Simons, J. Is TeF_8^{2-} the MX_n^{2-} Dianion with the Largest Electron Detachment Energy (5 EV). *J. Chem. Phys.*, 1992, *97* (4), 2826–2827. https://doi.org/10.1063/1.463025.
91. Christe, K. O.; Sanders, J. C. P.; Schrobilgen, G. J.; Wilson, W. W. High-Coordination Number Fluoro- and Oxofluoro-Anions; IF_6O^-, TeF_6O^{2-}, TeF_7^-, IF_8^- and TeF_8^{2-}. *J. Chem. Soc., Chem. Commun.*, 1991, 837–840. https://doi.org/10.1039/C39910000837.
92. Vasiliu, M.; Peterson, K. A.; Christe, K. O.; Dixon, D. A. Electronic Structure Predictions of the Energetic Properties of Tellurium Fluorides. *Inorg. Chem.*, 2019, *58* (13), 8279–8292. https://doi.org/10.1021/acs.inorgchem.8b03235.
93. Kieck, T.; Liu, Y.; Stracener, D. W.; Li, R.; Lassen, J.; Wendt, K. D. A. Resonance Laser Ionization Spectroscopy of Tellurium. *Spectrochim. Acta Part B: At. Spectrosc.*, 2019, *159*, 105645. https://doi.org/10.1016/j.sab.2019.105645.
94. Anusiewicz, I. Mg_2Cl_{5-} and $Mg_3Cl_7^-$ Superhalogen Anions. *Aust. J. Chem.*, 2008, *61* (9), 712. https://doi.org/10.1071/CH08212.
95. Kubel, F. The Crystal Structures of $SrAlF_5$ and $Ba_{0.43(1)}Sr_{0.57(1)}AlF_5$. *Z. Anorg. Allg. Chem.*, 1998, *624* (9), 1481–1486. https://doi.org/10.1002/(SICI)1521-3749(199809)624:9<1481::AID-ZAAC1481>3.0.CO;2-F.
96. Frazier, A. W.; Waerstad, K. R. The Phase System Alumina-Phosphorus Pentoxide-Fluorine-Water at 25 Degree C. *Ind. Eng. Chem. Res.*, 1993, *32* (8), 1760–1766. https://doi.org/10.1021/ie00020a033.
97. Kaufman, V. Wavelengths and Energy Level Classifications for the Spectra of Aluminum (All Through AI XIII). *JPCRD*, 2009, *20*, 775. https://doi.org/10.1063/1.555895.
98. Osorio, E.; Olson, J. K.; Tiznado, W.; Boldyrev, A. I. Analysis of Why Boron Avoids Sp^2 Hybridization and Classical Structures in the B_nH_{n+2} Series. *Chem. Eur. J.*, 2012, *18* (31), 9677–9681. https://doi.org/10.1002/chem.201200506.
99. Filippov, O. A.; Belkova, N. V.; Epstein, L. M.; Shubina, E. S. Chemistry of Boron Hydrides Orchestrated by Dihydrogen Bonds. *J. Organomet. Chem.*, 2013, *747*, 30–42. https://doi.org/10.1016/j.jorganchem.2013.04.025.
100. Moussa, G.; Moury, R.; Demirci, U. B.; Şener, T.; Miele, P. Boron-Based Hydrides for Chemical Hydrogen Storage: Boron-Based Hydrolytic and Thermolytic Hydrides. *Int. J. Energy Res.*, 2013, *37* (8), 825–842. https://doi.org/10.1002/er.3027.

101. Li, H.-W.; Miwa, K.; Ohba, N.; Fujita, T.; Sato, T.; Yan, Y.; Towata, S.; Chen, M. W.; Orimo, S. Formation of an Intermediate Compound with a $B_{12}H_{12}$ Cluster: Experimental and Theoretical Studies on Magnesium Borohydride $Mg(BH_4)_2$. *Nanotechnology*, 2009, *20* (20), 204013. https://doi.org/10.1088/0957-4484/20/20/204013.
102. Druzina, A. A.; Bregadze, V. I.; Mironov, A. F.; Semioshkin, A. A. Synthesis of Conjugates of Polyhedral Boron Hydrides with Nucleosides. *Russ. Chem. Rev.*, 2016, *85* (11), 1229–1254. https://doi.org/10.1070/RCR4644.
103. Grimes, R. N. Synthesis and Serendipity in Boron Chemistry: A 50 Year Perspective. *Journal of Organometallic Chemistry*, 2013, *747*, 4–15. https://doi.org/10.1016/j.jorganchem.2013.04.018.
104. Longuet-Higgins, H. C.; Roberts, M. D. V. The Electronic Structure of an Icosahedron of Boron Atoms. *Proc. R Soc. Lond. A.*, 1955, *230* (1180), 110–119. https://doi.org/10.1098/rspa.1955.0115.
105. Pitochelli, A. R.; Hawthorne, F. M. The Isolation of the Icosahedral $B_{12}H_{12-}^{2}$ Ion. *J. Am. Chem. Soc.*, 1960, *82*, 3228–3229. https://doi.org/10.1021/ja01497a069.
106. Sivaev, I. B.; Bregadze, V. I.; Sjöberg, S. Chemistry of Closo-Dodecaborate Anion $[B_{12}H_{12}]^{2-}$: A Review. *Collect. Czech. Chem. Commun.*, 2002, *67* (6), 679–727. https://doi.org/10.1135/cccc20020679.
107. Bhattacharyya, P.; Boustani, I.; Shukla, A. Why Does $B_{12}H_{12}^-$ Icosahedron Need Two Electrons to Be Stable: A First-Principles Electron-Correlated Investigation of $B_{12}H_n$ (n = 6, 12) Clusters. *J. Phys. Chem. A*, 2021, *125* (51), 10734–10741. https://doi.org/10.1021/acs.jpca.1c09167.
108. Ivanov, S. V.; Miller, S. M.; Anderson, O. P.; Solntsev, K. A.; Strauss, S. H. Synthesis and Stability of Reactive Salts of Dodecafluoro- *c Loso* -Dodecaborate(2–). *J. Am. Chem. Soc.*, 2003, *125* (16), 4694–4695. https://doi.org/10.1021/ja0296374.
109. Malischewski, M.; Bukovsky, E. V.; Strauss, S. H.; Seppelt, K. Jahn–Teller Effect in the $B_{12}F_{12}$ Radical Anion and Energetic Preference of an Octahedral $B_6(BF_2)_6$ Cluster Structure Over an Icosahedral Structure for the Elusive Neutral $B_{12}F_{12}$. *Inorg. Chem.*, 2015, *54* (23), 11563–11566. https://doi.org/10.1021/acs.inorgchem.5b02256.

3 Superhalogen Properties of Some Transition Metal Oxides

Abhishek Kumar, Ambrish Kumar Srivastava and Neeraj Misra

3.1 INTRODUCTION

The elements of the seventh group (halogen) in the periodic table have the highest electron affinity. In 1981, Gutsev and Boldyrev [1] predicted a molecular species with higher electron affinities (EAs) than halogen, called "superhalogen", and Wang et al. [2] experimentally confirmed it in 1999. The general formula for superhalogen is MX_{k+1}, where M is the main group or transition metal atom, X is a halogen atom, and k is the valency of M. Therefore, a typical superhalogen consists of a metal atom that may be surrounded peripherally by oxygen, halogen, pseudohalogen, *etc.*, such that the last electron is delocalized over peripherally attached atoms, and hence, electron affinity increases [3–7]. Such hypervalent species possess extraordinary oxidizing capability. As noticed by Bartlett and Lohmann during 1960s (much before the conceptualization of superhalogens) that PtF_6 can oxidize both O_2 molecule and Xe atom having very high ionisation potentials of about 12 eV. This is attributed to an enormous EA of PtF_6 which is 6.8 eV [8]. Not only PtF_6, a number of other transition metal fluorides have been investigated and their EAs are found to be higher than halogens as discussed elsewhere in the book. Even more interesting is the case of transition metal oxides. For instance, the EA of MnO_4 was found to be 5 eV that had also been verified experimentally [9]. Thus, MnO_4 belongs to the class of superhalogen and $KMnO_4$ is a widely known oxidizing agent. FeO_4 and CrO_4 also behave as superhalogen due to their higher EAs, 3.8 eV and 4.96 eV, respectively [10–12]. Not only as strong oxidizing agents, superhalogens find various applications in different fields such as in superacids [13–15], lithium ion batteries [16], hydrogen storage [17], organic superconductors [18], ionic liquids [19], *etc.* This led to the exploration of new superhalogens even today [20–23]. A recent progress on the design and application of superhalogens has been reported by Srivastava [24].

Transition metals possess variable oxidation states due to presence of *d*-orbital electrons. In other words, they can bind with a number of oxygen atoms due to

 DOI: 10.1201/9781003384205-3

participation of their d electrons in the bonding. To explore further the extent to which these metals can bind with O atoms, we have performed a number of studies on various transition metal elements [25–31]. In this chapter, we highlight the bonding of Ni, Pd, Pt, Re, Ru, Co and V with successive O atoms. We focus on the EAs of resulting MO_n species (M = Ni, Pd, Pt Re, Ru, Co and V; $n = 1-5$) and discuss their superhalogen properties.

3.2 COMPUTATIONAL NOTE

All computational results discussed in this chapter are obtained using density functional theory [32] as implemented in Gaussian 09 [33] program. The functional incorporating generalized gradient approximation for exchange and correlation potentials, B3LYP [34,35] in conjunction with the SDD basis set were used throughout these calculations. Various possible initial geometries were modelled with the help of GaussView 5.0 package [36] and optimized without any symmetry constraint in the potential energy surface. The vibrational frequency calculations were also performed at the same level of theory to ensure that the optimized geometries belong to at least some local minima. The calculations are repeated for higher spin states to recognize the ground state spin multiplicities.

The stability of transition metal oxides clusters has been analysed by calculating dissociation energy (D_e) against O atom and O_2 molecule:

$$D_e\left(MO_n^- \rightarrow MO_{n-1}^- + O\right) = E\left[MO_{n-1}^-\right] + E[O] - E\left[MO_n^-\right]$$

$$D_e\left(MO_n^- \rightarrow MO_{n-2}^- + O_2\right) = E\left[MO_{n-2}^-\right] + E[O_2] - E\left[MO_n^-\right]$$

where E [. . .] represents the total energy of respective species including zero-point energy correction. The adiabatic EA of MO_n species is calculated by the difference of energies between neutral MO_n and their anions both in their ground state configurations.

3.3 RESULT AND DISCUSSION

3.3.1 Nickel Oxides (NiO_n)

The outer electronic configuration of Nickel (Ni) is $3d^84s^2$. The minimum energy conformers of NiO_n (n = 1–5) neutrals and anions are shown in Figure 3.1. From Figure 3.1, it's clear that Ni binds atomically up to three O atoms in both neutral as well as anions. For $n = 2$, a bent NiO_2 triplet structure with an angle of 130° is minimum unlike linear structure suggested by literature [25]. The anionic structure of NiO_2 is found to be linear and the NiO_3 structure for both neutral and anion a trigonal planar structure. The bond-lengths Ni–O of neutral NiO_n lie in the range 1.63–1.77 Å that is less than those in their anionic structures up to $n = 2$ but tend to become greater beyond it. For n >3, NiO_n species exist in the form of $(NiO_{n-2})O_2$ complexes in which two O atoms interact molecularly to central Ni but

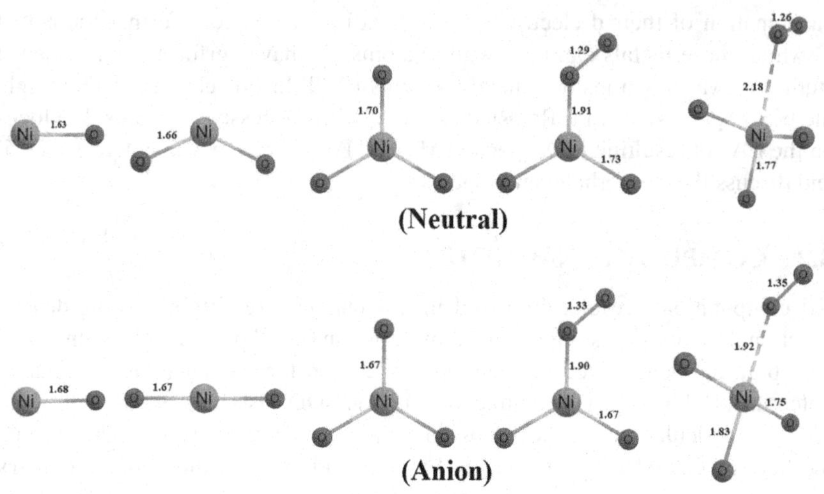

FIGURE 3.1 Optimised geometries of NiO_n ($n = 1–5$) with bond lengths (in Å) from Ref. [25].

remaining O bind atomically. In case of $(NiO_{n-2})O_2$ complexes, the distance n of O_2 moiety from Ni atom in neutral is greater than that in corresponding anions. In anionic complexes, molecule O_2 binds strongly, which makes anions more stable than neutrals.

The spin multiplicity (m) and partial NPA charges on Ni atom (q) of neutral and anionic NiO_n clusters are shown in Table 3.1. All neutral structures prefer higher spin states and same is true up to $n = 3$ in case of anions. For $n > 3$, NiO_n anionic clusters favour lower spin multiplicities. The corresponding dissociation energies (D_e) are also listed in Table 3.1. The positive D_e values imply that all NiO_n anions are stable against both dissociation channels. NiO_2^- possesses the highest dissociation energy against both channels among anions. It may indicate that NiO_2^- is the most stable anion and also supported by the fact that the dissociation of NiO_n^- for $n = 4$ favours $NiO_2^- + O_2$ against $NiO_3^- + O$ path by 2.91 eV. It is further established that NiO_4^- is more stable as compared to NiO_3^- for dissociation to O atom; thus, the lower multiplicity of NiO_4^- provides a little more stability as compared to NiO_3^-. For $n = 5$, NiO_n^- becomes $(NiO_3^-)O_2$ complex in which O_2 binds very weakly to Ni. Moreover, it is evident by its preferred dissociation to O_2 molecule by 2.63 eV.

Our systematic investigation reveals that Ni can bind successively up to three O atoms. For $n > 3$, NiO_n species exist in the form of complexes, for instance, as $(NiO_2)O_2$ when $n = 4$. Thus, Ni can expand its oxidation state to +6 at least in case of bonding with O atoms. The structures of NiO_n species are given in Figure 3.1 along with their bond lengths. The calculated EA of NiO, 1.47 eV is in perfect agreement with the anion photoelectron spectroscopic measurement [37]. The EAs of NiO_2 and NiO_3 are found to be larger than that of fluorine, which is 3.6 eV. Therefore, NiO_2 and NiO_3 belong to the class of superhalogens.

TABLE 3.1
Preferred spin multiplicity (*m*), electron affinity (EA) in eV and partial NPA charges on Ni atom (*Q*) of neutral and anionic NiO_n clusters for *n* = 1–5. Dissociation energies (D_e) of NiO_n anions are also given in eV from Ref. [25].

	NiO_n			NiO^-_n		$D_e(NiO^-_n)$	
n	*M*	*Q*	EA	*m*	*Q*	NiO^-_{n-1} + O	NiO^-_{n-2} + O_2
1	3	0.632	1.45	4	-0.103	8.25	—
2	3	0.854	3.90	4	0.508	7.46	7.75
3	3	0.796	4.45	4	0.643	5.05	4.54
4	3	0.821	4.25	2	0.675	5.33	2.42
5	3	0.840	4.20	2	0.751	5.39	1.76

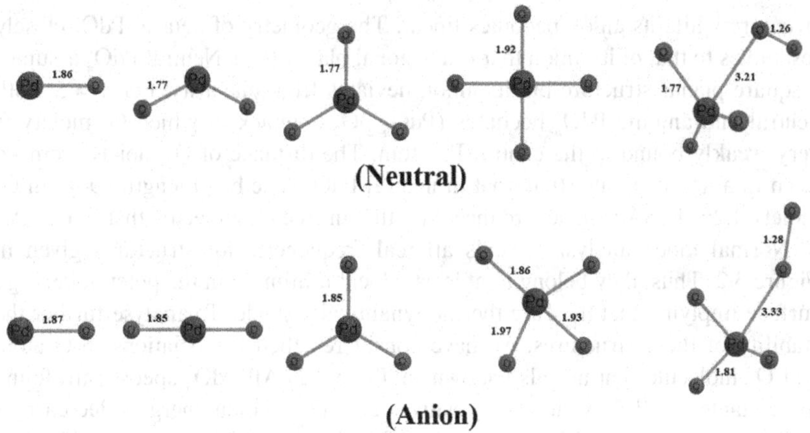

FIGURE 3.2 Optimised geometries of PdO_n ($n = 1$–5) with bond lengths (in Å) from Ref. [25,26].

3.3.2 Palladium Oxides (PdO_n)

Pd belongs to the same group as that of Ni with outer electronic configuration $5s^24d^8$. The optimized (ground state) structures of PdO_n (n = 1–5) neutral and anionic clusters are shown in Figure 3.2. The spin multiplicity, electron affinity and partial NPA charges on Pd atom of neutral and anionic PdO_n clusters are shown in Table 3.2. The bond length between Pd and O in neutral PdO_n cluster is smaller than that of corresponding anions. All PdO_n clusters energetically favour higher spin states except for n = 1 and 5 in anionic species. For n = 1 and 5, the higher spin states of PdO_2^- species are 0.49 and 0.42 eV higher in energy than corresponding lower spin states, respectively. Neutral PdO_2 takes a bent (C_{2v})

TABLE 3.2
Preferred spin multiplicity (*m*), electron affinity (EA) in eV and partial NPA charges on Pd atom (*Q*) of neutral and anionic PdO_n clusters for $n = 1–5$. Dissociation energies (D_e) of PdO_n anions are also given in eV from Ref. [25].

	PdO_n			PdO^-_n		$D_e(PdO^-_n)$	
n	*m*	*Q*	EA	*m*	*Q*	PdO^-_{n-1} + O	PdO^-_{n-2} + O_2
1	3	0.406	1.70	2	-0.329	6.07	—
2	3	0.915	3.60	4	0.524	6.63	4.73
3	3	0.947	3.75	4	0.693	5.08	3.74
4	3	0.744	4.57	4	0.634	4.30	1.41
5	3	0.939	3.80	2	0.773	5.04	1.38

structure while its anion becomes linear. The geometry of neutral PdO_3 closely resembles to that of its anion that is a trigonal planar (C_{2v}). Neutral PdO_4 assumes a square planar structure but its anion deviates from planarity. For $n = 5$, both neutral and anionic PdO_n becomes $(PdO_{n-2})O_2$ complex in which O_2 moiety is very weakly bound to the central Pd atom. The distance of O_2 moiety from Pd atom in anion is greater than that in neutral PdO_n. The bond length, O–O in O_2 moiety 1.26–1.28Å is in accordance with that in free O_2 molecule that is 1.27 Å.

Normal mode analysis reveals all real frequencies for structures given in Figure 3.2. Thus, they belong to at least a local minimum in the potential energy surface implying that they are thermodynamically stable. To analyse further the stability of these structures, we have considered their dissociations to O atom and O_2 molecule that are also shown in Table 3.2. All PdO_n species are found to be stable as all D_e values are positive. The dissociation energies decrease as the successive O atoms are attached to Pd. PdO_n anions are more stable than their neutrals due to high D_e values for both dissociation channels up to $n = 4$. Conversely for $n = 5$, dissociation energy of neutral PdO_n is higher than its anion against fragmentation to O atom but same in case of dissociation to O_2 molecule.

Evidently, Pd can bind with a maximum of four O atoms successively and form stable neutral and anionic PdO_n species up to $n = 4$. Thus, the maximum oxidation state of Pd can be as high as +8 at least in case of bonding with O atoms. Such a high oxidation state of Pd can be possible due to involvement of inner shell d electrons in bonding. The NBO analyses on these species to explore the participation of d electrons in bonding. We can see that the EA value increases remarkably as the successive O atoms are attached to Pd and reaches at its maximum of 4.57 eV for $n = 4$. This EA value is slightly lower than that of PdF_3 calculated with the same level of theory that is obvious due to relatively high electronegative nature of fluorine. The EA of PdO_2, 3.60 eV is also larger than fluorine atom and very large as compared to oxygen. The calculated EA values suggest that PdO_n species

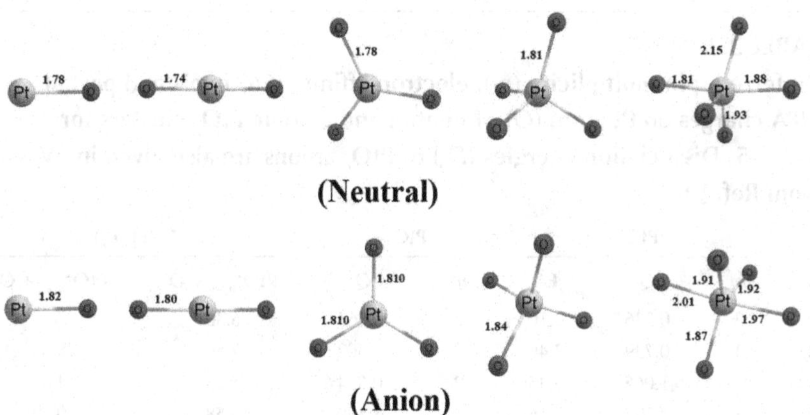

FIGURE 3.3 Optimised geometries of PtO_n (n = 1–5) with bond lengths (in Å) from Ref. [25].

behave as superhalogen for all $n \geq 2$. These large EAs result due to more positive charge localization on Pd atom.

All PdO_n species energetically favour higher spin states except for n = 1 and 5 in PdO^-_n. For n = 1 and 5, the higher spin states of PdO^-_n clusters are 0.49 eV and 0.42 eV higher than corresponding lower spin states, respectively. The dissociation energies (D_e) of PdO^-_n against fragmentation to PdO^-_{n-1} + O and PdO^-_{n-2} + O_2 are also listed in Table 3.2. All the PdO^-_n species are stable against both fragmentation channels. The dissociations of PdO^-_n follow almost the same trend as that of NiO^-_n up to n = 3. In general, PdO_n anions are more stable than NiO^-_n. For n = 4, the dissociation energies of PdO^-_n clusters are low as compared to NiO^-_n cluster, however, it can be recalled that unlike Pd, Ni does not bind atomically with four O atoms but forms a $(NiO^-_2)O_2$ complex that causes to increase its relative stability. In case of n = 5, PdO^-_n favours dissociation into $PdO^-_3 + O_2$ by 0.38 eV as compared to NiO^-_n that is consistent with an increase in the distance of O_2 moiety from metal atom in PdO^-_n as compared to NiO^-_n. Moreover, the dissociation energy of PdO^-_n (n = 5) against fragmentation to O atom is higher than that of NiO^-_n by 0.65 eV.

3.3.3 Platinum Oxides (PtO_n)

The optimized geometries of neutral as well as anionic PtO_n clusters have been shown in Figure 3.3. Like Pd, Pt binds stably with O atoms up to n = 4 in neutral and anionic PtO_n clusters. PtO_2 assumes a linear structure both in its neutral and anionic form, in contrast to NiO_2 and PdO_2, which is in accordance with some previous studies [38,39]. The PtO_3 closely resembles to that of PdO_3. On the other hand, PtO_4 takes a distorted tetrahedral geometry in contrast to planar geometry of PdO_4. For n = 5, PtO_n forms $(PtO_4)O$ complex instead of $(PtO_3)O_2$ as formed by Ni and Pd. However, in PtO_5 anion, all five O atoms bind atomically to central Pt

TABLE 3.3
Preferred spin multiplicity (*m*), electron affinity (EA) in eV and partial NPA charges on Pt atom (*Q*) of neutral and anionic PtO_n clusters for *n* = 1–5. Dissociation energies (D_e) of PtO_n anions are also given in eV from Ref. [25].

	PtO_n			PtO^-_n		$D_e(PtO^-_n)$	
n	*m*	*Q*	EA	*m*	*Q*	PtO^-_{n-1} + O	PtO^-_{n-2} + O_2
1	3	0.336	2.10	2	-0.363	6.30	—
2	1	0.739	3.40	2	0.302	7.50	5.83
3	3	1.065	3.45	2	0.854	5.13	4.66
4	3	1.297	4.26	2	1.199	3.58	0.73
5	3	1.079	5.08	4	0.903	3.88	-0.51

atom but we found that this structure becomes unstable and dissociates into PtO_3^- and O_2 fragments. The bond lengths Pt–O in neutral PtO_n are smaller than those in their anionic counterparts. Table 3.3 lists preferred spin multiplicity, electron affinity and partial NPA charges on Pt atom of neutral as well as anionic PtO_n clusters. All the neutral PtO_n favour the higher spin states except PtO_2 while its anions prefer lower spin multiplicity except PtO_5^-. The dissociation energies of PtO_n^- calculated by considering their fragmentation to PtO_{n-1}^- + O and PtO_{n-2}^- + O_2 are also listed in Table 3.3. All species are found to be stable against both dissociation paths except PtO_5^-. The dissociation energies also indicate that PtO_n anions are relatively more stable than PtO_n^- for $n \leq 3$. For $n \geq 4$, the PtO_{n-1}^- are relatively less stable as compared to PtO_n^-.

3.3.4 Rhenium Oxides (ReO_n)

The ground-state structures of ReO_n species from *n* = 1 to 5 in their neutral and anionic forms have been shown in Figure 3.4. We have considered all possible geometries in which O atoms bind atomically to the central Re atom. For example, linear and bent structures for *n* = 2, trigonal planar and T-shaped structures for *n* = 3, tetrahedral as well as square planar structures for *n* = 4, *etc.* are optimised to obtain corresponding minimum energy conformers. All ReO_n species energetically favour lower spin states except for *n* = 1. For *n* = 1, the ground states correspond to quartet and quintet in neutral and anionic species, respectively. The geometries of neutral ReO_n closely resemble those of corresponding anions. However, the bond length, Re–O in anionic ReO_n, is larger than that of the corresponding neutral counterparts due electronic repulsion created by modification in the charge distribution. To analyse the stability of these structures, we have considered their dissociations to O atom and O_2 molecule. All ReO_n species are found to be stable as all D_e values are positive. The dissociation energy for O

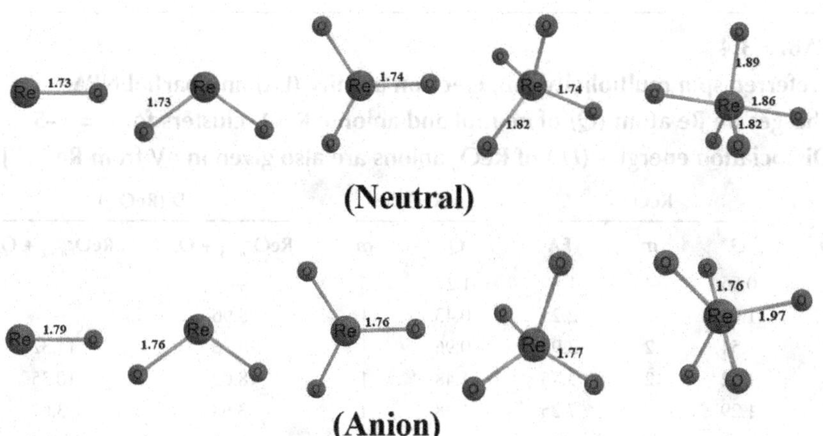

FIGURE 3.4 Optimised geometries of ReO_n ($n = 1–5$) with bond lengths (in Å) from Ref. [27].

atom increases up to $n = 3$, then starts to decrease as the successive O atoms are attached to Re. This suggests that ReO_3 is the most stable in neutral as well as anionic form among all ReO_n species. This is also reflected by its high D_e value against dissociation to O_2 and further supported by the fact that ReO_5 favours dissociation to O_2 as compared to O energetically by 1.75 eV. It should also be noted that ReO_n anions are more stable than their neutrals due to higher D_e values for both dissociation channels up to $n = 5$.

The Re can bind up to five O atoms successively and form stable neutral and anionic ReO_n species. Thus, the maximum oxidation state of Re can be as high as +9 at least in case of bonding with O atoms. However, a small D_e value (0.35 eV) for ReO_5 may suggest that the maximum oxidation state of Re is probably limited to +7. Note, however, that it is still possible to find a stable ReO_5 anion, at least theoretically. Such a high oxidation state of Re can be understood in terms of involvement of inner shell d electrons in bonding. We have performed NBO analyses on these species to explore the participation of d electrons in bonding. It is apparent that the number of 5d electrons participating in the bonding increases with the increase in O atoms up to $n = 5$.

The EA value increases remarkably as the successive O atoms are attached to Re up to $n = 5$. The EA of ReO_3, 3.91 eV, is larger than halogen and very large as compared to oxygen. Thus, calculated EA values suggest that ReO_n species behave as superhalogens for all $n \geq 3$. These large EAs result due to more positive charge localisation on Re atom. In ReO, charge contained by Re is +0.56 e. As the number of O atoms increases, charge on Re increases but saturates at +1.54 e in ReO_3. It is in accordance with the fact that ReO_3 possesses the highest dissociation energy and hence stability among all neutral species. Furthermore, in ReO^-, more than 70% of extra negative charge is located on Re. As successive O

TABLE 3.4
Preferred spin multiplicity (*m*), electron affinity (EA) and partial NPA charges on Re atom (*Q*) of neutral and anionic ReO_n clusters for n = 1–5. Dissociation energies (D_e) of ReO_n anions are also given in eV from Ref. [27].

	ReO_n			ReO_n^-		$D_e(ReO^-_n)$	
n	*Q*	*m*	EA	*Q*	*m*	ReO^-_{n-1} + O	ReO^-_{n-2} + O_2
1	0.55	4	1.80	-1.27	5	—	—
2	1.05	2	2.25	0.43	1	8.96	—
3	1.54	2	3.91	0.96	1	10.30	11.32
4	1.52	2	5.75	1.48	1	8.02	10.35
5	1.29	2	7.25	1.48	1	3.68	3.67

atoms are attached to ReO^-, extra charge starts to delocalise over several O atoms. Moreover, for n = 4, extra charge is almost completely delocalised over O atoms as NBO charges on Re in neutral and anionic ReO_4 are almost equal. This simply explains the reason for high EA values of ReO_n species with the increase in n. More strikingly, for n = 5, extra charge not only distributes itself to O atoms but also leads to electron transfer from Re to O atoms. This is what results in the very high EA of ReO_5, *viz.* 7.25 eV.

3.3.5 Ruthenium Oxides (RuO_n)

The optimised geometries of RuO_n species from n = 1 to 5 in their neutral and anion have been shown in Figure 3.5, in which O atom binds atomically to the central Ru atom. All RuO_n species energetically favour lower spin states except for n = 1 in neutral and n = 1, 2 in anions. The singlet and triplet states of RuO are, respectively, 1.81 eV and 1.25 eV higher in energy than the ground (quintet) state. In anionic species doublet states are 0.97 eV and 0.45 eV higher in energy than the quartet state for n = 1 and 2, respectively. The geometries of neutral RuO_n closely resemble to that of the corresponding anions up to n = 3. However, the bond length, Ru-O in the anionic RuO_n is larger than that of the corresponding neutral counterparts due to the electronic repulsion created by alteration in the charge distribution. Neutral RuO_4 assumes a tetrahedral T_d geometry that is deformed to C_1 in case of its anion. For n = 5, RuO_n takes the form of $(RuO_3)O_2$ complex in which O_2 moiety is at a distance of 1.89 Å from Ru (see Figure 5). The O-O distance in this moiety is 1.34 Å, which is slightly larger than the bond length found in O_2 molecule, *i.e.*, 1.27 Å. Thus, in case of RuO_5, two O atoms interact molecularly to the central Ru. Here it would be noteworthy to mention that we indeed find a stable C_s geometry of RuO_5 in which all O atoms bind atomically but 0.81 eV higher in energy than the $(RuO_3)O_2$ complex. To analyse the stability of these structures, we have considered their dissociations to the O atom. All

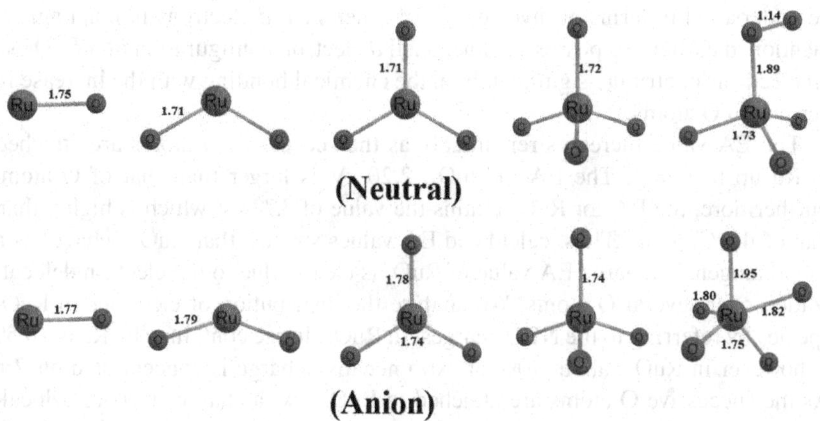

FIGURE 3.5 Optimised geometries of RuO_n (n = 1–5) with bond lengths (in Å) from Ref. [28].

TABLE 3.5

Preferred spin multiplicity (*m*), electron affinity (EA) and partial NPA charges on Ru atom (*Q*) of neutral and anionic RuO_n clusters for *n* = 1–5. Dissociation energies (D_e) of RuO_n anions are also given in eV from Ref. [28].

	RuO_n			RuO^-_n		$D_e(RuO^-_n)$	
n	*m*	*Q*	EA	*m*	*Q*	RuO^-_{n-1} + O	$RuO^-_{n-2} + O_2$
1	5	0.591	1.48	4	-0.299	7.22	—
2	1	0.993	2.20	4	0.478	7.20	7.90
3	1	0.942	3.60	2	0.671	6.80	8.12
4	1	1.022	3.72	2	0.987	6.00	6.20
5	1	1.068	3.80	2	0.828	3.18	3.28

RuO_n species are found to be stable as all D_e values are positive. The dissociation energy starts to decrease as the successive O atoms are attached to Ru, in general. RuO_2 possesses the highest dissociation energy of 7.19 eV among all neutral and RuO_3^- with the value of 8.11 eV, among its anions. It should also be noted that RuO_n anions are more stable than their neutral forms due to higher D_e values, up to n = 5. However, the difference in their D_e values becomes smaller with the increase in O atoms. For instance, RuO_3 favours dissociation than RuO_3^- by 1.32 eV but this difference is reduced to 0.21 eV for n = 4. Hence, it can be asserted that Ru can bind up to four O atoms successively and forms stable neutral and anionic RuO_n species. Thus, the maximum oxidation state of Ru can be as high as +7 at least in case of bonding with O atoms. Moreover, it is also possible to find a stable RuO_5 anion, at least theoretically. Such a high oxidation state of Ru can be understood in terms of involvement of inner shell *d* electrons in bonding. As mentioned earlier, Ru possesses inner shell *d* electron configuration of $4d^7$. These 4*d* electrons contribute significantly in the chemical bonding with the increase in peripheral O atoms.

The EA value increases remarkably as the successive O atoms are attached to Ru up to $n = 5$. The EA of RuO_2, 2.20 eV is larger than that of O atom. Furthermore, the EA for RuO_4 attains the value of 3.72 eV which is higher than that of the Cl atom. Thus, calculated EA values suggest that RuO_4 behaves as a superhalogen. The large EA value of RuO_4 is clearly due to the electron delocalization over several O atoms. We analyse the distribution of electrons in RuO_n species by referring to the NBO charges. In RuO, charge contained by Ru is +0.59 e, however in RuO^-, about 90% of extra negative charge is concentrated on Ru. As the successive O atoms are attached to RuO^-, extra charge starts to delocalize over several O atoms. Moreover, for $n = 4$, extra charge is almost completely delocalized over O atoms as NBO charges on Ru in neutral and anionic RuO_4 are almost equal. This simply explains the reason for high EA values of RuO_4. The electron delocalization in RuO_4 can also be explained on the basis of molecular orbital surfaces. The above discussion clearly reveals that RuO_4 belongs to the class of superhalogens.

3.3.6 Cobalt Oxides (CoO_n)

The ground-state equilibrium geometries of neutral and anion of CoO_n species are displayed in Figure 3.6. Except CoO_3, all CoO_n species tend to favour higher spin multiplicities energetically, irrespective of their charges. From Figure 3.6, it is also clear that the structures of CoO_n are independent of their charges, *i.e.*, approximately similar in neutral and anions. The equilibrium structure of CoO_2 is approximately linear unlike perfect linear structure of CoO_2^-. CoO_3 is a trigonal

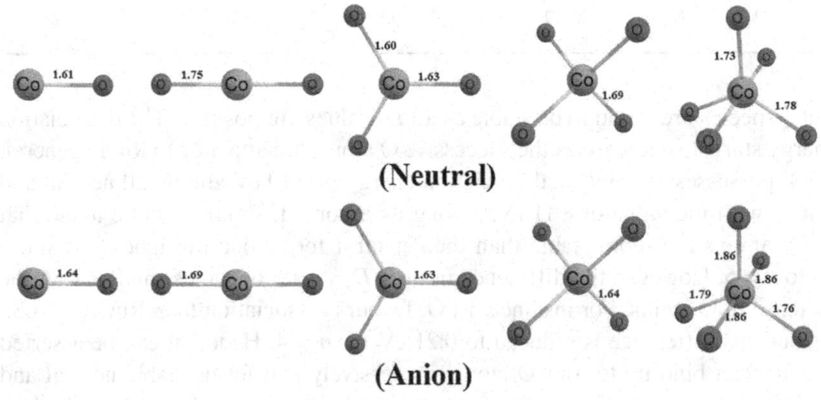

FIGURE 3.6 Optimised geometries of CoO_n ($n = 1$–5) with bond lengths (in Å) from Ref. [29].

planar in which one of the bond lengths is increased in contrast to equilateral CoO_3^-. CoO_4 and its anion assume tetrahedral structure. Neutral CoO_5 takes C_2 symmetry, whereas its anion takes C_1. The CoO_5 species and its anion are thermodynamically unstable as they dissociate into $CoO_3 + O_2$ fragments. To check whether the calculated ground states are spin contaminated, the expectation value of spin squared (⟨S2⟩) are listed in Table 3.6. All the calculated spin states are pure, having ⟨ S^2⟩ values with ±10 % of s (s + 1), where s is the total spin of unpaired electrons. However, a significant mixing of spin states can be seen in case of CoO_3 whose ⟨ S^2⟩ value is 1.38 in contrast to the expected value of 0.75. Therefore, it has significant mixing of its higher spin states that is consistent with the small difference in their energy values (Table 3.6). The frequency calculations performed over equilibrium geometries provide all real values that showing the structures belong to local minima in the potential energy surfaces. The stability of CoO_n species are also analysed by considering their dissociations into CoO_{n-1} + O and $CoO_{n-2} + O_2$ fragments.

The energies of O atom and O_2 molecule are calculated in their ground spin (triplet) states. The calculated dissociation energies for neutral and anionic CoO_n species are listed in Table 3.6. All CoO_n species are found to be stable against all possible dissociations, except CoO_5 and CoO_5^- that dissociate into $CoO_3 + O_2$ and $CoO_3^- + O_2$ fragments, respectively. This suggests that for $n \leq 4$, CoO_n species can be synthesized by using O atom or O_2 molecule in their synthesis. The stability of CoO_n species decreases with the increase in n due to the decrease in dissociation energy values. It is also interesting to note that CoO_3 favours dissociation into CoO_2 + O as compared to its anionic counterpart by 1.26 eV. This is in contrast to CoO_4 whose dissociation requires additional 0.21 eV as compared to its anion.

From Table 3.6, the EA increase successively with the increase in O atoms. Note that Co has outer orbital configuration of $3d^7 4s^2$. In the bonding of CoO, 4s orbital of Co interacts with 2p orbital of O atom, but not 3d orbital. For CoO_2, both

TABLE 3.6
Preferred spin multiplicity (*m*), electron affinity (EA) in eV and partial NPA charges on Co atom (*Q*) of neutral and anionic CoO_n clusters for n = 1–5. Dissociation energies (D_e) of CoO_n anions are also given in eV from Ref. [29].

	CoO_n			CoO_n^-		$D_e(CoO^-_n)$	
n	***m***	***Q***	**EA**	***m***	***Q***	CoO^-_{n-1} + O	$CoO^-_{n-2} + O_2$
1	2	0.72	1.64	1	-0.12	—	—
2	2	1.10	3.26	1	0.84	8.04	—
3	2	0.62	4.52	1	0.59	5.88	6.04
4	2	0.55	4.31	1	0.56	3.68	1.68
5	2	0.36	4.76	1	0.51	3.40	-0.80

4s and 3d electrons of Co participate in bonding with O atoms, leaving exactly half-filled 3d orbital. Note that +2 and +4 are well-established oxidation states of Co. For $n \geq 3$, the participation of 3d orbital in bonding of CoO_n increases. This causes increasing the coordination number of Co successively with the increase in O atoms. The oxidation state of Co can be as high as +7 due to the enhanced stability of CoO_4^- as compared to its neutral counterpart. The EA of CoO_n and VDE of CoO_n^- is much higher than the EA of Cl (3.62 eV) [40] for $n \geq 3$, allowing CoO_n species to behave as superhalogens. The EA of CoO_4, 4.31 eV, is higher than that of FeO_4, 3.8 eV [41]. The rise in EA of CoO_n with the increase in O atoms can be understood in terms of electronic charge distribution. In Table 3.6, we also list the NBO charges on Co atom in CoO_n and CoO_n^- species. The NBO charges on Co increase up to 1.1 *e* for CoO_2 and then decrease to 0.55 *e* for CoO_4. Thus, the addition of two more O atoms to CoO_2 reduces the charge transfer to O atoms by 50%. On the other hand, in CoO^-, about 85% of extra electron is contained by Co atom. The extra electron located on Co decreases to 25% in case of CoO_2^-, which results in the increase in the EA of CoO_2 and VDE of CoO_2^-. For $n \geq 3$, the extra electron is completely delocalized over O atoms, which leads to very high EA of CoO_n and VDE of its anions. In CoO_3^-species, more than 99 % of extra electron is contained by peripheral O atoms and merely 0.03 % is located on Co. Thus, CoO_3 and CoO_4 indeed belong to the class of superhalogens. CoO_n superhalogens can be used to form a new class of compounds having unusual oxidizing potentials.

3.3.7 Vanadium Oxides (VO_n)

The optimized structures of neutral VO_n (n = 1–5) clusters and their anions are shown in Figure 3.7. All neutral VO_n species favour lower spin state, *i.e.*, doublet except VO and VO_5. On the contrary, the ground states of VO_n^- are triplet except $VO_3^-$and VO_4^-, which are singlet. The ground state structures of VO_n clusters are approximately similar in neutral and anions. The geometry of VO_2 is more bent than that of VO_2^-, due to the fact that additional charge given to VO_2 goes to nonbonding orbitals of oxygen, which creates repulsion between O atoms. VO_3 is a trigonal planar in which one of the bond lengths is increased while VO_3^- shows equilateral structure. The structures of neutral VO_4 and its anion are distorted tetrahedral shape and exists in the form of $(VO_2)O_2$ complex in which V binds with one molecular oxygen and two oxygen atoms. However, in anionic form all O atoms bind dissociatively to V due to excess electron on O atoms. In case of VO_5, both neutral and anionic structures form $(VO_3)O_2$ complexes. The vibrational analysis also reveals that all frequencies are real for optimized structures; therefore, they belong to at least a local minimum in the potential energy surface.

The calculated dissociation energies (D_e) for anionic VO_n species are listed in Table 3.7. All VO_n species are found to be stable as all dissociation energy values are positive and the dissociation energy decreases as the successive O atoms are attached to V. The calculated EA of VO_n species are listed in Table 3.7. One can note that the EA increases as the successive O atoms are attached to V up to $n = 5$. The EA of VO_2, 2.04 eV is larger than that of O atom. Furthermore, the EA for

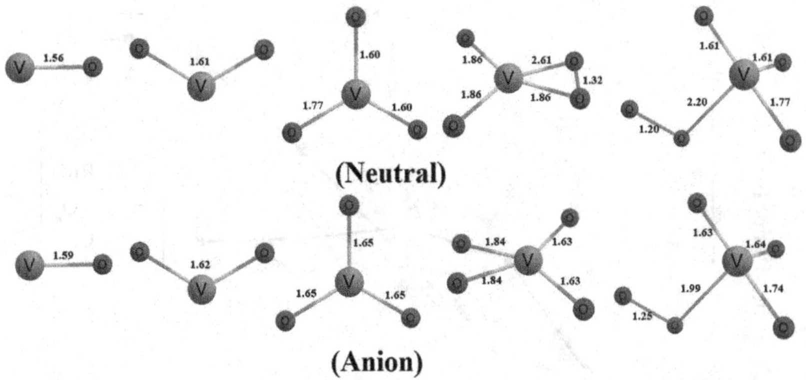

FIGURE 3.7 Optimised geometries of VO_n (n = 1–5) with bond lengths (in Å) from Ref. [30].

TABLE 3.7
Preferred spin multiplicity (m), electron affinity (EA) in eV and partial NPA charges on V atom (Q) of neutral and anionic VO_n clusters for n = 1–5. Dissociation energies (D_e) of VO_n anions are also given in eV from Ref. [30].

	VO_n			VO^-_n		$D_e(VO^-_n)$	
n	m	Q	EA	m	Q	VO^-_{n-1} + O	VO^-_{n-2} + O_2
1	4	0.65	1.09	3	-0.21	16.95	—
2	2	1.22	2.04	3	0.64	9.14	16.92
3	2	1.06	4.35	1	1.02	9.09	9.06
4	2	0.99	3.65	1	0.92	5.96	5.88
5	4	0.79	4.52	3	0.60	5.47	2.26

VO_3 attains the value of 4.35 eV, which exactly reproduces the experimental value 4.36 eV [42]. Although EA of VO_4 is smaller than that of VO_3, it is still higher than that of F atom. Therefore, EA values clearly show that VO_n (for $n \geq 3$) clusters behave as superhalogen.

3.3.8 Comparison

The calculated EAs of all these clusters are plotted in Figure 3.8 as a function of n. We can see that the EA value increases remarkably as the successive O atoms are attached to central metals. Thus, large EAs of MO_n clusters suggest that they behave as superhalogens for different values of n. These large EAs result due to more positive charge localisation on metal atoms.

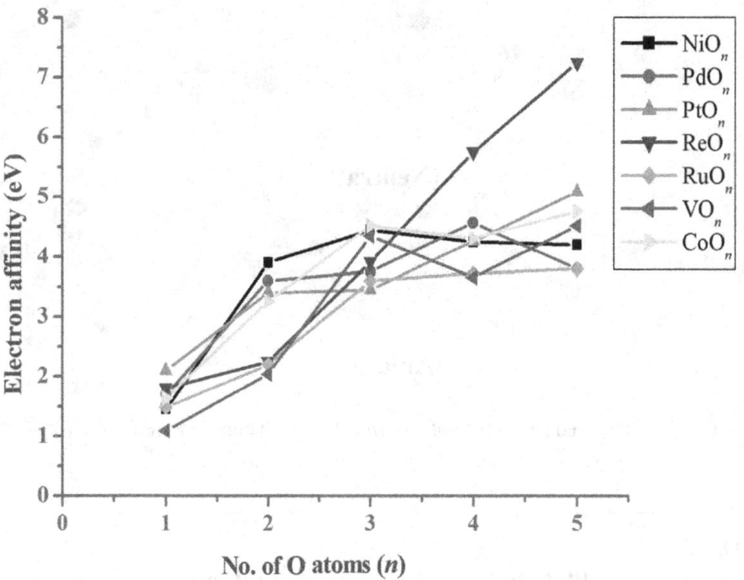

FIGURE 3.8 Electron affinity of neutral MO_n clusters as a function of n.

3.4 CONCLUSION

We have shown theoretically that Ni can bind up to three O atoms and Pd or Pt binds up to four O atomically. Thus, the maximum oxidation state of Ni may be as high as +6 that can jump further to +8 in case of Pd or Pt as far as the bonding with O atoms is concerned. This can be possible due to involvement of inner shell d electrons in bonding. The electron affinities of these species increase as the successive O atoms are attached. The large electron affinities of MO_n (M = Ni, Pd or Pt) for $n \geq 2$ suggest that these species can be regarded as superhalogens. In case of ReO_n species, EA value increases with the increase in n. More strikingly, for ReO_5 extra charge not only distributes itself to O atoms but also leads to electron transfer from Re to O atoms. In case of RuO_n, CoO_n and VO_n clusters, extra charge are almost completely delocalized over O atoms, explaining their high EA values. The stability of corresponding anions opens an opportunity to synthesize a new class of compounds by interaction of MO_n superhalogens with appropriate metal atoms having unusual chemical properties.

REFERENCES

1. Gutsev, G.L. and Boldyrev, A.I., 1981. DVM-Xα calculations on the ionization potentials of MXk+ 1− complex anions and the electron affinities of MXk+ 1 "superhalogens". *Chemical Physics*, 56(3), pp. 277–283.

2. Wang, X.B., Ding, C.F., Wang, L.S., Boldyrev, A.I. and Simons, J., 1999. First experimental photoelectron spectra of superhalogens and their theoretical interpretations. *The Journal of Chemical Physics*, 110(10), pp. 4763–4771.
3. Wen, H., Hou, G.L., Huang, W., Govind, N. and Wang, X.B., 2011. Photoelectron spectroscopy of higher bromine and iodine oxide anions: Electron affinities and electronic structures of BrO2, 3 and IO2–4 radicals. *The Journal of Chemical Physics*, 135(18), p. 184309.
4. Hou, G.L., Wu, M.M., Wen, H., Sun, Q., Wang, X.B. and Zheng, W.J., 2013. Photoelectron spectroscopy and theoretical study of M (IO3) 2–(M= H, Li, Na, K): Structural evolution, optical isomers, and hyperhalogen behavior. *The Journal of Chemical Physics*, 139(4), p. 044312.
5. Kandalam, A.K., Kiran, B., Jena, P., Pietsch, S. and Ganteför, G., 2015. Superhalogens beget superhalogens: A case study of $(BO_2)_n$ oligomers. *Physical Chemistry Chemical Physics*, 17(40), pp. 26589–26593.
6. Giri, S., Child, B.Z. and Jena, P., 2014. Organic superhalogens. *ChemPhysChem*, 15(14), pp. 2903–2908.
7. Smuczyńska, S. and Skurski, P., 2007. Is hydrogen capable of playing a central atom role in superhalogen anions? *Chemical Physics Letters*, 443(4–6), pp. 190–193.
8. Bartlett, N. and Lohmann, D.H., 1962. Dioxygenyl hexa fluoro platinate (V) O2+ [PtF6]-. *Proceedings of the Chemical Society of London*, (March), p. 115.
9. Hotop, H. and Lineberger, W.C., 1975. Binding energies in atomic negative ions. *Journal of Physical and Chemical Reference Data*, 4(3), pp. 539–576.
10. Rienstra-Kiracofe, J.C., Tschumper, G.S., Schaefer, H.F., Nandi, S. and Ellison, G.B., 2002. Atomic and molecular electron affinities: Photoelectron experiments and theoretical computations. *Chemical Reviews*, 102(1), pp. 231–282.
11. Gutsev, G.L., Khanna, S.N., Rao, B.K. and Jena, P., 1999. FeO 4: A unique example of a closed-shell cluster mimicking a superhalogen. *Physical Review A*, 59(5), p. 3681.
12. Gutsev, G.L., Rao, B.K., Jena, P., Wang, X.B. and Wang, L.S., 1999. Origin of the unusual stability of MnO4–. *Chemical Physics Letters*, 312(5–6), pp. 598–605.
13. Srivastava, A.K., Kumar, A. and Misra, N., 2017. A path to design stronger superacids by using superhalogens. *Journal of Fluorine Chemistry*, 197, pp. 59–62.
14. Srivastava, A.K. and Misra, N., 2015. Hydrogenated superhalogens behave as superacids. *Polyhedron*, 102, pp. 711–714.
15. Srivastava, A.K., Kumar, A. and Misra, N., 2017. Superhalogens as building blocks of a new series of superacids. *New Journal of Chemistry*, 41(13), pp. 5445–5449.
16. Srivastava, A.K. and Misra, N., 2016. Designing new electrolytic salts for Lithium Ion Batteries using superhalogen anions. *Polyhedron*, 117, pp. 422–426.
17. Srivastava, A.K. and Misra, N., 2016. Superhalogens as building blocks of complex hydrides for hydrogen storage. *Electrochemistry Communications*, 68, pp. 99–103.
18. Srivastava, A.K., Kumar, A., Tiwari, S.N. and Misra, N., 2017. Application of superhalogens in the design of organic superconductors. *New Journal of Chemistry*, 41(24), pp. 14847–14850.
19. Srivastava, A.K., Kumar, A. and Misra, N., 2021. Superhalogens as building blocks of ionic liquids. *The Journal of Physical Chemistry A*, 125(10), pp. 2146–2153.
20. Tripathi, J.K. and Srivastava, A.K., 2022. CF4-n (SO3) n (n= 1–4): A new series of organic superhalogens. *Molecular Physics*, 120(21), p. e2123748.
21. Xue, Q., Zhong, M., Zhou, J. and Jena, P., 2022. Rational design of endohedral superhalogens without using metal cations and electron counting rules. *The Journal of Physical Chemistry A*, 126(22), pp. 3536–3542.

22. Banjade, H., Fang, H. and Jena, P., 2022. Metallo-boranes: A class of unconventional superhalogens defying electron counting rules. *Nanoscale*, 14(5), pp. 1767–1778.
23. Sinha, S., Jena, P. and Giri, S., 2022. Functionalized nona-silicide [Si 9 R 3] Zintl clusters: A new class of superhalogens. *Physical Chemistry Chemical Physics*, 24(35), pp. 21105–21111.
24. Srivastava, A.K., 2023. Recent progress on the design and applications of superhalogens. *Chemical Communications*. DOI: 10.1039/d3cc00428g (Accepted).
25. Srivastava, A.K. and Misra, N., 2014. Unusual bonding and electron affinity of nickel group transition metal oxide clusters. *Molecular Physics*, 112(12), pp. 1639–1644.
26. Srivastava, A.K. and Misra, N., 2014. Theoretical investigations on the superhalogen properties and interaction of PdOn (n= 1–5) species. *International Journal of Quantum Chemistry*, 114(5), pp. 328–332.
27. Srivastava, A.K. and Misra, N., 2014. Superhalogen properties of ReO n (n= 1–5) species and their interactions with an alkali metal: An ab initio study. *Molecular Physics*, 112(15), pp. 1963–1968.
28. Srivastava, A.K. and Misra, N., 2014. First principle investigations on the superhalogen behaviour of RuO n (n= 1–5) species. *The European Physical Journal D*, 68, pp. 1–4.
29. Srivastava, A.K. and Misra, N., 2015. Superhalogen properties of CoO n (n≥ 3) species revealed by density functional theory. *Theoretical Chemistry Accounts*, 134, pp. 1–6.
30. Shukla, D.V., Srivastava, A.K. and Misra, N., 2017. Density functional study on the evolution of superhalogen properties in VO_n (n= 1–5) species. *Main Group Chemistry*, 16(2), pp. 141–150.
31. Srivastava, A.K. and Misra, N., 2014. Ab initio investigations on the stabilities of AuO_n^{q-}(q= 0 to 3; n= 1 to 4) species: Superhalogen behavior of AuOn (n≥ 2) and their interactions with an alkali metal. *International Journal of Quantum Chemistry*, 114(8), pp. 521–524.
32. Kohn, W., Becke, A.D. and Parr, R.G., 1996. Density functional theory of electronic structure. *The Journal of Physical Chemistry*, 100(31), pp. 12974–12980.
33. Frisch, M.J., Trucks, G.W., Schlegel, H.B., Scuseria, G.E., Robb, M.A., Cheeseman, J.R., Scalmani, G., Barone, V., Mennucci, B., Petersson, G.A., Nakatsuji, H., Caricato, M., Li, X., Hratchian, H.P., Izmaylov, A.F., Bloino, J., Zheng, G., Sonnenberg, J.L., Hada, M., Ehara, M., Toyota, K., Fukuda, R., Hasegawa, J., Ishida, M., Nakajima, T., Honda, Y., Kitao, O., Nakai, H., Vreven, T., Montgomery Jr., J.A., Peralta, J.E., Ogliaro, F., Bearpark, M., Heyd, J.J., Brothers, E., Kudin, K.N., Staroverov, V.N., Keith, T., Kobayashi, R., Normand, J., Raghavachari, K., Rendell, A., Burant, J.C., Iyengar, S.S., Tomasi, J., Cossi, M., Rega, N., Millam, J.M., Klene, M., Knox, J.E., Cross, J.B., Bakken, V., Adamo, C., Jaramillo, J., Gomperts, R., Stratmann, R.E., Yazyev, O., Austin, A.J., Cammi, R., Pomelli, C., Ochterski, J.W., Martin, R.L., Morokuma, K., Zakrzewski, V.G., Voth, G.A., Salvador, P., Dannenberg, J.J., Dapprich, S., Daniels, A.D., Farkas, O., Foresman, J.B., Ortiz, J.V., Cioslowski, J. and Fox, D.J., 2010. *Gaussian 09, revision B.01*, Gaussian Inc., Wallingford, CT.
34. Becke, A.D., 1993. A new mixing of Hartree–Fock and local density-functional theories. *The Journal of Chemical Physics*, 98(2), pp. 1372–1377.
35. Lee, C., Yang, W. and Parr, R.G., 1988. Development of the Colle-Salvetti correlation-energy formula into a functional of the electron density. *Physical Review B*, 37(2), p. 785.
36. Keith, T.A. and Millam, J.M., 2016. *GaussView, version 6.1, Roy Dennington*, Semichem Inc., Shawnee Mission, KS.

37. Wu, H. and Wang, L.S., 1997. A study of nickel monoxide (NiO), nickel dioxide (ONiO), and Ni (O2) complex by anion photoelectron spectroscopy. *The Journal of Chemical Physics*, 107(1), pp. 16–21.
38. Ramond, T.M., Davico, G.E., Hellberg, F., Svedberg, F., Salén, P., Söderqvist, P. and Lineberger, W.C., 2002. Photoelectron spectroscopy of nickel, palladium, and platinum oxide anions. *Journal of Molecular Spectroscopy*, 216(1), pp. 1–14.
39. Bare, W.D., Citra, A., Chertihin, G.V. and Andrews, L., 1999. Reactions of laser-ablated platinum and palladium atoms with dioxygen: Matrix infrared spectra and density functional calculations of platinum oxides and complexes and palladium complexes. *The Journal of Physical Chemistry A*, 103(28), pp. 5456–5462.
40. Weinhold, F. and Glendening, E.D., 2001. *NBO 5.0 program manual: Natural bond orbital analysis programs*, Theoretical Chemistry Institute and Department of Chemistry, University of Wisconsin, Madison, WI.
41. Gutsev, G.L., Khanna, S.N., Rao, B.K. and Jena, P., 1999. FeO 4: A unique example of a closed-shell cluster mimicking a superhalogen. *Physical Review A*, 59(5), p. 3681.
42. Wu, H. and Wang, L.S., 1998. A photoelectron spectroscopic study of mono vanadium oxide anions (VO x–, x= 1–4). *The Journal of Chemical Physics*, 108(13), pp. 5310–5318.

4 Transition Metal Fluorides as Superhalogens

Shamoon Ahmad Siddiqui, Ankit Kargeti and Tabish Rasheed

4.1 INTRODUCTION

Superhalogens are unique chemical species that have electron affinity (EA) and vertical detachment energy (VDE) values exceeding the corresponding values of halogens. These complexes have EA values greater [1] than those of even the most electronegative atom in periodic table, namely Cl (3.617 ± 0.003 eV) [2,3]. These molecular systems are of particular interest to scientists both from theoretical and practical viewpoints. The basic concept of superhalogens was initially introduced by Gutsev and Boldyrev in 1981 [4]. Since then, they have garnered wide attention amongst researchers over the past few decades. These highly electronegative species can be used to design novel complex compounds having quite interesting properties. The high oxidizing capability of superhalogens, which had been pointed out much earlier [5] has numerous potential industrial applications. Systems with high EA values play an important role in the synthesis of novel compounds as well. For example, the novel salts $LiAuF_6$ and $LiPtF_6$ have been synthesized by Graudejus et al. [6] using solutions of AuF_{4-} or PtF_6^{2-} as precursors. Superhalogens can also be used for developing organic superconductors [7] and their hydrogenation leads to the formation of superacids [8]. These superacids have useful applications in chemical industry. Srivastava et al. [9] and other research groups [10,11] have published articles on the usage of superhalogens as building blocks of superacids. Marcin Czapla et al. [12] have predicted the largest VDE complexes that are $[SbF_6(HF)_{12}]$-, $[AsF_6(HF)_{12}]$- and $[AlF_4(HF)_{12}]$- with values 14.03 eV, 14.03 eV and 13.96 eV, respectively, reported in literature so far. Freza and Skurski [13] calculated the electron binding energy of $[H_{12}F_{13}]$- species as 13.87 eV, which is almost three to four times higher than the chlorine atom with 3.62 eV. Recently, superhalogens have also been found to be useful for designing of safe electrolytic salts for lithium-ion batteries [14] and efficient materials for hydrogen storage [15].

The molecular structures of superhalogens generally consist of a metal atom at the centre surrounded by peripheral electronegative atoms such as fluorine (F), chlorine (Cl), oxygen (O), *etc.* Gutsev and Boldyrev [4] proposed the formula

 DOI: 10.1201/9781003384205-4

$MX_{(n+1)/m}$ for superhalogens, where n is the maximal formal valence of the central atom M, and m is the normal valence of electronegative atom X. As the number of electronegative atoms increases in superhalogens, the shared electron from the metal atom tends to get delocalized over them leading to increase in EA. An example of the same is LiF_2, which has superhalogen nature having EA of 5.45 eV [1]. The value is greater than the corresponding value of fluorine (3.399 ± 0.003 eV) [2,3]. Initial attempts by researchers (to investigate superhalogens) were mainly focused on chemical species containing *sp.* group of elements (of periodic table) as the central atom. Halogens were attached to the central atom forming the entire superhalogen complex. Formulae of these initial structures were MX_2 [1], MX_3 [16–18] and MX_4 [19,20], where M and X represent the central metal atom and halogen atom, respectively. Later, to increase the EA of superhalogens, scientists [21–25] started to focus on compounds comprising of *d*-group transition metals as the central atom. These transition metals have partially filled *d* orbitals in their ground or most stable oxidation state, which may be utilized for developing progressively more complex superhalogens by adding more halogen atoms. Due to increase in the number of halogen atoms, EA also increases, leading more effective superhalogens. Among halogens, fluorine has been the most preferred element in the formation of superhalogens, because it can react with all metallic elements. In fact, its reactivity surpasses even that of oxygen. EA of fluorine is second highest amongst all elements of periodic table second only to chlorine. Scientists have been attracted by fluorine to consider it as the electronegative component of superhalogens. Hence, superhalogens composed of transition metals and fluorine termed as transition metal fluorides (TMFs) are an important class of chemicals.

TMFs usually have high EA values, due to which they can be easily classified as superhalogens. Particularly, hexafluorides corresponding to third-row transition metals are regarded as prominent examples of superhalogens, displaying EA values in the range 5–9 eV [26]. However, it may be noted that fluoride-based superhalogens containing low atomic number metals as central atom, usually have low EA values. Since, TMFs have high EA values, they can act as oxidizers and when combined with cations, novel salts can be synthesized. Previous experimental investigations on TMFs have been limited in their scope due to scarcity of transition metals along with cumbersome laboratory procedures involved in chemical synthesis. Theoretical investigations on these systems facilitates the prediction of superhalogen nature of TMFs with high accuracy.

In the current chapter, feasibility of TMFs to act as superhalogens has been investigated through comparative analysis of theoretical results. These discussions are based on various TMFs studied earlier [27–33] theoretically by the chapter authors. The TMFs covered in this work are RhF_n (n = 1–7) [27], IrF_n (n = 1–7) [28], PtF_n (n = 1–6) [29], RuF_n (n = 1–7) [30], PdF_n (n = 1–7) [31], MnF_n (n = 1–6) [32] and FeF_n (n = 1–6) [33]. The properties discussed in this chapter of subject TMFs are fragmentation energy, average bond length, electron affinity, dimer formation of superhalogen complexes and their interaction with alkali atoms.

4.2 OBJECTIVE OF CHAPTER

To specifically focus on TMFs as superhalogen, some selected research articles [27–33] have been discussed here in detail that provides in depth understanding of these complexes. Comparative analysis of subject TMFs provides useful insights into the properties of these species useful for workers in this research area.

4.3 COMPUTATIONAL METHODOLOGY

All the calculations have been done by the self-consistent field technique using the linear combination of atomic orbital-molecular orbital approach. Total energies were calculated using Density Functional Theory [34,35] utilizing the hybrid functional B3LYP [36] along with the appropriate combination of basis set and pseudopotential. Various types of geometries were optimized using the *Gaussian 09W* [37] program package. Molecular modelling and visual inspection of the output was performed using the GaussView 6.1 [38] interface for all geometries. Specifically, Geometries of RhF_n (n = 1–7) complexes were calculated using DFT with B3PLY/SDD basis set level of theory. IrF_n (n = 1–7) complexes, PtF_n (n = 1–6) complexes and RuF_n (n = 1–7) complexes were also calculated at B3PLY/SDD basis set level of theory. PdF_n (n = 1–7) complexes and MnF_n (n = 1–6) complexes were also calculated at B3PLY/SDD And FeF_n (n = 1–6) complexes were calculated at B3PLY/6–311+G* basis set level of theory.

To investigate the effect of functionals and basis sets, test calculations have been carried out using B3LYP/SDD, B3LYP/LANL2DZ, B3PW91/SDD and MP2/SDD methods for selected complexes. After convergence most of the calculated values are well matched with experimental data. Such as the calculated ionization potential for Pd atom is found to be 8.692, 8.580, 8.619 and 7.982 eV for B3LYP/SDD, B3LYP/LANL2DZ, B3PW91/SDD and MP2/SDD methods, respectively, which are in excellent agreement with the experimental value 8.337 eV. Ir-F bond length is calculated as 1.944 Å by the B3LYP/SDD method, while it is calculated as 1.9347, 1.9315, 1.9344 and 1.935 Å by the B3PW91/SDD, MPW1PW91/SDD, B3PW91/LANL2DZ and MPW1PW91/LANL2DZ methods, respectively.

Therefore, based on these comparisons with different higher-level calculations and experimental data, we can say that the overall performance of B3LYP/SDD method and basis set combination is satisfactory for reported TMFs.

4.4 RESULTS AND ANALYSIS

Detailed discussion on superhalogen attributes of the following TMFs has been performed: RhF_n (n = 1–7), IrF_n (n = 1–7), PtF_n (n = 1–6), RuF_n (n = 1–7), PdF_n (n = 1–7), MnF_n (n = 1–6), FeF_n (n = 1–6). Theoretical calculations of subject TMFs have been carried out as per the scheme described in the computational methodology section. These calculations take into consideration the transition metal's variable valency, which leads to the multiple compounds containing them as central atom. The first superhalogen series RhF_n (n = 1–7) comprises of Rhodium (Rh) as the central atom. Rh belongs to the same group as cobalt (Co) that have ferromagnetic behaviour just like iron (Fe). With its outer electron configuration [Kr] $4d^8 5s^1$, Rh is known to possess a normal valence of 1, but this can exceed up to 4. However, Rh has partially filled 4d orbital because of which, it has variable coordination number. Hence, it able to bind with large numbers of F atoms. IrF_n (n = 1–7) is the second superhalogen investigated and analysed herewith containing Iridium (Ir) as central transition metal atom. Ir also belongs to the same group as Co having ferromagnetic behaviour. With its outer electronic configuration [Xe] $4f^{14} 5d^7 6s^2$, Ir is known to possess a normal valence of 2. Ir has partially filled 5d orbital because of which, it exhibits variable coordination number. Ir may also have oxidation numbers ranging from 1 to 7. Accordingly, this element has the ability to combine with many F atoms. Third superhalogen series are PtF_n (n = 1–6) with Platinum (Pt) as central atom. Pt has the electronic configuration [Xe] $6s^1 4f^{14} 5d^9$ and is known to possess a normal valence of 1, however, its oxidation number ranges from 1 to 6. The fourth TMF series discussed here is RuF_n (n = 1–7) containing Ruthenium (Ru) as central atom. Ru belongs to the same group as iron having ferromagnetic behaviour. Its electronic configuration is [Kr] $4d^7 5s^1$ and is known to possess a normal valence of 1. Ru has partially filled 4d orbital because of which, it has variable coordination number. Therefore, it can bind with high numbers of F atoms and has oxidation numbers ranging from 1 to 7. Fifth TMF series contains Palladium (Pd) as the central atom. Pd is known to possess normal valence of 1 and has filled 4d orbital with the oxidation numbers ranging from 1 to 7. The sixth superhalogen series studied are MnF_n (n = 1–6) composed of manganese (Mn) as central atom. Mn has an electronic configuration [Ar] $3d^5 4s^2$ and hence, has maximal valence of six. It is expected to form a series of superhalogen compounds with fluorine (F), having the general formula MnF_n (n = 1–6). The last TMF series discussed here is FeF_n (n = 1–6) containing Fe as central atom. Fe has electronic configuration [Ar] $3d^6 4s^2$, which shows that it can form a series of superhalogen compounds with F atoms. All subject TMFs have variable valency from +1 to +7. They were optimized in both their neutral and anionic states are shown below from Figure 4.1–4.7.

4.4.1 Bond Length Variation in Transition Metal Fluorides

Average bond lengths of the bonds between transition metal and fluorine atoms of subject TMFs were calculated and plotted. Figures 4.8–4.14 display the average bond lengths of subject TMFs *viz.*, RhF_n (n = 1–7), IrF_n (n = 1–7), PtF_n (n = 1–6), RuF_n (n = 1–7), PdF_n (n = 1–7), MnF_n (n = 1–6), FeF_n (n = 1–6) in both neutral and anionic forms. The values have been obtained from optimized geometrical parameters of these compounds. From the analysis of Figures 4.8–4.14, it is clear that the bond lengths in anionic forms are higher as compared to the corresponding neutral forms. The same can be attributed to delocalization of additional electron over the entire TMF in anionic form and consequent increase in electrostatic repulsion. It implies that the bond strength is weak in anionic forms as compared to the neutral forms. This observation agrees well with the earlier work carried

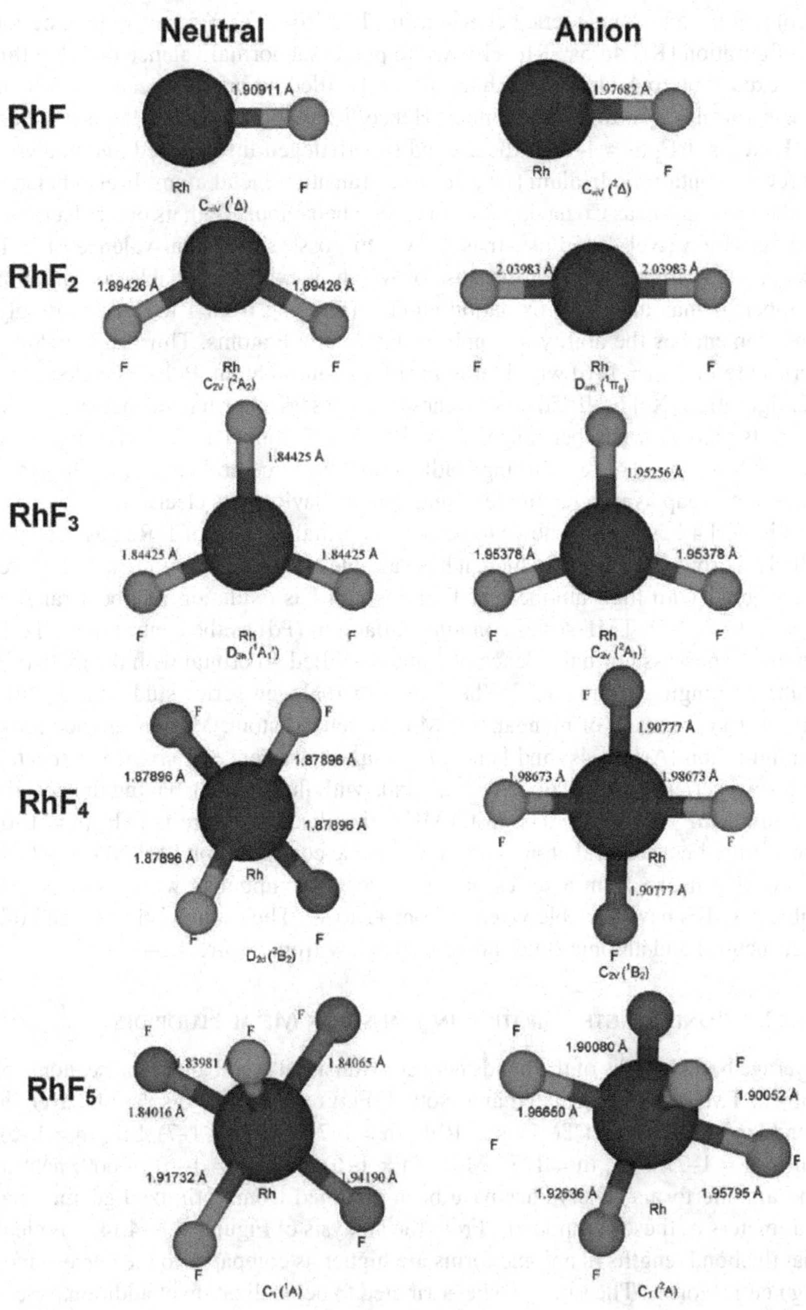

FIGURE 4.1 Optimised geometries of RhF_n Neutral and Anion clusters.

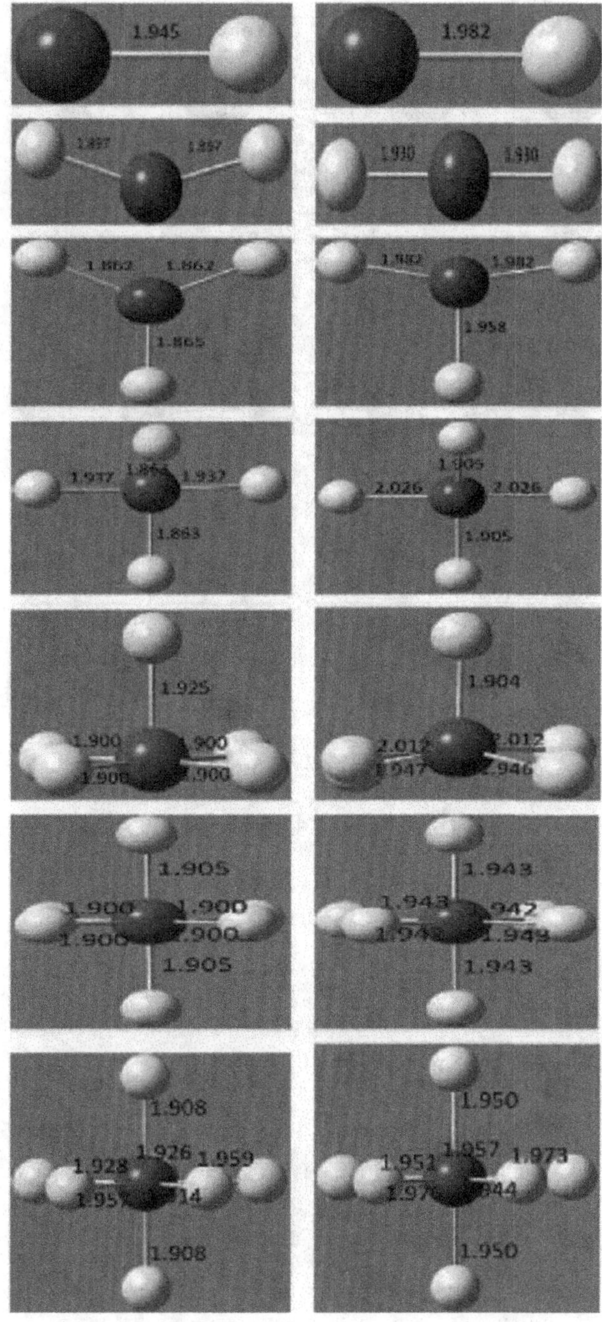

FIGURE 4.2 Optimised geometries of IrF_n Neutral and Anion clusters.

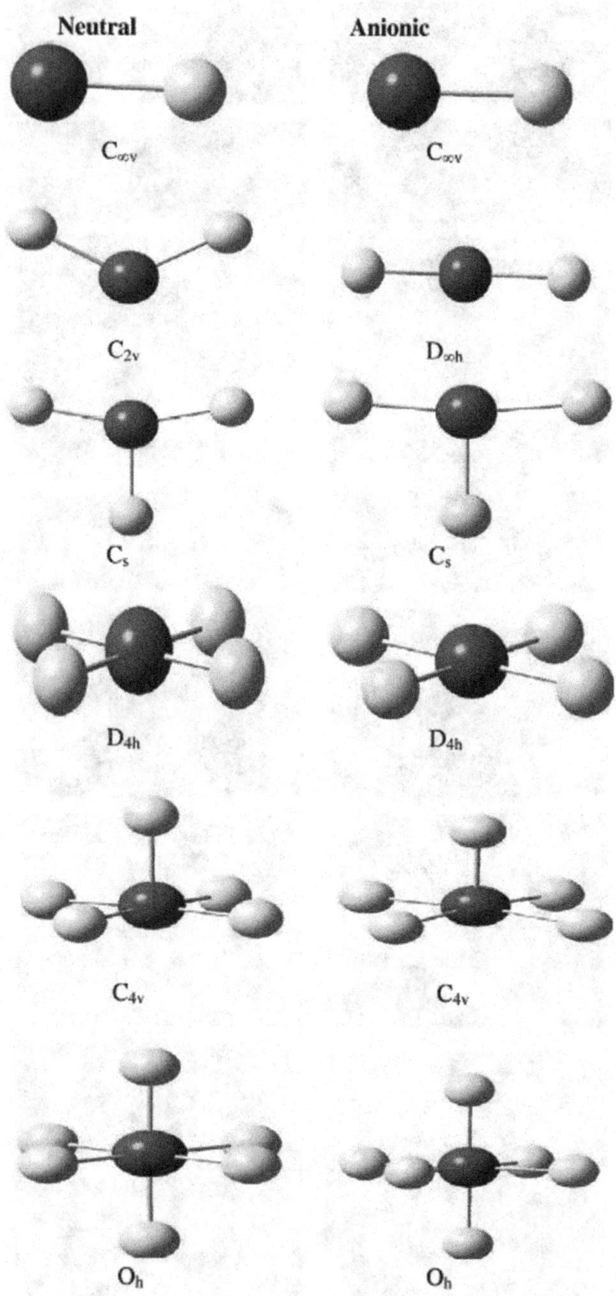

FIGURE 4.3 Optimised geometries of PtF_n Neutral and Anion clusters.

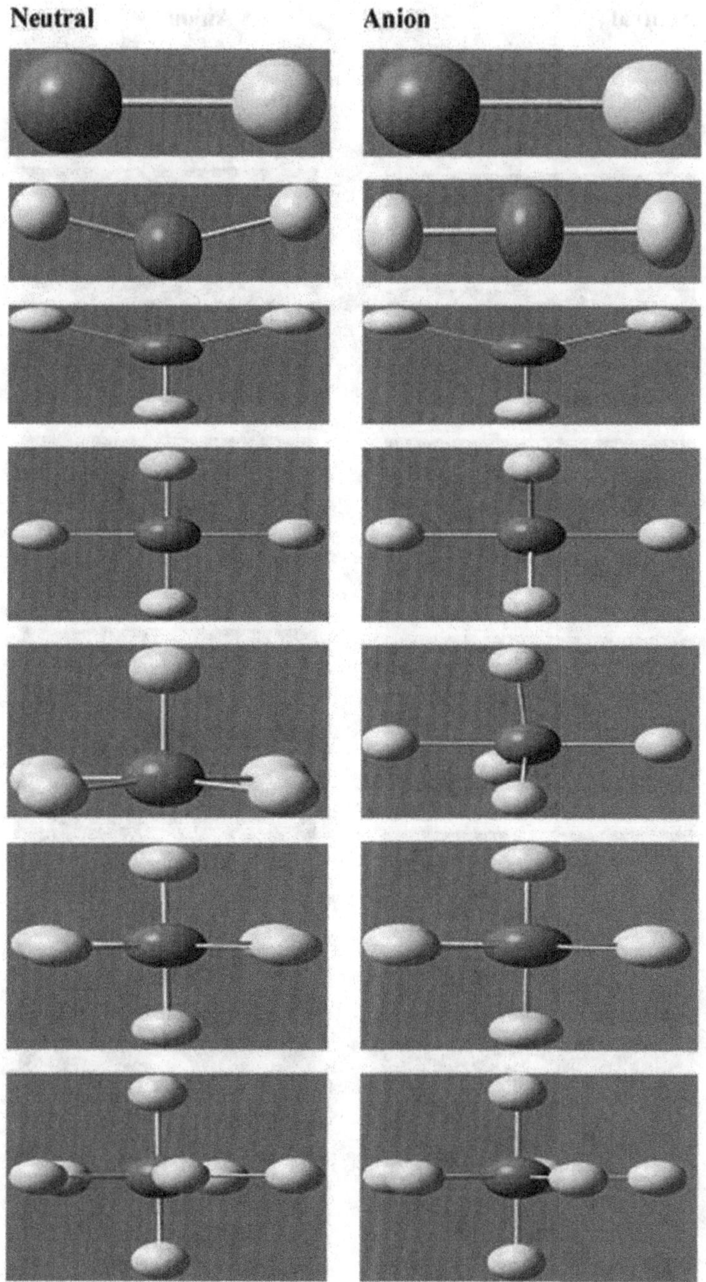

FIGURE 4.4 Optimised geometries of RuF_n Neutral and Anion clusters.

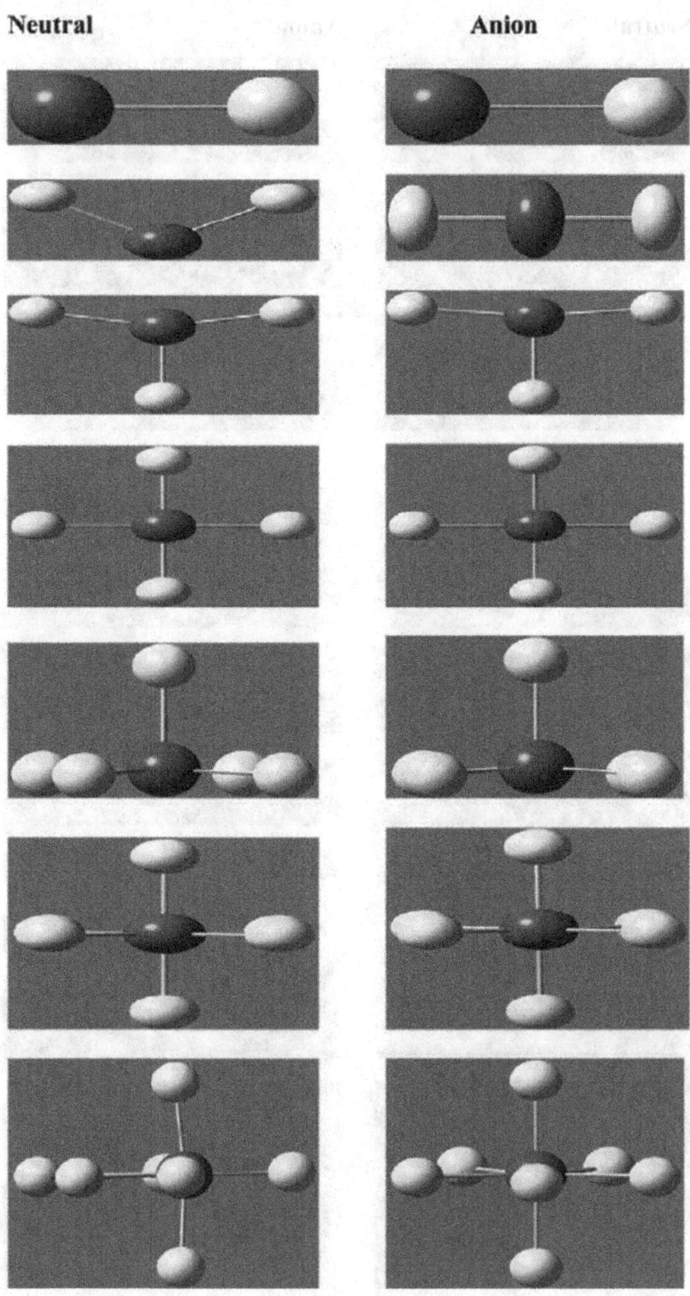

FIGURE 4.5 Optimised geometries of PdF_n Neutral and Anion clusters.

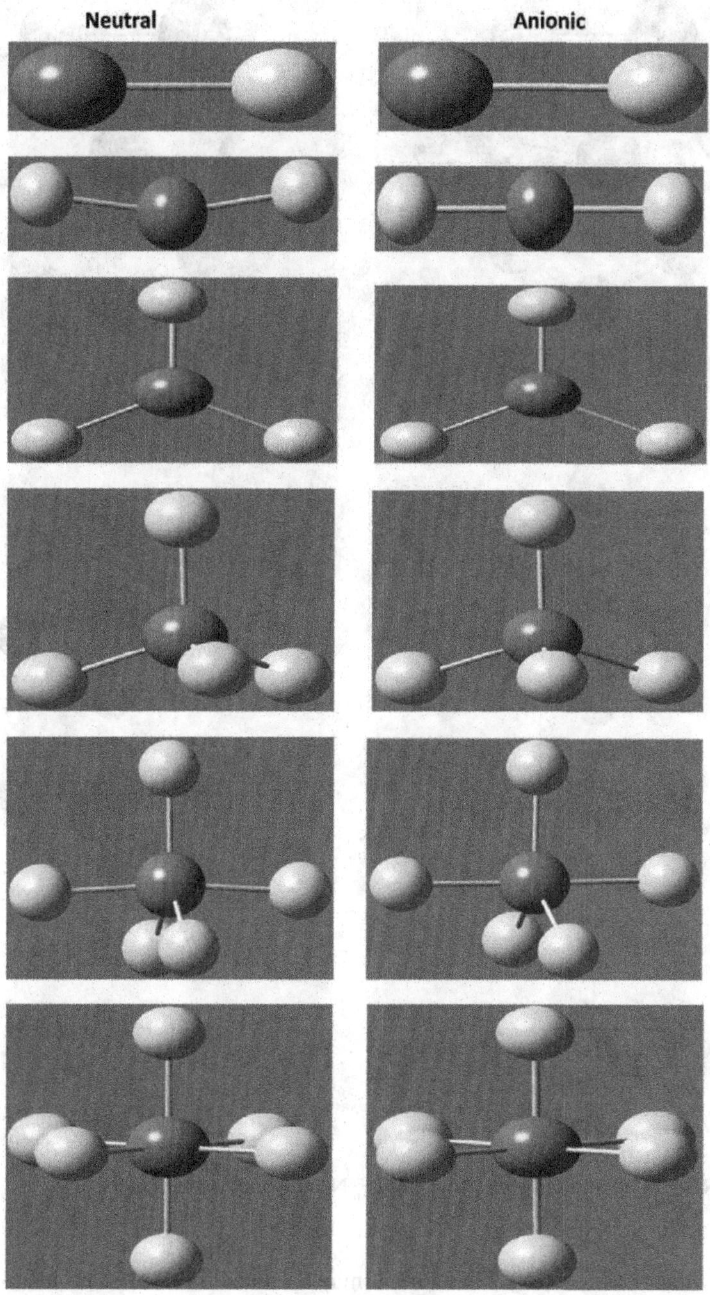

FIGURE 4.6 Optimised geometries of MnF_n Neutral and Anion clusters.

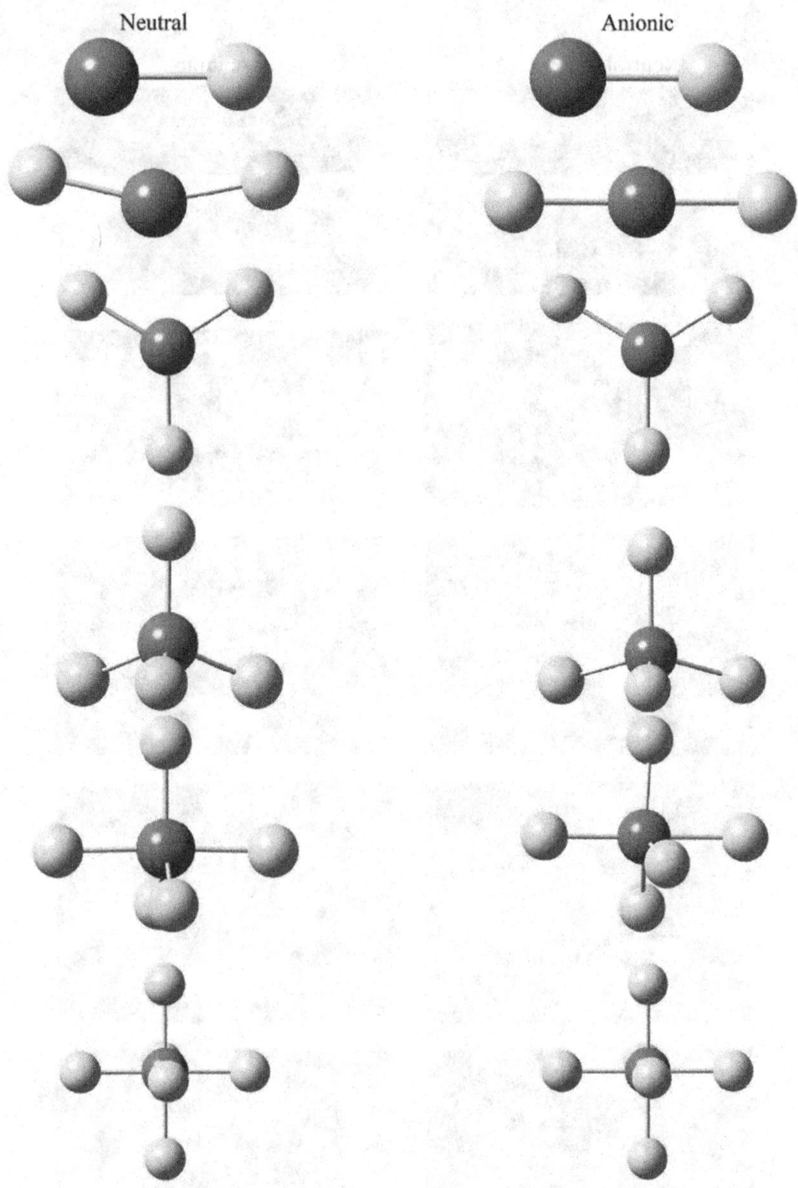

FIGURE 4.7 Optimised geometries of FeF_n Neutral and Anion clusters.

out by other workers [9,13]. Therefore, it may be stated that TMFs are much more likely to exist in neutral form rather than in the anionic form.

A particular interesting TMF for further discussion is of Figure 4.8 where average bond length of RhF_n (n = 1–7) cluster is shown. The transition metal Rh

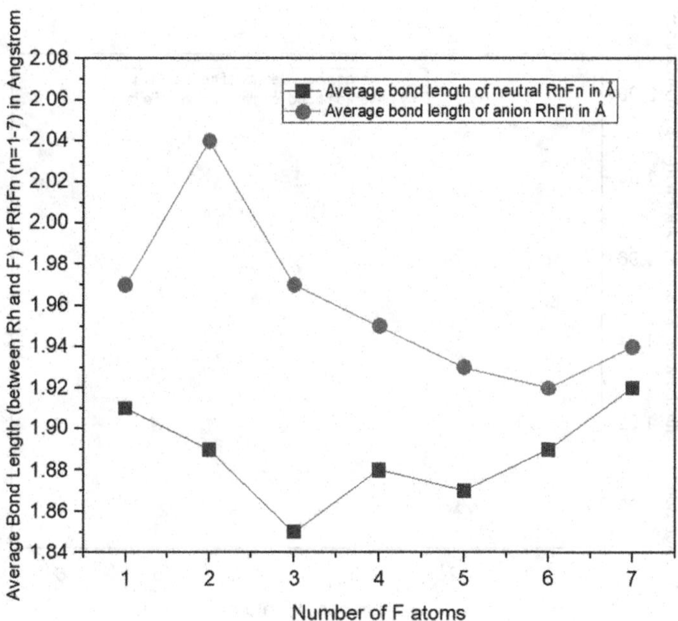

FIGURE 4.8 Avg. bond length of RhF_n (n = 1–7) cluster vs No. of F atoms.

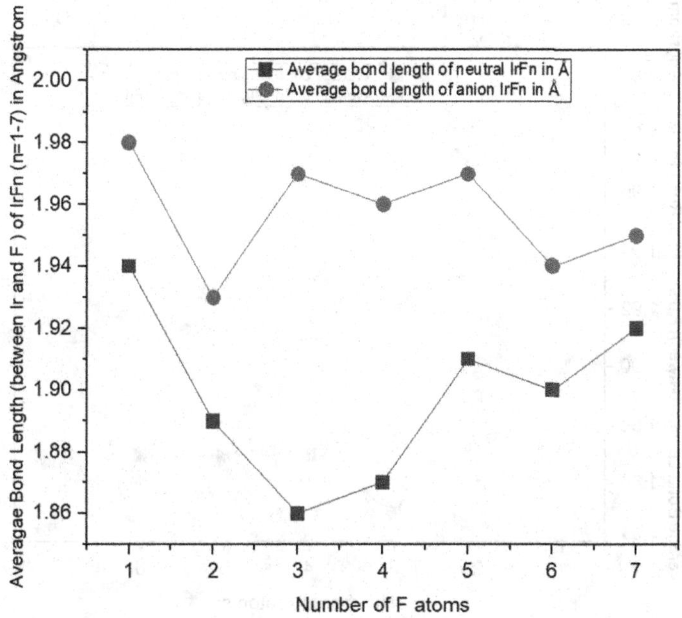

FIGURE 4.9 Avg. bond length of IrF_n (n = 1–7) cluster vs No. of F atoms.

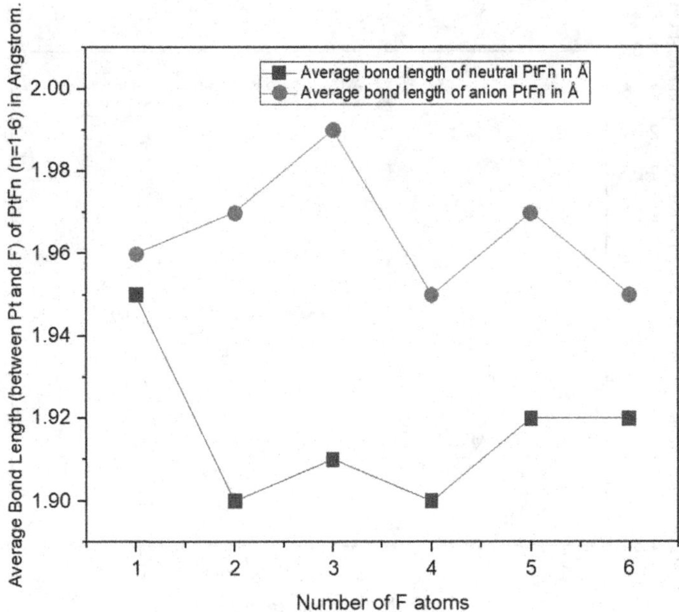

FIGURE 4.10 Avg. bond length of PtF_n (n = 1–6) cluster vs No. of F atoms.

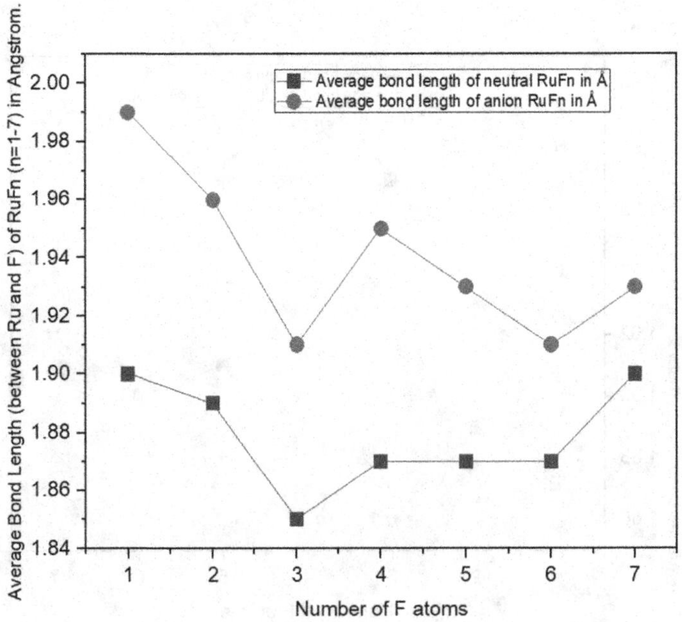

FIGURE 4.11 Avg. bond length of RuF_n (n = 1–7) cluster vs No. of F atoms.

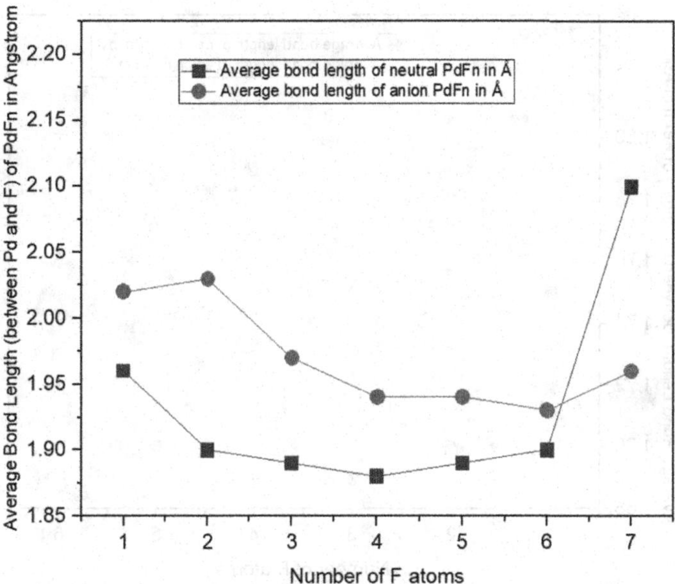

FIGURE 4.12 Avg. bond length of PdF_n (n = 1–7) cluster vs No. of F atoms.

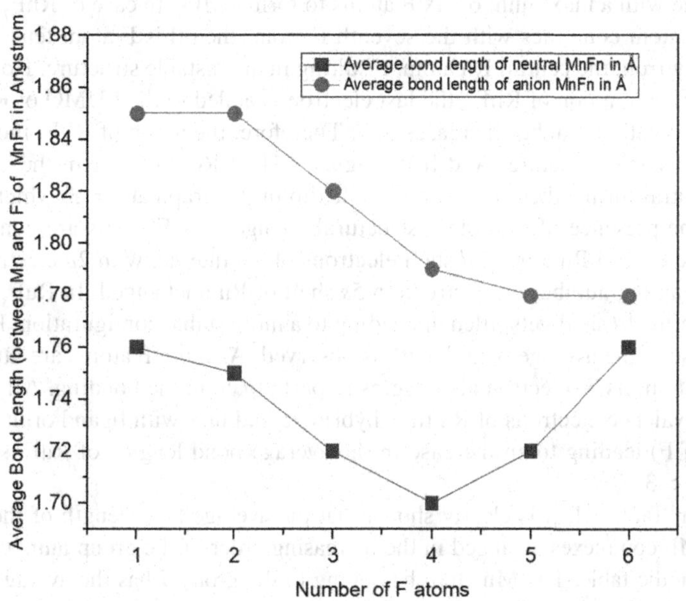

FIGURE 4.13 Avg. bond length of MnF_n (n = 1–6) cluster vs No. of F atoms.

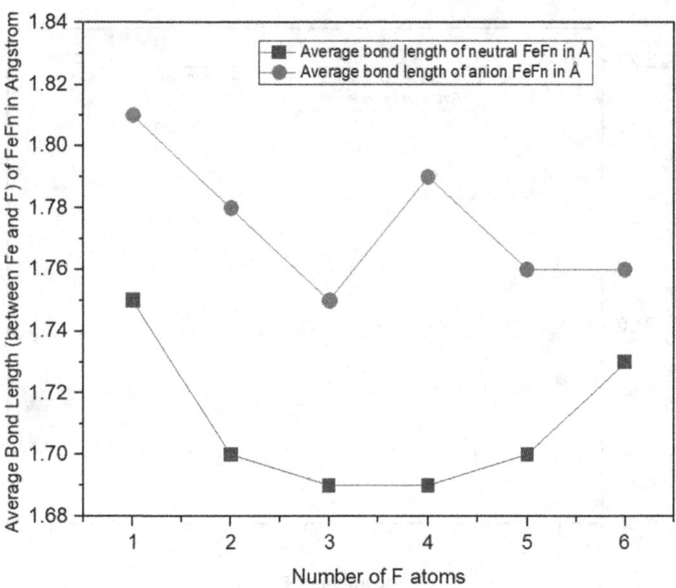

FIGURE 4.14 Avg. bond length of FeF_n (n = 1–6) cluster vs No. of F atoms.

(as discussed earlier) has the electronic configuration $[Kr]4d^85s^1$, which implies that the coordination number of Rh could not exceed beyond 6 and, hence, it can combine with a maximum of six F atoms to form TMFs. In case of RhF_7, when the Rh atom combines with the seventh F atom, then this F atom tends to go far away from the central Rh atom, resulting in an unstable structure. However, in the case of anion of RhF_7, the last electron is added to the LUMO of Rh and its coordination number increases to 7. Therefore, the anion of RhF_7 molecule exhibits stable structure. And from Figure 4.11 of RuF_n cluster in the anionic and neutral forms, there is an observable dip in the graph at n = 3, which indicates the presence of important structural changes. As F atoms are combined with the central Ru atom, 4*d* shell electrons of Ru interact with 2*p* electrons of F, leaving the number of electrons in 5*s* shell of Ru unchanged. In RuF_3 molecule, entire 4*d* shell gets filled up leading to a more stable configuration. Hence, decrease in the average bond length is observed. As more F atoms are attached to Ru atom, its 5*s* electron also begins to participate in the bonding. All the 4*d* and 5*s* valence electrons of Ru then hybridize and mix with ligand orbitals (2*p* shell of F) leading to an increase in the average bond lengths of RuF_n species when n > 3.

From Table 4.1, it is clearly shown that the average bond length of the various TMF complexes arranged in the increasing order of the group number from the periodic table. The Mn atom belonging to the group 7 has the average bond length of 1.735 Å while the Fe and Ru belongs to the group 8 of the periodic table and their average bond length is higher than the Mn atom complexes. The

TABLE 4.1
Structural parameter bond length comparison of TMFs central atom with Fluorine.

TMF complex	Structural parameter (average bond length of central metal atom with fluorine atom) in Å	Periodic table group No. of the central transition metal atom
$MnF_{(n)}$ [n = 1–6]	1.735	7
$FeF_{(n)}$ [n = 1–6]	1.710	8
$RuF_{(n)}$ [n = 1–7]	1.878	8
$RhF_{(n)}$ [n = 1–7]	1.887	9
$IrF_{(n)}$ [n = 1–7]	1.898	9
$PdF_{(n)}$ [n = 1–7]	1.891	10
$PtF_{(n)}$ [n = 1–6]	1.916	10

Rh and Ir belongs to the group 9 of the periodic table having average bond length higher than the group 8 complexes while Pd and Pt belongs to the group 10 of the periodic table showing the highest average bond length between the central atom to the fluorine atom showing 1.916 Å for Pt atom complex. This analysis of structural parameter bond length confirms that average bond length of the TMFs central atom with the fluorine atom increases as we move toward higher group number in the periodic table.

4.4.2 Fragmentation Energies of Transition Metal Fluorides

In Figures 4.15–4.21, the relative stabilities of various TMFs *viz.*, RhF_n (n = 1–7), IrF_n (n = 1–7), PtF_n (n = 1–6), RuF_n (n = 1–7), PdF_n (n = 1–7), MnF_n (n = 1–6), FeF_n (n = 1–6) in neutral and anionic forms have been plotted in the form of fragmentation energies vs number of F atoms graphs. Both neutral and anionic forms of subject TMFs have been studied by calculating the fragmentation energies needed to dissociate the TMFs into MX_{n-1}+F and MX_{n-2}+F_2 (where, M and X represent transition metal and F atom, respectively). Dissociation energies of neutral and anionic complexes are denoted by the terms ΔE_n and ΔE_n^-, respectively. They have been calculated by utilising the following expressions:

$$\Delta E_n = -\{E[MX_n] - E[MX_{n-m}] - E[F_m]\},\ m = 1,\ 2 \text{ and } n = 1,\ 2,\ \ldots v \quad (4.1)$$

$$\Delta E_n^- = -\left\{E\left[(MX_n)^-\right] - E\left[(MX_{n-m})^-\right] - E[F_m]\right\},\ m = 1, 2 \text{ and } n = 1, 2, \ldots v \quad (4.2)$$

In the above equations (4.1) and (4.2), v represents the variable valency that may be 6 or 7 as per the TMF chosen for investigation.

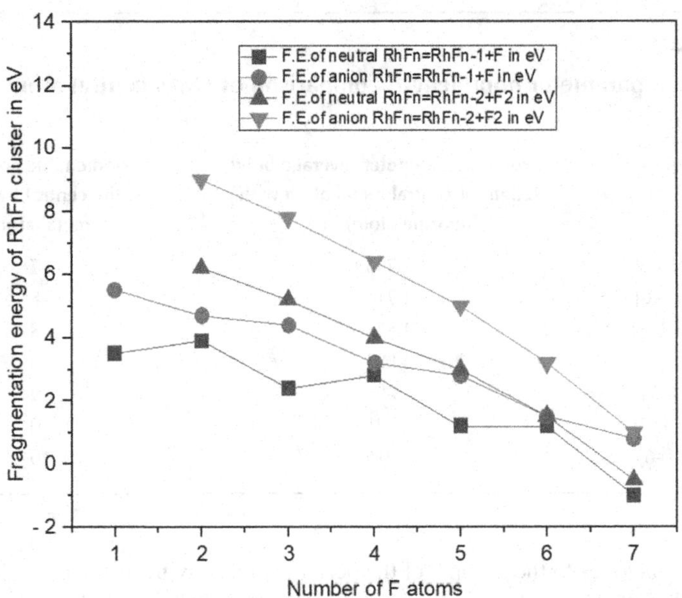

FIGURE 4.15 Fragmentation energy channel of RhF_n (n = 1–7) cluster vs no. of F atoms.

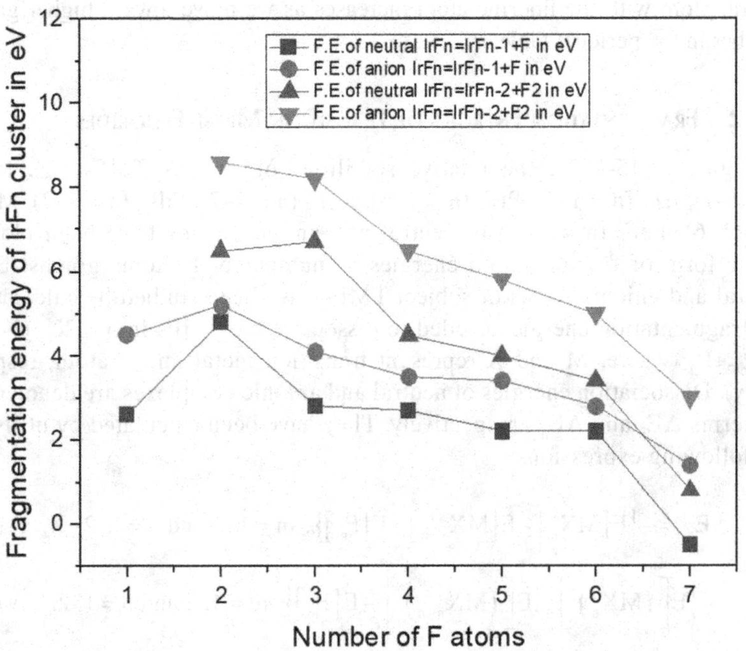

FIGURE 4.16 Fragmentation energy channel of IrF_n (n = 1–7) cluster vs no. of F atoms.

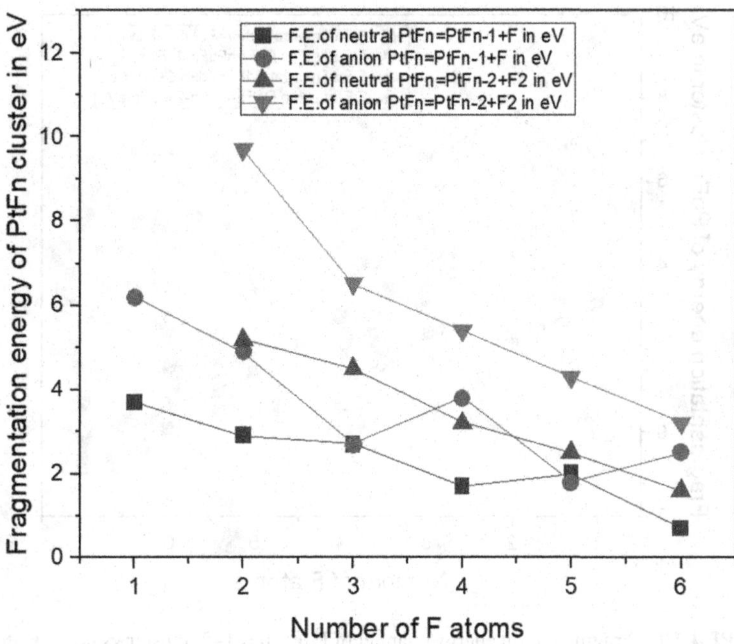

FIGURE 4.17 Fragmentation energy channel of PtF_n (n = 1–6) cluster vs no. of F atoms.

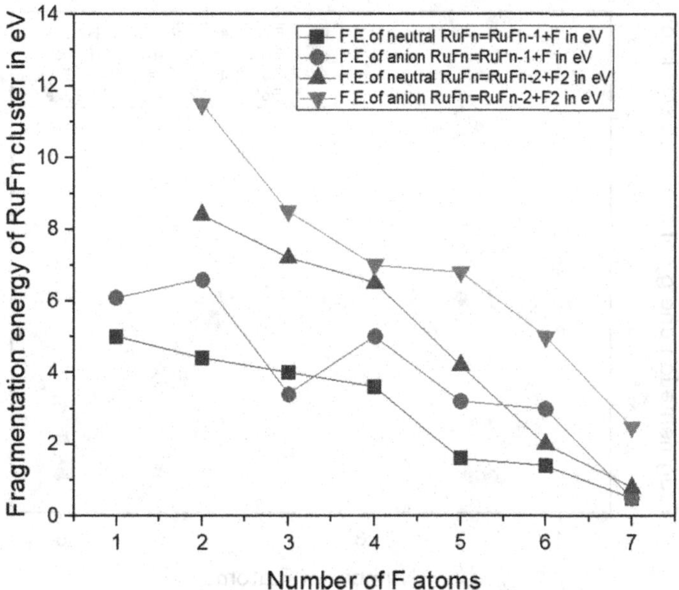

FIGURE 4.18 Fragmentation energy channel of RuF_n (n = 1–7) cluster vs no. of F atoms.

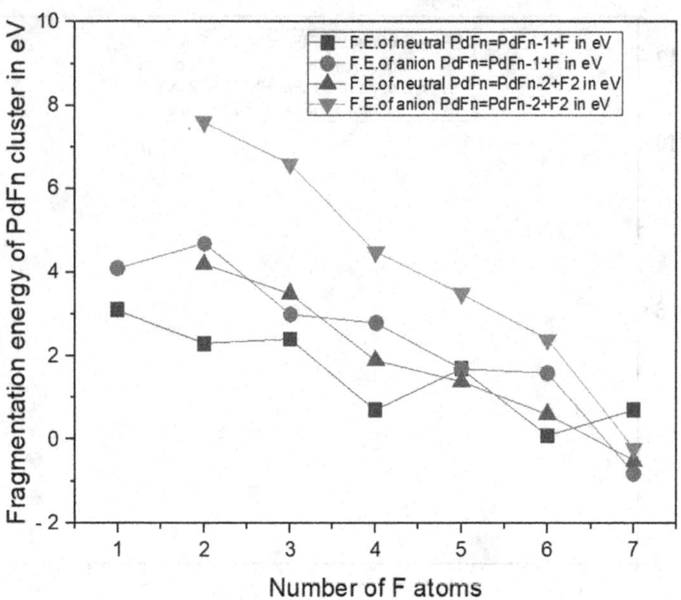

FIGURE 4.19 Fragmentation energy channel of PdF_n (n = 1–7) cluster vs no. of F atoms.

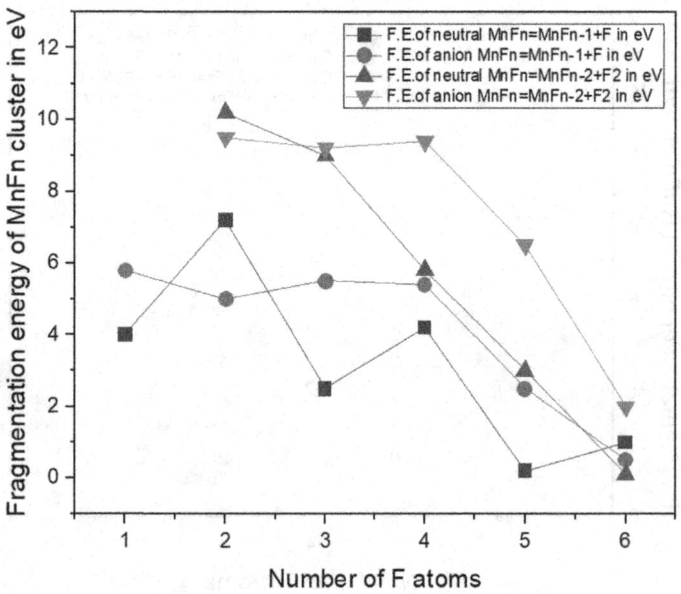

FIGURE 4.20 Fragmentation energy channel of MnF_n (n = 1–6) cluster vs no. of F atoms.

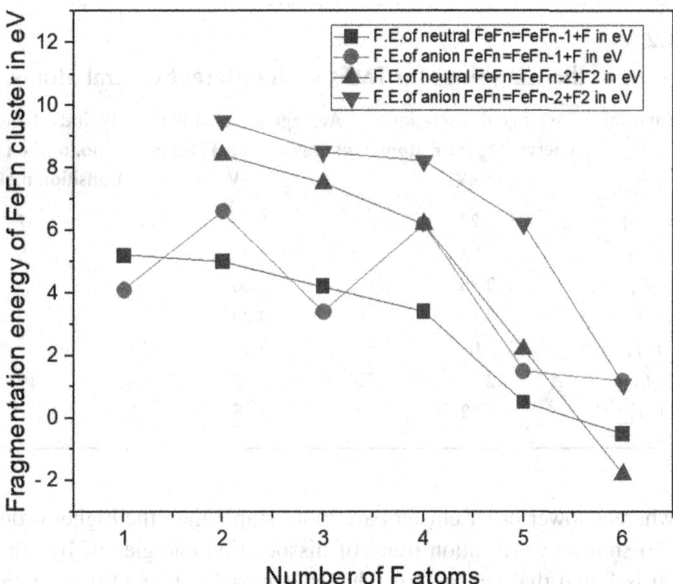

FIGURE 4.21 Fragmentation energy channel of FeF_n (n = 1–6) cluster vs no. of F atoms.

The dissociation energies of TMFs to F atom and F_2 molecule in neutral and anionic states are shown in Figures 4.15–4.21. It may be observed from the figures that there is a general trend of successive decrease in dissociation energies of both neutral and anionic forms of TMFs as the number of F atoms increase. This clearly indicates that the lower order clusters are more stable against dissociation (to F atom and F_2 molecule) as compared to higher order ones. Also, the dissociation energies of most neutral and anionic TMFs are positive, which clearly indicates that the binding energies are sufficient for protecting them against dissociation. Hence, it may be safely asserted that these molecules correspond to local minima on the potential energy surface (PES). Calculated vibrational frequencies of both the forms of subject TMFs have been found to be positive, which further supports this assertion. It may also be noted that, in general, the anionic forms of subject TMFs are more stable against dissociation to F atom and F_2 molecule as compared to their neutral forms. More specifically, these clusters are usually more stable against dissociation of F_2 molecule than F atom because dissociation energy of F_2 is higher as compared to the dissociation energy of the single F atom that shows that if F_2 is separated out from the complex it might be stable as compared to the single fluorine atom separation; therefore, F_2 molecules have higher probability to form rather than single fluorine atom. And F_2 has higher industrial application therefore this can be further utilized if dissociated from these superhalogens.

From Figure 4.15, it may be observed that the neutral RhF_7 cluster is unstable toward dissociation to both F and F_2 dissociation due to its negative fragmentation

TABLE 4.2
Comparison of the ΔE values for TMFs with different central atoms.

TMFs for (neutral system)	Average dissociation energy ΔE_n for F atom in eV	Average dissociation energy ΔE_n for F_2 atom in eV	Periodic table group no. of the central transition metal atom
$MnF_{(n)}$ [n = 1–6]	3.23	5.76	7
$RuF_{(n)}$ [n = 1–7]	2.98	4.80	8
$FeF_{(n)}$ [n = 1–6]	2.86	4.66	
$IrF_{(n)}$ [n = 1–7]	2.41	4.20	9
$RhF_{(n)}$ [n = 1–7]	2.0	3.10	
$PtF_{(n)}$ [n = 1–6]	2.2	3.42	10
$PdF_{(n)}$ [n = 1–7]	1.52	1.78	

energy, whereas lower order clusters are more stable than the higher order ones. Figure 4.16 shows the variation trend of dissociation energies of IrF_n (n = 1–7) clusters. It is found that neutral IrF_7 cluster is unstable toward dissociation to F atom due to its negative fragmentation energy while lower order clusters are more stable than higher order ones. The neutral and anionic PdF_7 cluster (Figure 4.19) is also found to be unstable against dissociation to F_2 molecule and F atom in neutral and anionic forms, respectively. The lower order TMFs containing Pd as central atom are found to be more stable than the higher order ones.

From above Table 4.2, Average dissociation energy (ΔE_n) of the fluorine atom and fluorine molecule is calculated. The results show that the ΔEn value for single F atom for Mn-F_n (n = 1–6) is 3.23 eV and for F_2 molecule is 5.76 eV which is higher for fluorine molecule as compared to the single fluorine atom. For all other TMFs when compared with their ΔE_n energy it is found that as we move from group 7 TMFs complexes to group 10 TMFs complexes there is a decline in the dissociation energy of both fragments like single fluorine atom and fluorine molecule. Single fluorine atom fragmentation energy or dissociation energy is found lowest for the PdF_n (n = 1–7) complex at 1.52 eV and similarly for F_2 molecule that is lowest when compared with other calculated TMFs at 1.78 eV. From this comparison analysis of dissociation energy with various TMFs it is seen that dissociation energy of F_2 is higher as compared to the single fluorine atom. This shows that dissociation of F_2 is more stable as compared to the single fluorine atom among various TMFs. And value of dissociation energy decreases as we move from group 7 to group 10 complexes due to increase of the atomic nuclei of the various central transition metal atoms that results in reduction of the dissociation energy of the fragments of single fluorine atom and fluorine molecule of various TMFs.

4.4.3 Adiabatic Electron Affinities of TMFs

Figure 4.22 shows the graph of variation trend of EA values (vs number of F atoms) of various TMFs *viz.*, RhF_n (n = 1–7), IrF_n (n = 1–7), PtF_n (n = 1–6), RuF_n

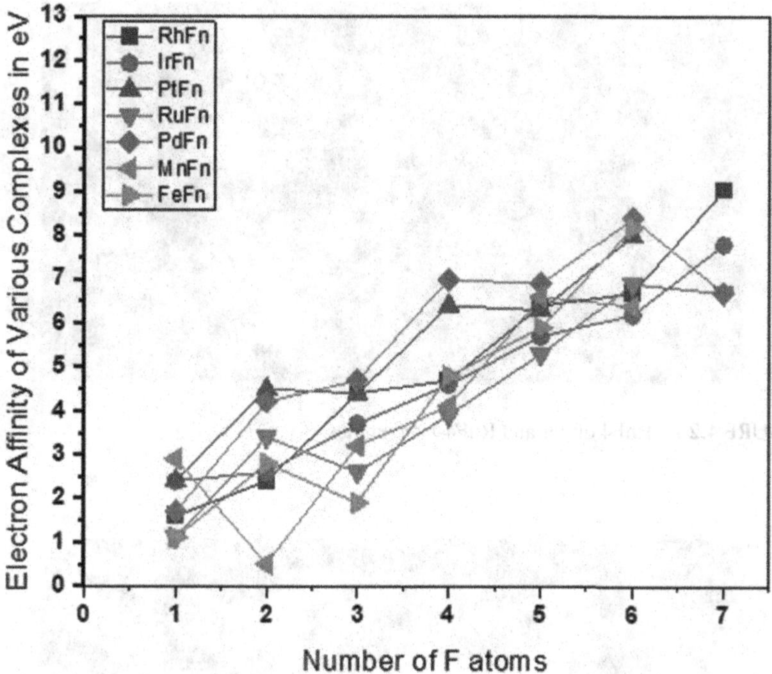

FIGURE 4.22 Electron affinities of all seven TMFs vs number of F atoms.

(n = 1–7), PdF_n (n = 1–7), MnF_n (n = 1–6), FeF_n (n = 1–6). These EA values have been calculated by taking the energy difference between neutral and corresponding anionic forms of TMFs, both in their ground state configurations. Among all the subject TMFs, the highest and second highest EA values are obtained as 9.05 eV and 8.4 eV for RhF_7 and PdF_6, respectively. FeF_6 and PtF_6 have the third and fourth highest EA values of 8.2 eV and 8.0 eV, respectively. This implies that TMFs having n > 4 exhibit very high EA values that are almost two times higher as compared to the EA values of Cl or F atoms. In general, from our results most of the TMFs show higher EA than the conventional halogens. Hence, due to their high structural similarity with the formula predicted by Gutsev and Boldyrev [4] and high EA, TMFs can be safely predicted to represent a new class of efficient superhalogens.

4.4.4 Optimised Structures of Selected Dimers and their Interaction with Alkali Atoms to Form the Superhalogen-Alkali Complex

From Figure 4.23 to 4.29, it is clearly shown that all the reported superhalogen forms the dimer that proves that superhalogen mimics completely halogen character of forming the dimers. The best suitable dimer configuration for all the shown TMFs belongs to their n = 4 cluster that suggests that higher order superhalogen will be favourable for producing dimers. Stability of these dimers is

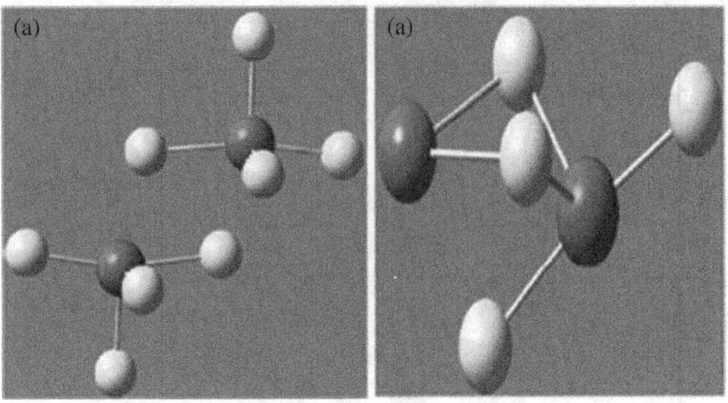

FIGURE 4.23 RhF4 dimer and RhF4-Na complex

FIGURE 4.24 IrF4 dimer and IrF4-Na complex

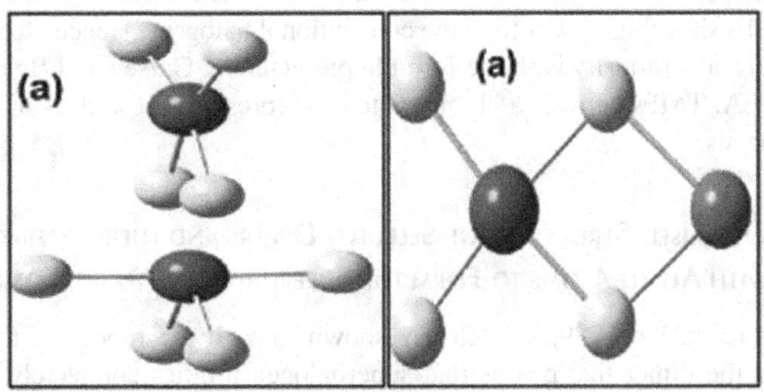

FIGURE 4.25 PtF4 dimer and PtF4-Na complex

to suffer from high staff turnover and low levels of staff motivation. Previous studies in this area have also shown that owner/managerial incompetence is the key factor for small firm failure (Panagiotakopoulos, 2020b).

The personal and wider social benefits, which result from increased training activity, should be added to the preceding list around the significance of staff learning. These benefits include increased job security, enhanced social status, improved employee morale, greater job satisfaction and a higher standard of living (Harrison, 2002; Panagiotakopoulos, 2020a). For example, low pay is an important part of the explanation for in-work poverty. Low pay is mostly concentrated in a few labour-intensive sectors of the economy and is especially prevalent in relatively small, private-sector service provider organizations operating in either very price-competitive markets (Panagiotakopoulos, 2016, 2019).

Recent research by the author has revealed that low-paying firms can become 'anti-poverty employers' by adopting employee-friendly HR practices including workplace training and development, flexible working and the provision of various non-wage fringe benefits that could improve the overall quality of working life for low-paid employees. The evidence shows that those small firms that can provide increased opportunities for staff learning do help their employees to secure advanced job posts with a better salary. In return, the organizations get back increased productivity and less absenteeism. Overall, the existing studies indicate that all the small enterprises, which can support the working poor using various training interventions seem to enhance significantly staff motivation and reduce employee stress, which leads to improved firm performance (Panagiotakopoulos, 2020a).

The preceding analysis is closely linked to the concept of 'good work' (Isles, 2010). For employees, 'good work' concerns the development of skills; choice, flexibility and control over working hours and the pace of work; trust, communication and the ability to have a say in decisions that affect them; and a balance between effort and reward. The concept of 'good work' is closely related to business ethics (corporate social responsibility), a term popularly equated with business philanthropy but more accurately described as all of a company's impacts on society and the need to deal responsibly with the impacts on each group of stakeholders. A strong record of corporate social responsibility improves an organization's image in both product and labour markets – aiding recruitment, retention and customer loyalty (Panagiotakopoulos, 2019).

Managing change in the small business context and the key role of learning

This last section examines the process of change in the small business context and provides guidance to firm owners on how to manage uncertainty based on effective learning provisions. It is undeniable that continuous change has become the norm in the contemporary business world, so small firm owners need to be ready to lead change effectively. Several variables may force small firms to introduce a number of operational or/and strategic changes in order to survive and remain competitive, such as unexpected external events (e.g. pandemic, environmental degradation), competitor behaviour, changing customer preferences, technological advances, regulations on product/service quality and so on. Normally, most owners adopt a responsive approach when it comes to managing change (i.e. they take action when certain problems arise) due to a lack of knowledge and resources to do proper forecasting. However, it is important to scan regularly both the external and internal business environment in order to adopt a proactive approach. Learning has a critical role to play in this respect since it can prepare employees for any forthcoming changes (i.e. it makes workers adaptable). Although the list of theoretical models in the existing change management literature is quite big, most models focus on the following four core stages (Carnall and Todnem, 2014):

1 *Identify the need for change*. In this initial stage, small firm owners need to identify any performance gaps, generate scenarios for anticipated changes in the internal and external environment of the organization and vision the future desired state. During that stage, small firm owners should review the planned strategic goals and monitor the actual organizational performance to see if the organization met its goals. At this stage, several quantitative and qualitative indicators should be monitored (e.g. productivity, sales, staff absenteeism rate, error rates, number of faulty products, customer satisfaction, etc.) to evaluate if there is any significant discrepancy from the initially planned objectives. It is important for small firm owners to place equal emphasis on qualitative performance indicators (i.e. *how* a goal has been achieved). Furthermore, an extensive analysis should be conducted on the external and internal environment of the organization to identify any key factors that may act as drivers for change.
2 *Explore resources for change*. In this stage, small firm owners/managers should attempt to identify various obstacles to change and examine ways to overcome them. For example, a common barrier to organizational change is employee resistance, which is normally caused by the fear of the unknown, the fear of financial loss, a threat to job security and the loss of existing social status, as well as loss of job autonomy. At this stage, it is also crucial for owners/managers to evaluate whether the firm has the necessary resources (physical, financial, human) or not to implement the required change, as well as assess the employee learning implications of the proposed change.

3 *Implement change*. In this stage, owners/managers should choose proper learning interventions to facilitate organizational changes. It is the stage during which owners/managers should provide a strong motive for change and help employees abandon old behaviours and adopt new ones. This stage normally involves the following steps: (a) scheduled staff meetings of the owner with existing employees to inform them about the nature of change, reasons for change and impact of change on their jobs in order to reduce their fear of the unknown; (b) increased informal staff learning provision to enable employees to develop the new gamut of skills that may be required by the organizational changes; (c) employee participation and involvement in change to feel that they are actively involved and go along with it; and (d) provision of suitable rewards (either monetary or non-monetary) to encourage employees to support with much enthusiasm the proposed changes (e.g. the provision of a flexible work schedule in return for changing work patterns). At this point of analysis, it is important to point out that those firms that managed to survive during COVID-19 were the firms that had an open-communication policy with their staff (i.e. regular updates on the pandemic, continuous advice, support for mental well-being, updates on company performance, etc.) and provided extensive training to their employees in order to cope with the emerging new working practices (e.g. information technology [IT] training to support extensive teleworking).

4 *Evaluate the outcome of change*. In this final change, small firm owners should monitor a number of key performance indicators (e.g. productivity, rate of errors, absenteeism, customer complaints, faulty products, etc.) before and after any organizational change to determine if the change actually had a positive impact on the overall performance of the firm.

The preceding brief analysis indicates that staff resistance to change is a common characteristic of most employees owing to the fear of the unknown. In this context, effective communication, training and negotiation through the provision of rewards are vital elements in the process of managing change effectively. The major challenge for small firm owners is to gain employee commitment during organizational changes instead of forcing through the proposed change by virtue of power. This requires from owners the ability to understand all the employee concerns. Although there are cases where a 'forcing' strategy might look like a desirable approach to implement change (e.g. when speed to organizational change is key such as, during a sudden financial crisis), still it can be risky in the long term because it can make employees feel frustrated and disappointed about owner/management attitude, thus leaving the firm as soon as alternative job opportunities become available.

Reflective Case Study

'*Microland - A*' is a small high-tech enterprise in the U.S. that produces payroll software programs for large corporations. It is run by a very enthusiastic and experienced owner. The company employs a general director, 10 full-time software design engineers (i.e. they design the programs working individually at their offices) and 3 full-time IT consultants (i.e. they sell the programs to the end users, they train them on how to use the programs and they provide after-sales support and advice). All design engineers and consultants are well educated, whereas all consultants have a strong background and expertise in software engineering. The rewards are very generous within the firm (above the market average) with consultants getting a higher wage than software design engineers (i.e. $400 per month extra). In terms of leadership, the owner has adopted a democratic style and staff participation in decision-making is highly encouraged. In terms of staff training, there are frequent formal seminars for both consultants and designers in order to further develop their technical skills, whereas informal learning is facilitated through extensive daily employee collaboration. Performance appraisal is conducted once a year by the general director based on a list of predetermined quantitative criteria (number of sales, number of programs developed, etc.). Each year, the best software designer (according to performance evaluation) is given the opportunity to get promoted to a consultant role (thus earning a higher salary), whereas the best consultant gets a holiday package and a generous bonus.

Problem

During the last year, designers' productivity has declined significantly, and this has caused serious problems for the firm.

Task

Based on case study data, discuss the impact of employee learning on company performance. Is staff learning at the heart of the business problems that the company is currently facing? Are there any other key factors that may affect organizational performance?

Key points of the chapter

This chapter has explored extensively the benefits of employee learning for individuals, organizations and society in general. The increasing interest that firms have been showing over recent years in employees and in practices related to their management, especially training, can be explained by the general acceptance of the fact that human resources and organizational knowledge are, at present, two of the main sources of sustainable competitive advantage for the company.

Although the contribution of staff learning to goal achievement is difficult to pinpoint due to the large number of variables affecting business performance, a number of research efforts have concluded that increased learning activity positively affects productivity, service quality and labour turnover. It further leads to reduced errors in production, reduced absenteeism, enhanced motivation and job satisfaction, increased staff confidence and job security.

The importance of learning is not confined to financial terms. A series of empirical studies have shown that increased employee learning in the small business context has a critical social role to play since it can be a core variable in the fight against in-work poverty. Moreover, the role of learning is crucial when it comes to managing change since it can ensure that employees have the set of new skills required to support the proposed change, as well as the required confidence to develop new behaviours and adapt to new circumstances.

References

Bakhshi, H., Downing, J., Osborne, M. and Schneider, P. (2017) *The future of skills: employment in 2030*. London: Pearson and Nesta.

Carnall, C. and R. Todnem (2014) *Managing change in organizations*. 6th ed. Essex: Pearson Education.

DeSimone, R. L., Werner, J. M. and Harris D. M. (2002) *Human resource development*. 3rd ed. San Diego.: Harcourt College Publishers.

Grugulis, I., Warhurst, C. and Keep, E. (2004) 'What's happening to skill?', in Warhurst, C., Grugulis, I. and Keep, E. (eds.) *The skills that matter*, pp. 1–18. Hampshire: Palgrave Macmillan.

Harrison, R. (2002) *Learning and development*. 3rd ed. London: Chartered Institute of Personnel and Development (CIPD).

Isles, N. (2010) *The good work guide: how to make organizations fairer and more effective*. London: Taylor and Francis.

Johnson, S. (2002) 'Lifelong learning and SMEs: issues for research and policy', *Journal of Small Business and Enterprise Development*, vol. 9, no. 3, pp. 285–295.

Kerr, A. and M. McDougall (1999) 'The small business of developing people', *International Small Business Journal*, vol. 17, no. 2., pp. 65–74.

Noe, R. (2019) *Employee training and development*. 8th ed. New York: McGraw-Hill.

Panagiotakopoulos, A. (2009) 'An empirical investigation of employee training and development in Greek manufacturing SMES', PhD thesis, University of Leeds.

Panagiotakopoulos, A. (2011) 'Workplace learning and its organizational benefits for small enterprises: evidence from Greek industrial firms', *The Learning Organization*, vol. 18, no. 5, pp. 364–374.

Panagiotakopoulos, A. (2016) *A short guide to people management for HR and line managers*. London: Routledge Publications.

Panagiotakopoulos, A. (2019) 'In-work poverty: reversing a trend through business commitment', *Journal of Business Strategy*, vol. 40, no. 5, pp. 3–11.

Panagiotakopoulos, A. (2020a) 'Exploring the link between management training and organizational performance in the small business context', *Journal of Workplace Learning*, vol. 32, no. 4, pp. 245–257.

Panagiotakopoulos, A. (2020b) *Effective workforce development: a concise guide for HR and line managers*. London: Routledge Publications.

Patton, D., Marlow, S. and Hannon, P. (2000) 'The relationship between training and small firm performance; research frameworks and lost quests', *International Small Business Journal*, vol. 19, no. 1, pp. 11–27.

Rainbird, H. (2000) 'Training in the workplace and workplace learning: introduction', in Rainbird, H. (ed.) *Training in the Workplace*. Basingstoke: MacMillan Press.

3 The process of learning in small firms

Nature of training in small firms

In the HRD literature, two major types of training have been identified in order to describe the ways organizations choose to develop employee skills, that is *formal* and *informal* training. *Formal* training involves all the planned learning interventions that have predetermined learning objectives, scheduled delivery, systematic evaluation of learning outcomes and accreditation of learning. By contrast, *informal* training involves all the learning interventions that are usually not planned and take place on an *ad hoc* basis.

A series of research studies conducted during the last two decades have confirmed that small firms are less likely than larger organizations to provide any sort of *formal* training intervention for their staff (Kitching and Blackburn, 2002; Panagiotakopoulos, 2011). All these studies have revealed that much learning in small enterprises takes on an *informal* character, with an emphasis on subtle forms of influence and guidance from employers and others at the workplace rather than planned forms of instruction. Training is often incorporated within routine working practices and is not an additional activity. Moreover, the findings have shown that most training is undertaken at the workplace because it is cost-effective, as well as can be tailored to suit the specific needs of worker and employer (i.e. it can be delivered at times and in formats that are convenient to both sides).

A common view in the literature suggests that training provision is positively correlated with firm size. However, this reflects a view that training is defined as purely *formal* and mainly provided by external training vendors. However, it should be pointed out that this reflects a very narrow definition of training. Widening the definition to include less-formal ways of learning reduces the gap between small and larger businesses. As Johnson (2002) argued several years ago, the extent to which there is a problem of limited learning activity in small firms depends on the definitions and

DOI: 10.4324/9781003381815-4

measures that are being used. The more formal and quantitative the measure, the more negative is the picture painted. Conversely, where the definition is broadened to include more informal learning activities, learning that takes place on the job and that does not necessarily lead to formal qualifications such as, learning by doing, teamwork, visits to trade fairs, dialogue with customers and suppliers, personal development meetings and staff meetings, the gap between small firms and larger employers is reduced. In this context, the notion of informal learning seems to be central to understanding the training strategies of small firms (Eraut, 2004; Panagiotakopoulos, 2009).

However, although employee training is important, it should be pointed out that without small firm owner support, the major focus of an organization is likely to be on activities other than training. Top management and employee cooperation and mutual contribution are prerequisites for the training interventions in order to benefit the organization and the individuals. In this context, employees must be well informed by the firm owner about the importance and benefits of the training activity in order to be motivated to participate actively in this process. Employees need to be ready to learn, whereas small firm owners need to know the factors affecting employee behaviour so that they can encourage and motivate their employees continuously. Such factors involve the general state of the economy, technological changes, the work environment, employee motivation to learn, knowledge and skills already possessed by employees and so on. For example, technological changes impact employee behaviour since technology creates redundancies and employee obsolescence, requiring the ongoing training of employees to acquire new, marketable skills.

Furthermore, it is of the utmost importance to be stressed that the whole process of employee learning is strongly inter-related with all the major human resource management activities including staff planning, recruitment and selection, performance appraisal and rewards (as discussed in Chapter 4). For example, poor human resources forecasting (e.g. labour shortage) may result in the provision of more extensive training so that existing staff may cover all the business needs. Similarly, poor selection (i.e. owners not following a rigorous process) may lead to an increased need for further training for newcomers. Also, a lack of performance evaluation may lead to the design of learning activities that are not oriented towards filling the required employee learning gaps (i.e. learning efforts that lack intent). Moreover, the absence of a reward package that can motivate employees to acquire new skills may prove detrimental towards human resource development. In short, employee learning should be a part of a coordinated effort to improve the productive contribution of people in achieving the firm's objectives. Changes in just one part of the human resource management (HRM) function are unlikely to bring the desired organizational outcomes.

The process of employee learning in small firms

Initial training (i.e. employee orientation) is essential for any organization regardless of its size since it enables new employees to become familiar with their work tasks, get to know their colleagues and become aware of the firm's core cultural values. This, in turn, helps the new employee feel bonded to the organization very quickly, reduces their stress caused by the fear of the unknown, boosts their confidence and gives them a sense of belonging to the company. Several research studies have shown that a rigorous employee orientation program has a radical impact on staff turnover (Stone, 2002; Panagiotakopoulos, 2013, 2014). In the small business context, employee induction can be done by the firm owner or some other experienced colleague. It may involve a range of learning activities ranging from self-learning audio-visual material to observational learning (i.e. shadowing). The preceding learning process concerns the induction of all newcomers.

For the existing staff, the usual learning process in small firms may take place in 3 main stages.

The learning needs analysis stage

This stage is the first step in developing employee skills in any business context. Learning needs analysis is a process by which an organization's learning needs are identified and translated into learning objectives. According to Armstrong (1999), training needs analysis is essentially concerned with defining the performance gap at work. This initial phase of any learning process establishes what kind of training is needed, by whom, when and where. Without determining the need for employee training, there is no guarantee that the right training will be provided for the right employees (i.e. training may end up being a waste of time and effort). Training needs assessment can also reveal if staff training is the appropriate solution to a performance problem, since all performance issues are not training issues. Other variables at work, including tools and equipment, factory space, work processes and poor leadership, can impact employee performance and may not have any training implications at all. Also, this stage can be used also as a diagnostic process since it can reveal shortcomings in several other core HRM activities (as noted in the previous section).

In the small business context, the major tool used to identify employee learning gaps is the performance evaluation conducted by the firm owner along with an overall assessment of the external business environment (i.e. macro-environment scanning). More specifically, the owner should perform a careful and detailed performance evaluation (using relevant tools as described in Chapter 4) to identify any weaknesses in employee performance that training interventions might improve. Although a rigorous performance appraisal can be a time-consuming process, yet the size of small enterprises provides the owner with the chance to conduct a meaningful performance evaluation process without spending too much time and money. For example, a behaviour rating scale tool can be used to assess staff performance in specific job-related criteria that will give the owner the opportunity to identify any major employee learning gaps. Direct observation (by the owner) of employee performance is just one method of identifying major learning gaps. Customer feedback and peer observations could also form the basis for some valuable input, so they should not be ignored. Both customers and colleagues are significant sources of information when it comes to employee learning. Hence, the small firm owner should pay careful attention to these raters of staff performance.

It should be stressed that not all organizational problems can be solved through training. There is a range of issues at work where training may have minimal or no contribution (e.g. a lack of proper equipment/machinery, poor rewards, etc.). Therefore, it is important for owners to identify the root of the organizational issue (i.e. diagnosis of a problem) before designing any learning interventions.

The action stage

Informal learning is the predominant skills development practice within the small business sector due to various resource constraints that such organizations face resulting in employees being much less likely than those in larger ones to have the opportunity to participate in off-the-job (formal) learning interventions. Furthermore, informal learning is especially valuable for small enterprises because they occupy increasingly specialist markets, thus having a need for highly *specific* skills rather than 'generic and transferable' skills mostly developed through formal training. For the purposes of our analysis, informal learning is used to denote any learning activity (either planned or incidental) that takes place at the workplace and does not involve any sort of external training provider or formal accreditation of learning.

Before exploring the informal learning practices in small firms, it is crucial to explain how individuals learn using the popular learning cycle developed by David Kolb in 1984. The learning cycle reveals that learning is a continuous process, the direction of learning is determined by individuals' needs and goals and it is affected by different styles of learning. The learning cycle consists of 4 main stages. The initial stage is the one in which the learner faces a particular experience (i.e. deals with a social situation). The experience is then followed by the individual's effort to understand and clarify the situation (i.e. the learner reflects on their new experience in the light of their current knowledge). This stage leads to the crucial cognitive one where the learner attempts to solve the problem and draw conclusions (i.e. reflection gives rise to a new idea). In the final stage, these conclusions are put into practice (i.e. the learner applies their ideas to their contexts).

It is important for small firm managers to understand that employees have different learning styles. There are people who learn through active experimentation (i.e. doing) and people who are abstract thinkers (i.e. theorists) and are keen to develop theoretical models. Therefore, the learning interventions designed should take into account that sort of learning diversity. Looking at Kolb's learning cycle, it becomes evident that informal learning comes with its limitations. For example, employees do not have much opportunity to reflect on their experiences since they learn while performing tasks (i.e. there is frequent interruption of learning, etc.). Therefore, it is crucial that the small firm owner provides sufficient time for employees to reflect on their daily experiences and develop a sound theoretical understanding that can facilitate behavioural change.

The main training techniques used for Informal learning interventions involve the following:

Coaching: Essentially, this concerns the guidance that employees get by their immediate supervisors (e.g. small firm owner or shop manager) in relation to their job tasks. The supervisor provides a specific example of what is to be achieved and then examines the employee's performance and offers counselling on how to maintain effective performance and correct any performance problems. The process of coaching is complex and requires a range of qualities to be possessed by the supervisor. In particular, the supervisor needs to keep a steady pace of guidance, as well as make clear the relevance of the particular learning intervention to individual employee needs and organizational objectives.

Furthermore, as mentioned in the previous paragraphs, it is equally important to accommodate different learning styles. Numerous research studies in the fields of education, psychology and neuroscience have concluded that every individual learns differently (Kolb, 1984; Paine, 2019). In simple terms, this means that an employee's learning style refers to the preferred way in which they absorb, process, comprehend and retain information. Some employees may understand a job task by receiving proper verbal instructions, whereas others may need to see it in practice. Employee learning styles depend on cognitive, emotional and environmental factors, as well as one's previous experiences. It is therefore important for a supervisor to understand the differences in their employees' learning styles so that they can implement best practice strategies into their daily activities when doing coaching.

Moreover, constructive feedback provision is of the utmost importance. Detailed, timely and accurate feedback clarifies expectations, helps employees learn from their mistakes and builds confidence. Feedback should be clear and specific, as well as include suggestions on how employees could do things differently using 'positive' language (e.g. instead of saying, 'You shouldn't do this because ...', the supervisor may say, 'You could try doing this since ...'). It is important to avoid using personal critique towards the employee. When providing feedback, the focus should be on job behaviour not on the employee's personal attributes (i.e. personality traits).

Another core element of effective coaching is careful listening to employee concerns. Showing empathy to learners is important. The supervisor is vital to understand the emotional state of employees while they participate in the learning process (e.g. identify any signs of resistance to learning), as well as adopt a positive stance in difficult situations. Other key coaching elements include the ability to encourage employees during learning, allow them gradually to undertake further responsibilities, build trust and set specific and challenging goals to be achieved.

Shadowing (observational learning): This is a process whereby the employee observes a task which is done by others (e.g. manager, coworker) who serve as role models and then the employee performs the observed act. In other words, this is a learning process through which employees develop their skills by simulating other people's behaviour. During shadowing, it is important for employees to have a clear overview of the tasks to be performed by the supervisor, as well as have the time to reflect on the supervisor's behaviour. A follow-up meeting after each shadowing session is essential so that the supervisor can clarify the lessons to be learned.

Communities of practice: This could also be a form of both planned and unplanned learning since employees get to develop their knowledge and skills through extensive interaction with their counterparts (i.e. the process of social learning). The theory of social learning indicates that learners are active in the learning process since they observe their environment and become motivated to reproduce a behaviour from their social context. In other words, social learning theory is based on the axiom that people may also learn from their interactions with others in a social context (Bandura, 1976). Communities of practice are social structures of employees who share an interest and passion in a specific work domain, and they learn how to perform their jobs more effectively as they interact with each other. Although that form of learning has its own limitations (e.g. the quality of learning may be questioned since the person who shares their knowledge around a specific subject area may not be an 'expert'), yet it provides employees with various benefits, including extensive learning and improved retention of information and satisfaction of their social needs (Panagiotakopoulos, 2020). Social learning at work can be enhanced through various ways including team-working tasks and projects and proper arrangement of the physical environment, as well as online employee social forums. It should be stressed that trust, familiarity and mutual understanding among employees are necessary preconditions for the successful sharing of tacit knowledge since members of '*communities of practice*' are normally less willing to share knowledge and expertise in situations where trust is lacking.

Learning by doing: This type of incidental learning is based on the wider concept of experiential learning where employees learn by experience (e.g. trial and error). This learning technique can be enhanced at work through job enrichment. In particular, as employees get the chance to perform a wide variety of tasks or participate in new projects they are exposed to a range of challenges and problems. This problem-solving process enables them to develop their subject-matter expertise and develop a vast array of skills. At the heart of experiential learning is the opportunity to engage in a range of new tasks, have sufficient job autonomy and a good network of colleagues to stimulate knowledge sharing. Therefore, job design becomes central to this type of learning. It is suggested that the small firm owner should avoid creating jobs that are highly monotonous and restrictive in terms of skills acquisition and application. As is pointed out in Chapter 4, an interesting job with various and challenging tasks can prove to be a powerful motivator for employees.

The evaluation phase

Employee learning evaluation is defined as the collection of information necessary to measure how well a learning intervention met its objectives (Stone, 2002). Unfortunately, the vast majority of small enterprises do not include an evaluation phase as part of their learning efforts due to misconceptions that undermine this stage of the learning process, such as the perception that learning evaluation adds cost to the learning process, as well as it is very time-consuming. However, learning evaluation is a key stage in the workforce skills development process that cannot be ignored since it can help small firm owners determine whether a learning intervention met its objectives or not (i.e. to understand if the learning activity was suitable or needs further improvement), as well as identify if employees actually benefitted from the learning activity and improved their job performance. It is important that the evaluation is conducted before, during and at the end of each learning intervention. This enables small firm owners to understand the extent of knowledge and skills possessed by employees before they start the learning activity, evaluate the effectiveness of the learning intervention by assessing employees' understanding during the learning event and evaluate whether the learning intervention met its objectives (Panagiotakopoulos, 2020). To measure the effectiveness of employee learning, small firm owners should monitor the following: the extent to which employees' enjoyed participating in the learning intervention, the extent to which employees expanded their knowledge after a learning intervention and were able to apply their learning to their current job tasks and the extent to which the learning intervention benefitted both the enterprise and employees (e.g. measurement of various performance indicators such as error rates, customer complaints, etc.). Direct owner observation, peer observation and customer feedback could be valuable sources of information in this process.

Barriers to employee learning in small enterprises

The existing studies around this area have revealed that there are a host of factors that affect the propensity of small firms to train (Panagiotakopoulos, 2011). In particular, small firms suffer from a number of specific obstacles that make it difficult for them to engage in learning activities, including time and cost constraints; 'poaching' concerns (i.e. fear of skilled labour being 'poached' by larger companies) due to the absence of internal labour markets and poor financial rewards; managerial ignorance on the long-term impact of training on firm performance (i.e. low managerial commitment); problematic job design and work organization that provide limited opportunities for learning (e.g. narrow range of tasks, repetitive work, low employee autonomy and involvement, etc.); absence of a systematic and well-organized performance appraisal process to identify skill gaps; low employee commitment due to the lack of promotional opportunities; a lack of in-house expertise; owner fear of losing bargaining power; competitive pressures; and available skills in the external labour marker. For example, small firm owners usually are not willing to invest time and money in workforce skills development interventions (especially formal) bearing in mind the possibility of well-trained labour being 'poached' by larger competitors who may be able to offer more generous financial rewards and working conditions. However, one of the major factors in any discussion about training in the small business context appears to be the small firm owners' attitude towards employee development since firm owners determine the overall learning culture for the company. Existing evidence shows that some owners in certain sectors (e.g. high-tech) are willing to invest in staff development arguing that developing their talents will help them gain a sustainable competitive advantage in the market, whereas several others view employee learning as an operational expense rather than an investment (i.e. owners who are either not aware or yet to be convinced of the benefits of employee learning).

Overcoming the major barriers to learning

As mentioned in the previous section, one of the key barriers to employee learning is owners' ignorance about the several benefits of skills development for company performance. The previous chapter (i.e. Chapter 2) gives a good overview of the main research findings around this area and lays the ground for the provision of more frequent learning opportunities at work. For those firm owners who do acknowledge the importance of stall learning, the next sections provide useful guidance on how to overcome a series of other core barriers to learning. In short, the subsequent analysis indicates that small firm owners can take a number of steps in order to create a conducive learning environment, in particular the following:

1 They need to adopt a learning approach to the overall business strategy. This means that they should be willing to introduce small operational changes that are aligned with the general firm strategy, which they treat subsequently as experiments for performance improvement.
2 It is also helpful on several occasions to adopt a participative leadership style, where valuable employee input is provided on all business activities. In this way, owners get the chance to identify the root of any organizational deficiencies (i.e. employee knowledge gained through experience lays the ground for future learning interventions), whereas employees feel motivated and bonded to the company since they realize that they form a key part of the organization (i.e. their own learning is at the heart of future developmental activities). More information on this process is provided in Chapter 5.
3 They should conduct regular performance appraisals to identify employee learning needs. This ensures that any learning intervention does not take place for its own sake (i.e. it is not a waste of time and money) and that employees do broaden their knowledge base in areas that do matter to them and the organization itself.
4 They should provide a range of financial or/and non-financial rewards to employees for the exchange of ideas, as well as for the application of new skills. For example, the provision of an extra day off or restaurant vouchers may be given to those employees who promote learning at work significantly (e.g. an experienced worker providing extensive guidance and advice to a newcomer).
5 They should design jobs in such a way that they enhance workplace learning (e.g. encouragement of team working through proper task allocation). They should also focus on providing their staff with frequent informal learning interventions that do not require extensive time and budget (e.g. extensive coaching by experienced colleagues, as well as shadowing).

6 They should allow staff to reflect on their behaviour, as well as question ideas and actions. It is important for the owner to view employee mistakes as learning opportunities. For example, a formal meeting at the end of each month among employees and their immediate supervisor (or small firm owner) may provide enough room for reflection. Also, the arrangement of frequent informal social events outside working hours may promote this sort of social learning.
7 To avoid 'poaching' of skilled labour, much emphasis should be placed by the small firm owner on the provision of various non-monetary rewards, which do not require many resources and are highly valued by most employees according to several research studies. For example, the following rewards have been identified in the literature as having a radical impact on the intention of employees to stay loyal to their employer: interesting job (varied tasks and responsibilities that lead to increased job satisfaction), flexible working arrangements, supportive colleagues (chosen through a careful staff selection process), competent supervisor, tolerance of mistakes, learning opportunities, health and safety policies (increased welfare policies) and fair treatment at work (absence of discrimination; Panagiotakopoulos, 2014, 2016).

Reflective Case Study

'*PhoneplanU*' is a small mobile phone retailer in the U.K., which is owned by a very experienced IT specialist. The company sells high-quality pay-as-you-go mobile phones and mobile phones with fixed-term usage contracts for all networks through its 4 branches in London. It operates in a highly competitive marketplace, competing with 5 very large mobile phone retail companies and a host of smaller independent retailers.

Leadership and work organization

Each branch employs 5 full-time employees plus a general branch manager. Each branch sells a vast array of technologically advanced mobile phones. In terms of marketing promotion practices, the company produces a monthly brochure summarizing its best deals, which is widely distributed to consumers. As for the selling techniques, sales staff always use a computer program that prompts them to ask questions about customer requirements, which then produces a list of recommended products and deals to choose from. The firm owner has adopted an autocratic supervisory style in order to monitor closely all staff and business processes with the help of the general branch directors. Employee participation in decision-making is non-existent, and all strategic decisions are made by the company owner. Employee job autonomy is very limited (i.e. all staff should follow a specific process around selling) so that the owner could ensure consistency in selling, whereas there is no task variety (i.e. each employee has a very restrictive set of tasks to perform in order to develop a strict specialization in their jobs).

Staff selection, performance appraisal, training and rewards

Store managers have responsibility for hiring sales staff, although they receive much guidance and support from the company owner. The hiring process involves the submission of a formal application form by the prospective candidate and the attendance of a formal structured interview with the respective branch manager. The selected candidates normally possess higher education qualifications in the area of 'business and management'. Almost all employees are under 26 years of age. The company currently pays its sales staff at an hourly rate of 9£ an hour (i.e. the national minimum wage), with an extra bonus for meeting sales targets. The maximum bonus an individual is likely to receive will add 10% to monthly wages. Only the top 20% of sales staff receive any bonus at all. Top-performing sales staff and shops are praised in the company's weekly email newsletter to all staff. The overall performance of each branch is closely monitored by the company owner using strictly a number of financial indicators (e.g. sales, profitability). Stores are also set very strict quantitative sales targets, and employee performance is measured purely against those targets through monthly performance appraisal reports (e.g. the company's software system records the number of enquiries handled by each member of staff, the proportion of enquiries that result in sales and the value of each sale). The company owner deals with under-performing store managers and store managers deal with under-performing employees. The attitude of the company owner towards under-performing managers and employees is very much one of 'get improved or you'll get fired'. In terms of staff training and development, all newly hired employees receive a 3-day intensive on-the-job training around their main tasks, whereas existing staff attend once a year a formal training seminar around health and safety.

Overall working environment within the firm

Internal communication takes place predominantly in a formal way through scheduled personal meetings and emails. All internal communication is made available to the owner of the company as part of the quality monitoring process. As for the informal communication among employees in each shop (i.e. horizontal communication), this is generally discouraged in the company. Team working is rather limited in the company because all employees work individually at their own workstations, which are equipped with all the necessary IT facilities. In terms of organizational culture, the official company values state that friendly customer service and the provision of high quality are at the heart of business success.

In practice, the firm owner makes all the key strategic business decisions and keeps store managers and employees under tight control. The working environment is competitive as poor staff performance is not tolerated and superior performance is rewarded.

Problems

The firm owner plans to open 2 more new stores in the next 3 years in different regions across the U.K. However, over the last 6 months, sales in existing stores have begun to stagnate, whereas the company is not performing as well as its closest rivals. Also, staff turnover is high, with 50% of staff having left the company during that period.

Tasks

Critically analyze the roles of ***leadership, motivation, communication*** and ***culture*** at PhoneplanU using relevant theory and academic literature, as well as provide specific suggestions on how problems could be resolved. Consider the role of ***employee learning*** and discuss its contribution towards business success in the specific case-study scenario.

Key points of the chapter

Employee learning activities begin when a newcomer joins the organization (i.e. induction), usually in the form of employee orientation. Newcomers must learn new behaviours, facts, procedures, values and must establish relationships in order to be successful in a new position. The next steps in the learning process involve the learning needs analysis stage (i.e. to identify what training is needed and by whom), the action stage (i.e. design and implementation of learning interventions) and the evaluation stage (i.e. measuring the impact of a learning activity on firm performance).

The major informal learning techniques involve coaching, shadowing, communities of practice and learning by doing. Each one has its own strengths and limitations. Coaching is the most popular informal learning technique in the small business context, which can help employees develop a range of job-related skills. Effective coaching requires from the firm owner an in-depth understanding of the different employee learning styles, the timely provision of clear and constructive feedback, careful listening of employee concerns and understanding of their needs (i.e. empathy), mutual trust with the learners and continuous encouragement during the learning process.

Several barriers to learning have been identified in the literature including ignorance on the benefits of staff learning, resource constraints, the chosen business strategy, limited management commitment, poaching concerns and many more. Small firm owners can take various steps to overcome the main obstacles to learning, including the implementation of regular, well-designed performance evaluations to identify any employee learning gaps; encouragement of employee input on the business practices (i.e. employee involvement in decision-making); and the provision of suitable rewards that can motivate staff to learn (e.g. interesting job, flexible work practices, etc.).

References

Armstrong, M. (1999) *A handbook of human resource management practice*. 7th ed. London: Kogan Page.

Bandura, A. (1976) *Social learning theory*. Hoboken: Prentice Hall Publications.

Eraut, M. (2004) 'Informal learning in the workplace', *Studies in Continuing Education*, vol. 26, no. 2, pp. 247–273.

Johnson, S. (2002) 'Lifelong learning and SMEs: issues for research and policy', *Journal of Small Business and Enterprise Development*, vol. 9, no. 3, pp. 285–295.

Kitching, J. and R. Blackburn (2002) *The nature of training and motivation to train in small firms*. Department for Education and Skills (DfES), Research report No 330. Nottingham: DfES publications.

Kolb, D. A. (1984) *Experiential learning: experience as the source of learning and development*. Hoboken: Prentice Hall Publications.

Paine, N. (2019) *Workplace learning: how to build a culture of continuous employee development*. London: Kogan Page.

Panagiotakopoulos, A (2009) 'An empirical investigation of employee training and development in Greek manufacturing SMES', PhD thesis, University of Leeds.

Panagiotakopoulos, A. (2011) 'What drives training in industrial micro-firms? Evidence from Greece', *Industrial and Commercial Training*, vol. 43, no. 2, pp. 113–120.

Panagiotakopoulos, A. (2013) 'The impact of employee learning on staff motivation in Greek small firms: the employees' perspective', *Development and Learning in Organizations*, vol. 27, no. 2, pp. 13–15.

Panagiotakopoulos, A. (2014) 'Enhancing staff motivation in tough periods: implications for business leaders', *Strategic Direction*, vol. 30, no. 6, pp. 35–36.

Panagiotakopoulos, A. (2016) 'Staff poaching', in Wilkinson, A. and S. Johnstone (eds.) *Encyclopedia of human resource management*, pp. 412–413. Northampton: Edward Elgar Publications.

Panagiotakopoulos, A. (2020) *Effective workforce development: a concise guide for HR and line managers*. London: Routledge Publications.

Stone, R. J. (2002) *Human Resource Management*. 4th ed. Milton: John Wiley and Sons.

4 The interlink of HRD and other HR activities

HR planning

HR planning is a key activity in the people management domain since it can ensure that the firm has the right number of employees, in the right jobs and at the right time (i.e. there is no labour surplus or skill shortages). The planning process should take into account both the internal (i.e. culture, strategy, budget, etc.) and external environment (i.e. state of the economy, industry competition, etc.) of an organization so that proper labour forecasting can be done.

A common mistake for the small firm owner is to focus on short-term resource needs rather than on the firm's long-term people requirements. Such a non-strategic and reactive approach can cause several problems for the organization, such as an inability to meet increased product/service demand that could eventually lead to loss of profit and harm the firm's brand image. The main approach in the HR planning activity that can be used by small firm owners is the qualitative one. The qualitative approach uses expert opinion (usually the opinion of line managers) to predict future staffing requirements. In this approach, line managers make predictions on labour demand and supply using their experience, as well as various sources of information (e.g. governmental financial reports, sectoral research studies, national statistical data on demographics and education, etc.). Moreover, small firm owners should frequently monitor the staff turnover rate, as well as conduct exit interviews to identify any major organizational weaknesses that affect staff retention. HR planning has significant implications for employee learning since an organization that is understaffed (or employees lack the required skills) is in greater need of extensive learning interventions.

Another key element of HR planning that is closely linked to employee learning is the process of job design. In particular, the way that jobs are organized within a small enterprise is of critical importance for staff learning. Job enrichment is a popular process that is characterized by adding different tasks and responsibilities to existing jobs to make them more

DOI: 10.4324/9781003381815-5

interesting and challenging. Job enlargement is another process that is characterized by adding more tasks (of the same nature) to existing jobs. The core difference between job enrichment and job enlargement lies in the quality and quantity of tasks. More specifically, job enrichment involves task variety (qualitative dimension), whereas job enlargement involves more duties (quantitative dimension). Small firm owners should strive to enrich employee jobs (instead of focusing on job enlargement and an increased workload) in order to enable staff to develop further their knowledge and skills through experiential learning. As employees engage in different tasks, they get the chance of gaining rich experience and enhancing their knowledge base through the process of learning by doing (Mullins and McLean, 2019).

HR recruitment

Future job vacancies may occur either through someone leaving the organization or as a result of expansion. Recruiting a new staff member may be the most obvious tactic when a job vacancy occurs but it is not necessarily the most appropriate. Some of the options that small firm owners need to consider are the following: Re-organize the work so that the job tasks can be performed by the remaining employees, use overtime and subcontract the work. Hence, the firm owner should conduct a detailed micro-environmental analysis (e.g. business strategy, corporate culture, nature of tasks, available budget, etc.) to determine what option suits best their business context.

If the decision is that the firm should recruit new employees, then an accurate job description should be produced involving the job title and context (e.g. location of work, background information of the firm), the job summary (e.g. duties, working conditions, performance standards) and the necessary, as well as desirable, attributes needed by potential applicants to perform successfully in the job. During recruitment, it is crucial for small firm owners not to discriminate against potential employees on the basis of unrelated job characteristics including gender, age, disability, ethnicity, sexual orientation, religious beliefs and so on. This is important not only for legal and ethical reasons but also for financial ones since the firm may lose valuable talent if there is discrimination against potential applicants.

The available recruitment techniques can be divided into two main categories depending on whether the organization fills the vacancies internally (i.e. using existing employees) or externally (i.e. using candidates from the external labour market). Each approach has its own strengths and weaknesses. For example, internal recruitment leads to enhanced staff motivation since employees realize that they are heavily valued by the company (i.e. intense employer interest in staff development). However, there is a lack of new ideas in the firm (i.e. employees are affected by the existing firm culture and cannot easily think 'outside the box'). By comparison, external recruitment provides firms with a greater pool of applicants, so more skilled employees may be attracted compared to the existing ones. Yet, external candidates need much more time to become familiar with the organizational context (i.e. employee orientation may become a time-consuming process).

Normally, the great majority of small firm owners choose to fill any job vacancies from the external labour market owing to the small size of the organization (i.e. absence of internal labour markets). In terms of recruitment techniques, the most popular one in the case of internal recruitment is the word of mouth, whereas for external recruitment, most owners use existing employee referrals and online recruitment (i.e. cyber-recruiting including social media advertisement, vacancies advertised on the corporate website, etc.). All the various techniques have benefits and drawbacks and the choice of a technique has to be made in relation to the particular job vacancy, the type of labour market in which the job falls, the business objectives and so on. It is important for the small firm owner to place much emphasis on staff recruitment because this has also a radical impact on employee learning. For example, if the 'wrong' candidate is being attracted in terms of knowledge, skills and attributes, then more extensive training may be needed for existing staff to cover organizational inadequacies (Panagiotakopoulos, 2016).

HR selection

After the recruitment process, there is the short-listing process where the small firm owner needs to identify the candidates most likely to perform successfully in the job. During that process, owners/managers need to look at the essential and desirable skills and attributes of the individual candidates and rank them. It is quite unusual for one selection method to be used alone. A combination of two or more methods should generally be used to strengthen the validity of the selection decision. The choice of the methods is dependent on various factors, including the selection criteria for the post to be filled, the abilities of the small firm owner and time and cost considerations. Various selection techniques are available, including interviews, psychometric tests, trial days and references from previous employers, with the most popular one being discussed in the next section.

Interviewing is the most popular selection technique for small firm owners. It is a planned conversation for the purpose of gathering rich data about an individual. The format may vary (i.e. structured vs unstructured interviews) depending on the post with more skilled jobs favouring an unstructured approach to interview so that the owner can explore both the 'hard' and 'soft' skills of a candidate. For more low-skilled posts, structured interviews (i.e. predetermined list of questions in a standardized format) are normally used. Although interviews suffer from several weaknesses (as is the case with every selection tool), yet they can provide small firm owners with lots of information on the individual's skills and personality. In particular, the owner has the opportunity to observe the body language of candidates and determine if they have the necessary social skills and motives required for the job, evaluate the candidate's subject-matter expertise and assess if the candidate fits with the overall team and firm culture. Arguably, the main disadvantage of interviews is the interviewer's bias (since it is based on subjective evaluations that are affected by variables such as gender, age, educational level, previous experience, emotional intelligence, etc.) that may affect their final decision.

It is important for small firm owners to design an effective interview process to ensure that the right person is being chosen for the job vacancy. It is advised prior to the interview to prepare a list of the key personal attributes and skills required for the post to be filled. Then, it is equally important to design a list of suitable open-ended, job-related questions that will enable them to gather rich information on the applicant's ability to perform the job successfully and contribute towards the firm's objectives. Frequent probing should be used during the interview in order to learn as much as they can about the applicant. Moreover, it is important for the owner to elaborate on the key tasks and responsibilities associated with the particular job vacancy (Panagiotakopoulos, 2016).

Employee selection is also another critical HR activity, which is closely linked to workforce skills development. Therefore, mistakes cannot be tolerated during that process. For example, hiring the 'wrong' person (e.g. lack of person–job fit) will create additional requirements in terms of employee learning. An employee that lacks the required skills or does not match the prevalent firm culture, will need more extensive coaching to be able to perform effectively. Also, if the owner places much emphasis on the 'hard skills' of the candidate (e.g. IT skills, numerical skills, etc.) while ignoring other vital 'soft skills' (e.g. willingness to share knowledge with colleagues, interpersonal skills, etc.), then this may have a radical impact on staff learning since this limits the extent of social learning at the workplace.

Performance appraisal

Performance appraisal (PA) is defined as the systematic process of reviewing employee performance. This normally requires the owner/manager and employee to take part in a performance review meeting. During the performance evaluation process, job behaviour/performance should be measured using both quantitative and qualitative indicators. In the small business context, this process may take place in a less formal manner. However, it is important to be implemented regularly since it provides a dynamic link to employee selection, training and rewards and lays the ground for performance improvement.

The first stage in the PA process is for owners to set proper indicators (both quantitative *and* qualitative) of employee performance in order to be able to identify learning gaps and areas of poor performance that could be addressed through corrective action. It is important for small firm owners/managers to place equal emphasis on the qualitative criteria of performance evaluation (i.e. kind of behaviour exhibited by employees while trying to achieve the desired organizational objectives). For example, the process through which a sale of a product/service is completed does have an impact on corporate brand image, employee well-being and customer satisfaction. In other words, instead of focusing solely on measuring *what* has been achieved, it is equally important for owners to evaluate *how* it was achieved (i.e. the process). For example, some organizations manage to have increased profitability over a prolonged period by making their employees work very hard (i.e. extensive overtime work), thus causing them to suffer from 'burnout' after some time. Such long-term implications can be detrimental to the organization both in ethical and financial terms (e.g. poor brand image and problems in attracting talented staff, reduced sales, increased staff turnover and absenteeism, etc.).

In the small business context, performance appraisal is normally done informally by the small firm owner (or line manager). However, there can be other raters of employee performance, including colleagues, customers and employees themselves. The major techniques of performance appraisal include *ranking* (the manager ranks each employee from best to worst according to their overall job behaviour), *grading* (employees' overall performance is matched with a specific grade definition), *behaviour rating scales* (rating scale that evaluates employee performance using specific job-related criteria), *essay description* (a written statement prepared by the rater describing an employee's strengths and weaknesses) and *management by objectives* (setting of measurable goals and then reviewing the progress made; Panagiotakopoulos, 2016).

Arguably, *behaviour rating scales* are one of the tools that can offer much detailed information to the employee about their performance since it may outline learning gaps in specific areas of the job holder. Various job-related dimensions can be used including job knowledge, customer relations, safety, creativity, punctuality, team working and so on to measure staff performance. In particular, a series of job-related criteria are being developed and employees get a specific grade in each one of them. Although this may look a time-consuming tool to be used by small firm owners (since it requires some considerable effort to prepare PA forms with detailed job-related criteria for each post), yet it is a very useful tool that can bring many positive organizational benefits in the longer term since it is closely linked to staff selection, training and rewards. A careful PA is of the utmost importance for employee learning since it provides insights into the employee's key weaknesses (i.e. learning gaps). Without a sound PA process, the process of staff learning may lack focus.

On completion of the PA process, a performance review meeting should take place so that owners can help underperforming staff get improved, as well as praise high flyers in order to maintain their performance. The following are some of the key steps that small firm owners/managers should take in order to ensure that this meeting has a developmental character: (a) They should try to describe events and job behaviour, not judging employee attitudes (e.g. several studies have stressed that offensive criticism has a negative impact on employee morale and goal achievement); (b) they should provide positive feedback (i.e. praise staff for their strengths) in addition to areas for improvement; (c) employees should be encouraged to talk, explain job behaviour and share ideas on performance improvement. The review meeting should bring mutual benefits to employees and superiors. Any developmental intervention should be mutually agreed on so that employees can be more committed to actively participating in any learning interventions; (d) specific performance improvement objectives should be set, as well as changes in job-related behaviour (i.e. abstract comments and goals should be avoided).

Employee rewards

The purpose of the next section is to look at the concept of employee compensation, which directly affects staff motivation and workforce skills development. In simple terms, motivation is an internal state that drives human behaviour. It is concerned with the reasons that make employees decide to 'go the extra mile' at work, as well as develop themselves (Stone, 2002). It is therefore of critical importance for small firm owners to boost employee motivation in order to improve productivity, advance product/service quality, enhance staff knowledge and cope with any macro-environmental challenges (i.e. manage change effectively).

Several theories of motivation have been developed throughout the years trying to outline the main factors and processes affecting employee behaviour (e.g. Maslow's needs hierarchy theory, Herzberg's two-factor theory, Vroom's expectancy model, Locke's goal-setting theory, Adams's equity theory, etc.). The findings from numerous studies on employee motivation have revealed that the following core variables do have a major impact on employee behaviour at work: employee perception, values, personal needs, personality characteristics, innate and acquired abilities, job design (e.g. increased task variety, job autonomy), the culture of the organization, relationships with colleagues and managers, financial rewards

2.2 Pandemic and Portuguese institutional

"Winning the war against the new coronavirus" was how the President of Portugal, Marcelo Rebelo de Sousa, referred to the crisis generated worldwide by the SARS-CoV-2 coronavirus pandemic, which causes Covid-19 disease. In late January 2020, the World Health Organization (WHO) declared the outbreak of this virus a public health emergency of international concern (WHO, 2020a). On March 11 and after the ongoing assessment of the outbreak and the alarming levels of spread, severity, and inaction, the WHO declared Covid-19 a pandemic (WHO, 2020b). The globalization of the disease became a crisis of worldwide magnitude, affecting all spheres of society: health, economy, education, and everyday life. And, consequently, it challenged the communication strategies of this crisis, put in place by different institutions with great social relevance.

During a phase of order disturbance, especially in the first moments, it is very difficult to keep calm and give the appropriate indications, especially when there are a multitude of risk variables. The health crisis caused by the pandemic has acquired an unprecedented scale, with frames of uncertainty that make any exercise of predictability difficult. However, it is possible to make a preparation to face different crises and, in this way, overcome them with more chances of success.

A crisis, from the communication point of view, is not an event, it's a process that requires previous preparation from all those who surround a moment of attention. One of the most important figures in all the processes is that of the communication manager, who must transmit information to the preferred audiences, having as a mission to anticipate and safeguard the reports presented by the spokesperson of the different institutions.

In a period of turbulence, the search for information is a constant for the various publics, and the social media play a crucial role in this communication process. The social media space can work as a mechanism of multiplication, discussion, and position-taking of governmental and institutional measures, because, being an exceptional situation, the publics can directly enter in dialogue with the institutions. In the case of the University of Minho in Portugal, the crisis caused by Covid-19 implied, precisely, this communication with several "layers of information", considering the diversity of the institution's preferential publics and the new publics. The crisis in Portuguese universities began with the closure of the University of Minho (UMinho) on March 7, 2020, when the Portuguese Minister of Health announced in a press conference the closure of some schools and national institutions due to the outbreak of coronavirus, including "the building of the History course of the University of Minho".[1] Part of its facilities were, thus, abruptly closed and, with this, a set of concerns emerged, about the dimension of the health problem, about the conditions of provision of its services, and about its economic and financial sustainability, among others. In the following days, the Rector's team, with the support of the Commission for the Preparation and Management of the Internal Contingency Plan COVID-19 (appointed by Rector's dispatch of March 3), defined a set of measures to respond to the crisis. Gradually, the institution was closing buildings and transferring activities to the online work model, and on March 10 all teaching activities were definitively suspended.[2] The crisis situation was in place, but the magnitude of its consequences was unpredictable.

In this context, communication with internal audiences became urgent, as well as the relationship with the external community, to share information, reduce anxieties, and control the risk of worsening the critical situation. The University communicated predominantly through online media, with a low degree of segmentation and following a four-pronged strategy:

1 Continuous communication (including Rector's orders that decreed the closure of the teaching complexes, restrictions in the operation of the university residences, suspension of teaching activities, closure of libraries and canteens, and regulation of telework; documents that were shared in the online media and announced in press conferences);
2 Severity of the crisis (through the announcements of the "Committee for the Preparation and Management of the Internal Contingency Plan COVID-19 of the University of Minho" and the advice of public health authorities);
3 Use of digital media (with the use of the institution's website and social networks, the creation of the Covid-19 page on the UMinho website, the publication of videos of the Rector's messages, and the sharing of all dispatches and communiqués);
4 Unity (by developing activities in the digital space with its own contents, in dialogue with the different publics of the institution).

These four guiding axes aimed at sending warning signals to the community and showed the vulnerable situation of an institution that was struggling against an unexpected (though more or less expected) crisis. With the informative notes, the rectorial dispatches, and the official communiqués about the new pandemic, the University sought to enter dialogue with its publics to reduce the risks of the situation worsening.

Founded in 1973, the University of Minho is currently a public foundation of private law, organized in 11 schools and hosts about 20,000 students, operating in two campuses (in the cities of Braga and Guimarães). UMinho has a Communication and Image Office that cooperates with the Rectory "in the definition of policies and strategies of Communication and Image".[3] Having been one of the first academic institutions to suspend face-to-face activities, the University of Minho was faced, from the very first moment, with the challenge of managing the effects of the pandemic at the higher education level, a sector where, unlike others, the decision process required combining the governmental guidelines with the institution's own sphere of autonomy.

2.3 The Case of the University of Minho

The closure of the University of Minho's facilities to prevent an outbreak of Covid-19 (starting March 7, 2020), constitutes one of the most remarkable events in the history of the institution. However, this episode was experienced identically by many higher education institutions (HEIs) around the world, as Knight (2020) portrays:

> … we are now in the fourth week of a worldwide lockdown… With special risks resulting from students working together in close quarters, institutions of higher education were faced with the problem of figuring out what to do with very little notice. Residential campuses had to decide when and how to close dormitories. Professors needed to learn how to move their subjects online very quickly to preserve some semblance of continuity. Some colleges and universities already had plans in place to deal with a crisis, although the magnitude of this was unprecedented…with the Covid-19 pandemic, the challenges were gigantic.
>
> (Knight, 2020, p.131)

Seeking to analyze the communicative behavior of this university at the moment the crisis erupted, a case study was carried out on the relevance of dialogue in communication with publics in the process of managing the Covid-19 crisis. The case study methodology (Yin, 2014) has dominated crisis communication research (Coombs & Holladay, 2010) and is widely used in research on HEIs (Ruão, 2008). Crisis communication studies seem to have started with the analysis of concrete cases in the professional world, which were then explored academically with the expectation of developing theory.

Although crisis communication studies have focused on the analysis of the for-profit sector, we believe in the particular relevance of analyzing the performance of public institutions in times of crisis. In fact, when faced with a threat, citizens tend to look to the public sector for the responsibility of the initial response to the crisis, so the vitality of public sector communication networks is essential to the fulfillment of its social mission (Gainey, 2010). Hence, we consider it appropriate to study the risk management campaign carried out by a public university.

2.4 "UMinho online" campaign

In view of the suspension of face-to-face teaching activities in HEIs in Portugal, as of March 2020 and in the context of the Covid-19 pandemic, the Ministry of Science, Technology and Higher Education suggested that Portuguese universities and polytechnics develop "efforts to stimulate distance teaching-learning processes" and "digital interaction between students and teachers".[4] Since the first days of the closure of its facilities and in the same line of understanding, the University of Minho "tried to find solutions and bet on maximizing the success of the transition from face-to-face teaching to technology-mediated teaching".[5] In this context, the university rectory contacted the Institute of Social Sciences (school where the area of specialization and knowledge in Communication Sciences is located), requesting the preparation of a campaign for internal audiences, aimed at promoting adherence to distance learning and telework.

It was developed a Risk Communication campaign to be carried out in the context of the critical situation of the institution. As we have seen in the literature review, Risk Communication can be used as a response to crises, as a containment mechanism that seeks to sensitize the public about the need to adopt safe behaviors (Sturges, 1994; Coombs, 2010). Following these assumptions, the University decided to go ahead with a campaign, with the following purposes: (first) to inform its internal publics about the rules of behavior in a pandemic situation and (second) to motivate them for the necessary behavioral change (to work online), to protect the internal community and avoid the worsening of the crisis situation.

To meet the needs of the University's rectory, the Communication team started by conducting a benchmarking study. This is an analysis aimed at identifying and understanding good practices carried out by organizations around the world, to rethink the way companies/institutions act based on already tested models (Anand & Kodali, 2008). The study included the analysis of other universities' communication campaigns within the Covid-19 outbreak, evaluating their patterns of action. The famous Xerox benchmarking model (Anand & Kodali, 2008) was adapted, and the following research steps were carried out: (1) identification of the cases to be studied, taking into account the degree of proximity to UMinho's environment and looking for countries with a diversity of pandemic situations (in March, 2020); (2) definition of the data collection methods, which included the analysis of the universities' websites and social networks (due to the ease and speed of access); (3) categorization and interpretation of the collected data, whose results we present below; and (4) incorporation of the results in a strategic proposal.

This research was organized by countries. We started with the case of Italy, a country where the pandemic was already reaching extremely serious levels, and the results of the observation allowed us to conclude that universities mainly sent information messages on how distance learning was being set up and how some activities were being restructured according to the crisis situation (such as graduation ceremonies). Next, we studied the communication of universities in France, where the pandemic was beginning to show alarm signals. Here, institutions were limiting themselves to informing the public about the closure of facilities and about distance learning operating models. In Spain, universities were providing guides to the use of the new distance learning platforms, with some relying on videos to engage their students and inform about Covid-19 and the implications for the university. In the UK, where the outbreak of Covid-19 was in its early stages, universities were maintaining their classic communication model, reporting on the pandemic situation and showing what research was being undertaken at the institution to help communities, with particular emphasis on the discovery of the vaccine for the new coronavirus. In the Netherlands, universities showed an apparent normality, putting information about the pandemic alongside the dissemination of planned activities, in a "low profile" communication model (Caponigro, 2000). In Belgium, institutions took advantage of their researchers' involvement in pandemic-related research, covering areas such as medicine, biology, or the social sciences. In the US, in a period when contagion was still low, universities showed awareness of the pandemic situation, but their information was still incipient. And in Portugal, HEIs combined campaigns in favor of social isolation with the provision of information on distance learning.

In summary, we can say that the messages were mostly informative and low engagement. However, there were also some good examples of appealing, engaging, and/or emotional content, as in the University of Bologna's networks[6] or the Autonomous University of Madrid,[7] and on the MIT website.[8] The preferred channels were email, website, and social networks. Media were perceived as supporting channels (for communication with external audiences). The institutional websites presented Covid-19 as the predominant subject and contained links to epidemiological information, rules and regulations, distance learning platforms, (internal) expert testimonials, pandemic-related research projects, and solidarity practices, among other regular subjects in the life of an academy. The social networks were extensions of the website, replicating information about COVID-19 related rules, practices and research, as well as general information about teaching and research activities. They also included campaigns calling for confinement and social distance and for engagement in new teaching and research practices. As preferred media, written statements (informative texts and open letters), dispatches and regulations, videos (with statements, interviews, personal statements, and motivational messages), posters (informative and engagement), and photographs (of the campus, people, or new work environments) stood out.

The main strategic communication trend consisted in associating the call for social isolation – with hashtags like "#stayhome" – to the need for distance learning – having as communication axis the idea that telecommuting was a necessity in times of emergency and not an option. This message was shared in various pieces of communication, such as texts, posters, flyers, or guides/tutorials, through online channels.

2.5 Branding and communication strategy

After the benchmarking report was prepared and shared, a communication strategy was defined for the University of Minho, which had as its primary target the internal public – students, faculty, and staff – and, marginally, the local and national community. As communication objectives a set of intentions was identified: (1) to improve the understanding of the extraordinary circumstances in which the academic community found itself; (2) to motivate for the adoption of technologically mediated teaching-learning processes; (3) to appeal for participation in the joint construction of the best distance learning solutions; and (4) to overcome resistance to the adoption of telework.

Aiming at a deeper understanding of the ideas, attitudes, and behavior of the target publics towards the extraordinary situation they were experiencing, we carried out an analysis of the interaction on UMinho's social networks (Facebook and Instagram). We intended to find perceptual barriers to overcome and locate communication opportunities. In this framework, we arrived at a set of insights or mental representations about the situation by the public, which we synthesized in Table 2.1.

Based on the benchmarking study, on the communication model already in place at UMinho (traditionally corporate), and on the evaluation of our

Table 2.1 Study of the perceptions and attitudes of internal publics

The faculty consider that...
"Distance learning is a lot of work"
"Distance learning implies technical knowledge that I don't have and that takes a lot of time and effort to acquire"
"Distance learning doesn't work"
"We live in a temporary emergency situation that should not require so much effort on our part"
"I want to help. I want to be supportive. I want to protect and be protected"…
The students consider that...
"Distance learning is boring"
"Distance learning requires difficult technical learning"
"Distance learning focuses on individual work"
"We are experiencing a temporary emergency situation that should not require so much effort on our part"
"I didn't join this University to take distance learning classes"
"I want to help. I want to be supportive. I want to protect and be protected"…
The staff considers that...
"Telecommuting will interfere with my family life"
"Telecommuting will transfer operating costs to my family budget"
"There are tasks that cannot be done in telework"
"We are experiencing a temporary emergency situation that should not require so much effort on our part"
"I want to help. I want to be supportive. I want to protect and be protected"…

Andrade et al. (2020).

audiences, we defined that the campaign should have an institutional tone, integrate the different audiences, and convey the following key message:

Communication Axis –

> Exceptional circumstances require exceptional measures, on the part of the University and Society as a whole. In this context, telework and distance learning are not an option of the institution, but a necessity in times of emergency. The understanding, collaboration, participation and solidarity of all is fundamental to cope with the new needs. We are all part of the solution.
>
> (Andrade et al., 2020)

From the identification of this axis, a communication concept (or a creative idea for the campaign) resulted:

> University united. University in solidarity.

This concept was further broken down into different signatures in the six phases of the campaign (Table 2.2), which answered the AIDA model. This model defines the cognitive stages that can be motivated by a persuasive message when it acts in the minds of audiences and were identified as follows: Attention, Interest, Desire, and Action (Joannis, 1998.)

These sentences signed daily institutional messages – with a close, emotional, and warm tone – written in the font defined in the University's Identity Manual – NewsGotT – and inserted in posters and banners with images representing the internal community and work and distance learning environments, where the colors of UMinho's logo and its schools were predominant. The communication pieces were placed on the University's website and networks

Table 2.2 Campaign unfoldings

Stage/effect	*Publics*	*Signature*
First stage attention	University community	United University. Solidarity university
Second stage interest	University community	Distance university. Solidarity university
Third stage interest	Students teachers	Distance learning. Solidarity education
Fourth phase action	Staff	University at home. University in solidarity
Fifth stage reinforcement	Schools	United schools. University in solidarity
Sixth phase reinforcement	Services	Present services. Solidarity University

Andrade et al. (2020).

(Facebook and Instagram) and were also shared on the social networks of the organic units. The campaign also had an English version, to reach international students, teachers, and researchers working in the campi. On the networks, the posters took the form of gifs so that the two language versions – Portuguese and English – were more visible. The campaign ran from March 30 to May 13, 2020, included two daily publications (one in the morning and the other in the afternoon), in a cross-sharing strategy between the University's channels and those of its Schools.

The initial phases of the campaign corresponded to moments of development of visibility of the theme (ATTENTION) – with statements like "Social isolation is a sign of respect and responsibility" or "Preventing contagion is a duty of citizenship"; of promoting familiarity (INTEREST) with the idea of "distance in proximity" – with statements like "To work at a distance is not to take unnecessary risks" or "Far can be close in multiple ways" –; of reducing prejudice and stimulating new practices (DESIRE) – with propositions like "Teaching is an exercise in sharing. Even at a distance" or "Teaching at a distance is sharing screens, sharing life"; and of promoting the clear involvement of the public – with testimonials like "UMinho at your home. Safe teaching" or an excerpt from the institution's hymn that appeals to the idea of "common home".

Phases 5 and 6 promoted the memorization of the message and sought to reinforce responsible attitudes and behaviors, calling for unity among the University's schools and recognition of the role of services. If the initial posters used purchased images, due to the need for quick reaction, the images of phases 5 and 6 correspond to real photographs of the internal public's workplaces (including their own homes), in an exercise of personalization and closeness.

As for the campaign's effects, these can only be assessed in the long term, since it was intended to act on perceptions, attitudes, and behaviors, i.e., factors that cannot be detected by counting likes, shares, and views. Still, the campaign sought to comply with the rules of crisis communication and pandemic communication, given the preparation of informative and empathetic messages sent through channels close to the priority audiences (Knight, 2020).

2.6 Concluding remarques

How Covid-19 changed how consumers relate to brands? In the case study the "UMinho Online" campaign corresponded to what Sturges (1994) calls adjustment communication. It included advisory messages or expressions of sympathy, which intended to develop stakeholders' sensitivity to risk and lead them to behavioral adjustment. Understanding the issues seems to improve the stakeholders' reaction to risk or crisis situations, but it is not enough. It is important to generate the emotional motivation that underlies all human relationships, and this seems to be the key point of a dialogue that promotes changes in knowledge, behaviors, and practices. Besides being this involvement that generates the transfer of the stakeholders' status to "our publics and new publics", to which the digital channels try to respond, either in the (re)knowledge of interactions (likes and shares) or in the need to respond in a more agile way, as in the comments of the publications that were presented here. In this sense, we understand emotional motivation in crisis communication as a form of civic citizenship, fundamental in the life of public institutions. It presents itself as a social brand in moments of crisis.

Although it is defined as a health crisis, impacting public health, the Covid-19 pandemic also amplified the crisis of the university institution itself. The response that universities have found to continue their activities has sharpened the public debate about the nature and mission of the academy. In the public space, the universities' actions were analyzed more through the prism of the solutions implemented to overcome the effects of the suspension of classroom activity than through the strategies of communication of the actions developed. The adoption of a technologically mediated teaching model, with signs of extending beyond the official confinement period, has aggravated the perception that universities give in to a kind of fascination with virtualization.

The communication strategy of the University of Minho in the context of the pandemic crisis was, to a large extent, oriented towards the promotion of distance learning and teleworking as exceptional measures for an equally exceptional time. Determined by the specific goal of mobilizing the academic community for a temporary change of paradigm, the communication campaign that we have reported on in this chapter aimed at raising awareness among different audiences – students, faculty, and staff – to the need for a universal commitment. With a strong focus on the immediate, it sought to be essentially an exercise in exhortation to the imperatives of the moment. Presented as the only way to correspond to the duty of continuing its mission in safe conditions, the implementation of a "virtual university" plan opened, however, the horizon to a crisis that, not being new, will extend beyond the pandemic crisis itself, the identity crisis of the academy, which should keep the University of Minho, and universities in general, on alert and particularly active behind the scenes of Risk Communication.

Notes

1 https://www.dn.pt/edicao-do-dia/08-mar-2020/18-pessoas-infectadas-em-portugal-11899337.html.
2 RT-25/2020.
3 https://www.uminho.pt/PT/uminho/Unidades/Servicos.
4 Clarification Note from the Ministry of Science Technology and Higher Education, March 13.
5 Note from the Communication, Information and Image Office of the University of Minho, March 18th.
6 https://www.inst agram.com/unibo/.
7 https://www.instagram.com/p/B91LbBxADAi/.
8 http://www.mit.edu/.

References

Anand, G., & Kodali, R. (2008). Benchmarking the benchmarking models. *Benchmarking: An International Journal, 15*(3), 257–291.

Andrade, J. G., Ruão, T., & Oliveira, M. (2020). Os bastidores da comunicação de risco: A UMinho em tempos de pandemia. In M. Martins & E. Rodrigues (Eds.), *A universidade do Minho em tempos de pandemia: Tomo II: (Re)Ações* (pp. 127–157). Braga: UMinho Editora. ISBN: 978-989-8974-28-0. https://10.21814/uminho.ed.24.6

Botan, C., & Soto, F. (1998). A semiotic approach to the international functioning of publics: Implications for strategic communication and public relations. *Public Relations Review, 24*(1), 21–44. https://doi.org/10.1016/S0363-8111(98)80018-0

Caponigro, J. R. (2000). *The crisis counselor: A step-by-step guide to managing a business crisis*. London: McGraw-Hill Companies.

Coombs, W. T., & Holladay, S. J. (Eds.) (2010). *The handbook of crisis communication*. Singapore: John Wiley & Sons.

Eiró-Gomes, M., & Duarte, J. (2005). Que públicos para as relações públicas. *Actas Do III Sopcom, VI Lusocom e II Ibérico, II*, 453–461.

Gainey, B. S. (2010). Educational crisis management practices tentatively embrace the new media. In W. T. Coombs & S. J. Holladay (Eds.), *The handbook of crisis communication* (pp. 301–318). Singapore: John Wiley & Sons.

Grunig, J., & Repper, F. (1992). Strategic management, publics and issues. In J. Grunig (Ed.), *Excellence in public relations and communications management* (pp. 131–172). Mahwah NJ: Lawrence Erlbaum Associates.

Joannis, H. (1998). *O processo de criação publicitária*. Lisboa: Edições Cetop.

Knight, M. (2020). Pandemic communication: A new challenge for higher education. *Business and Professional Communication Quarterly, 83*(2), 131–132. https://doi.org/10.1177/2329490620925418

Livingstone, S. (2005). *On the relation between audiences and publics. Audiences and publics*. Bristol and Portland, OR: Intellect Books.

Ruão, T. (2008). *A comunicação organizacional e os fenómenos de identidade: A aventura comunicativa da formação da universidade do Minho, 1974–2006*. Tese de Doutoramento. Braga: Universidade do Minho.

Sturges, D. L. (1994). Communicating through crisis: A strategy for organizational survival. *Management Communication Quarterly, 7*(3), 297–316. https://doi.org/10.1177/0893318994007003004

WHO. (2020a). *2019-nCoV outbreak is an emergency of internacional concern*. World Health Organization. http://www.euro.who.int/en/health-topics/health-emergencies/coronavirus-covid-19/news/news/2020/01/2019-ncov-outbreak-is-an-emergency-of-international-concern

WHO. (2020b). *WHO Director-General's opening remarks at the media briefing on COVID-19-11 March 2020 11 de março de 2020*. World Health Organization. https://www.who.int/es/dg/speeches/detail/who-director-general-s-opening-remarks-at-the-media-brie ng-on-covid-19-11-march-2020

Yin, R. K. (2014). *Case study research: Design and methods*. Thousand Oaks, CA: Sage Publications.

3 How brands coped successfully with Covid-19

Inês Teixeira-Botelho
COO Cata Vassalo

Where were you on September 11, 2001? This is a question that most people living today know how to answer. Some know precisely what they were doing at that time and even what day of the week it was. More than a collective memory or flashbulb memories (Brown & Kulik, 1977), this event transformed our global consciousness and modus vivendi forever. Covid-19 and moments of confinement too.

The pandemic has changed us as a society and as people and, therefore, consumers. The results of the EY Future Consumer Index (2022) surveys demonstrate that the sense of awareness about uncertainty and risks, both in relationships, health, personal finances, work, and the future, led to the emergence of a new mindset in that the sense of urgency – and emergency – prevails. Allied to an unstable and unpredictable geopolitical and economic scenario, which leads to increased inflation and the fragility of supply chains, consumers have changed the way they consume – reducing impulse purchase – as well as prioritizing and using their money and time.

The same study indicates that most consumers want to increase their savings and that younger generations, with a focus on Generation Z, are focused on this goal. In this sense, there is a growing interest in products and services with greater durability and meaning, both because this avoids unnecessary expenses and because it is positive for environmental concerns, namely not only through the preference for subscriptions and rentals but also through the growth of more circular economic models, based on concepts such as resale and/or upcycling.

As a response to constraints on mobility, reduced hours, and social distancing protocols in public spaces resulting from the Covid-19 pandemic, people began to adopt e-commerce as a necessity. This channel registered, according to eMarketer data (2021), a growth of 27.6% worldwide during the year 2020.

This transformation directly impacted the consumer journey (Think with Google, 2021), with search engines becoming the most used tool along this journey, both in the discovery stage and in the following stages, by allowing and contributing to other elements such as the benefits, ratings, comparison, and price that can come under closer and more informed scrutiny.

DOI: 10.4324/9781003382331-4

Studies by Google (2022) demonstrate that consumers no longer only use digital channels not only to buy but also to search before physically visiting a store. In 2021s Christmas (Think with Google, 2022) consumers returned to physical stores, with the results of in-person purchases reaching almost pre-pandemic levels. However, the use of digital channels continued to rise – online research on a smartphone inside a store has tripled since the beginning of 2022.

To survive Covid-19, brands, both B2B and B2C, were forced not only to restructure their businesses from a logistical and operational point of view but also to adjust on-the-go, during confinement and post-confinement, the experience provided throughout the consumption journey.

3.1 A new paradigm: take advantage or embrace opportunities consequently

The Covid-19 pandemic, with the consequent restrictions on mobility and the confinement of people at home, implied a restructuring of society and companies.

From an operational point of view, there had to be a paradigm shift. If remote work was a condition of a niche of knowledge workers, during the pandemic it became not only the main way to keep companies running but also the basic structures of society, such as education and, in part, issues of non-urgent health.

Remote work, both in EU countries and in the US, increased exponentially in March and April 2020 and continues at higher levels than before the pandemic (Marcus, 2022). This change not only opened several discussions about the impact on productivity, cost reduction, improvements in quality of life, boundaries between personal and professional life, and reduction of the environmental impact of commuting to work but also forced a technological acceleration and investment by companies and telecommunications infrastructures, so that it could become a reality.

Remote work can have negative and positive effects on productivity and innovation within companies, since on the one hand it can have a negative impact in terms of motivation, interaction, and knowledge exchange. However, on the other hand, it alleviates and reduces costs and resources that can be applied to the business (OECD, 2020) and avoids non-productive time such as commuting to work. Barrero et al. (2021) estimated a general increase in the productivity of 4.6% in the US, of which 1% is due to the reduction in travel times to and from workplaces.

As a result of all this, the OECD estimates that internet traffic worldwide increased by up to 60% between September 2019 (pre-pandemic) and March 2020. Sandvine (2020) refers to a 40% increase in traffic from February 1, 2020, to April 19 of the same year – the period of the first confinement. In his study, he distinguishes between downstream and upstream traffic, the first of which includes an exponential increase in video consumption through services such as Netflix and Amazon, which have grown during the pandemic. Netflix, for example, saw its subscribers grow by 16 million during the first quarter of 2020 and its shares increased more than 60% through October 2020, compared to the S&P500 which reach just 6.5% in 2020 (Forbes, 2020). This growth ended up not materializing in the post-pandemic scenario, with Netflix recording a loss of 200,000 subscribers in the first quarter of 2022 and around 1 million subscribers in the second.

To face uncertainty, many brands have tried to reinvent themselves by taking advantage of the needs created by the pandemic or just following the course of change. The famous brand, pioneer in bagless vacuum cleaners, Dyson was one of the manufacturers that tried to respond to the appeal of the UK government and produce ventilators and PPE. However, as founder Sir James Dyson mentioned, demand was lower than initially anticipated, with the government backing out of its order and the brand following the pre-pandemic course of its business.

The Californian company Zoom Video Communications, created in 2013, took advantage of the changes introduced by both remote work and the need for interaction between family and friends, to ensure the rapid adoption of its software as the preferred one globally. With an audience as diverse as executives, work teams, fitness and yoga instructors, or children in tele school, Zoom has become a new way of communicating and existing both during the pandemic – where its billing results increased by 355%, in the second quarter of 2020 – as in the post-pandemic period with remote work being part of the organizational context.

Uber Eats also knew how to read and keep up with the opportunities, as it launched a campaign with free delivery fees to support local restaurants and encourage orders for transporting meals. In addition, the company also made available in their app the delivery of other products, food, consumption, and/or first necessities, with the collection in supermarkets, stores, and pharmacies. The brand's results went from 1.9 ($bn) in 2019 to 4.8 and 9.3 ($bn), respectively, in 2020 and 2021. Also, Amazon and Alibaba, market leader in B2B e-commerce, started to have in product purchases food a considerable part of its business. Data provided by Alibaba show that in October 2020 food & beverage was the segment with the highest demand on the platform.

In parallel with these brands, which grew as a direct result of the changes resulting from the Covid-19 pandemic, other industries had the challenge of readjusting and reinventing themselves to improve not only their offer of products and services but also the shopping experience.

3.2 Cata vassalo: essence digitalization

The fashion industry is responsible for around 10% of total carbon emissions, which is greater than the sum of the impact caused by international flights and maritime transport (UNEP, 2019). If not contradicted, all studies indicate that this value will tend to increase along with the exponential growth that the industry has had in recent decades. Thus, more and more brands and stakeholders have favored the transformation of fast fashion approaches, typical of a consumer society, to slow fashion.

Sophisticated algorithms, the vast amount of data collected on consumers and their behavior (Dias & Teixeira-Botelho, 2020) and the filter bubbles we live in (Pariser, 2012) combined with the new global awareness and concerns about uncertainty and instability, cause the, intrinsic and extrinsic, approach on individualization to increase. As such, brands must focus on both products and services that become individualized, interactive, and meaningful experiences to establish humanized relationships with their consumers.

Cata Vassalo is a Portuguese jewelry brand that today operates strongly in the Iberian market but sells globally. It has two main consumer segments: brides looking for unique and custom pieces for their weddings; and women, in general, who want jewelry for everyday use and special moments. The brand, which was always based on a slow fashion positioning, until the pandemic sold preferably at a physical store – the atelier in the Greater Lisbon area (Portugal). Online sales represented around 37% of the business until the beginning of 2020, but, at the end of 2022, sales at physical points represent only 37% of turnover – a 180-degree change in the operation.

Alongside the operational digitalization of the team and processes, which until the beginning of the confinement were mainly based on offline tools and systems, the brand focused on an integrated strategy of rebuilding all contact points of the consumer journey.

Choosing jewelry is always a high-involvement purchase, especially in the case of pieces to be worn on special occasions such as weddings. The fact that it is not an impulsive purchase implies a scrutiny that is rigorous in itself and depends not only on objective elements (design, materials, price, and reviews) but also on subjective ones that go beyond the product itself, such as location, decoration and appearance of the point of sale, service/usability and general experience, packaging and even sensory elements. All this affects the perception that one has of quality and, therefore, the sensitivity to price.

Thus, the brand's first objective was to bring the online experience closer to the offline experience, working on all these details, since during confinement online was the only way to exist.

In the first stage, brides' appointments, which were always physical, started to be carried out by Zoom. The pieces could not be tried on by the customers and, therefore, the Wow moment! was lost but they were used by the salespeople on the online meetings, and enthusiasm and empathy began to overcome this restriction. The online conversations were scheduled in a software that made it possible to collect information from the customers and upload inspirational images of what they like, how they want to look, their dresses and hairstyles, etc. This made it possible for the meetings to be prepared in advance.

The fact that the clients were in their homes and in their surroundings gave them a kind of emotional comfort and allowed them to invite other people they trust to watch and participate online. The services became more intimate and emotional and, therefore, the relationship with the brand as well.

At the same time, to work on awareness, and because the wedding industry itself was paralyzed by restrictions, the brand focused on communicating hope. Cata Vassalo invited public figures, who had been recently engaged, to give testimonials on video – without editing to be emotional and real– about how their marriages had been and to leave a message of comfort. Words like cancellation were replaced by others like postpone, and the notion of "soon" was always introduced. The statement "keep dreaming" was subtly placed in all brand's communication on social media.

Instead of a wave of fear and uncertainty, hope and the notion that this extra time would allow for better preparation and self-knowledge to improve the outcome of each bride's dream day were worked on.

For the female audience in general, the same emotions coated with empowerment and attitude were communicated. The brand carried out a campaign for a new collection, completely photographed at home – by the model with her own cell phone – wearing comfortable clothes complemented with jewelry. Under the general fashion industry motto of the "comfortable" style, combined with the e-sport trend, not only the need to take care of ourselves and our image at home was communicated but also the hope that the special days would soon be reality. This positioning helped transform consumer awareness that jewelry does not need to be worn only on special days but also in everyday life as a complement, in addition to working again on the idea that soon the pieces would be used on special occasions.

Cata Vassalo developed the online movement #Umasómarca (which means "just one brand"), where they invited other brands, including competing brands, to show their protagonists, from CEOs to team members, through videos with testimonials. Business experiences, expectations, and fears were shared on a day-to-day basis, digital workshops were held, and conversations were held together, via streaming, humanizing each brand, and making its backstage known to the final consumer.

The concept of slow fashion was worked on with the filming of content that demonstrated the steps of making Cata Vassalo jewelry and accessories, thus making customers an integral part of the process. The concept of upcycling was also highlighted. Cata Vassalo launched a collection and campaign, in partnership with another brand in the textile fashion industry, in which accessories were created with waste fabrics.

Alongside this "perpetual contact" (Katz & Aakhus, 2002) with the consumer and, therefore, the close relationship, the website experience was redesigned, making it more mobile, clean, and intuitive. There was a commitment to create a clear and objective communication – all the technical details of a piece were provided so that the consumer knows its dimensions, weight, and materials – as well as providing images of the jewels worn at different angles and approximations and in the white background. The offer of different methods of payment has been increased; a private user area with login was created to work on repurchases with purchase history; the cost of shipping has been reduced; automated emails with status change notifications were improved to empower the customer (purchase, preparation, shipping, and tracking numbers information); the homepage now has elements of social proof through links to articles and press content; based on data collected, cross-selling recommendations and suggestions were inserted in line with customer tastes; integrations were added so that the pieces can be shared via email, Instagram, WhatsApp, and Facebook – enhancing word-of-mouth; and, finally, to work on the need and feeling of *fomo* and a sense of urgency, a back-in-stock alert system was made available when products were out of stock. An SEO optimization strategy was also developed, with the brand always favoring organic communication.

The approach to communication and sales has become increasingly omnichannel, moving from email, video call, telephone, social media messaging systems, and WhatsApp – now equipped with a product catalog with prices – and SMS being all of them active parts of the consumer journey. The experience of the website and Instagram was also approximated and integrated, through the placement of tags of products from the online store in Instagram content and the use of tags on the website.

The personalization of the consumer experience continued to be carried out in every detail as it had been previously done – all shipments are accompanied by handwritten postcards signed by the designer and founder of the brand – and the use of a CRM system was expanded to get to know the brand's consumers in depth – names, relationships with other consumers, tastes, preferred contact points, purchase history, and interactions.

After the confinement period, the brand continued to adopt emotional communication, the individualized relationship, and the omnichannel approach, also turning physical services into hybrid processes, using tablets where the team's salespeople access customer files and image repository of pieces and collections. In addition, the possibility of making online purchases with a physical collect in the atelier was created.

As mentioned, the brand's strategy has drastically transformed the weight of online sales in the business (63%) and means that today 62% of consumers are new purchases and 38% are recurring customers, with an average of around three online orders per customer.

3.3 Spotify: from music to podcasts and… to books

The Covid-19 pandemic has brought a new awareness to consumers, making them to prefer subscriptions, which are flexible and affordable, to owning physical goods. Thus, streaming businesses, such as Netflix or Spotify, gained enormous relevance in the modus vivendi. This type of value proposition was found in moments of confinement at home, in the change in the consumer's mindset, who now prefer the digital consumption of products, without commitment and with flexibility, as well as in the involvement that they put in purchases to save money and spend only on what they need or fill it, the right context to evolve.

October 2022 data provided by Spotify show that the platform has 456 million active users, of which 195 million subscribe to the paid service. Despite direct competition from Apple Music, YouTube Music, and Amazon Music, Spotify continues to remain at the top of consumer preferences and maintain pace in subscription growth.

With mobility restrictions and people confined to their homes, Spotify understood that consumers no longer listened to music on mobile devices with headphones or in cars on the way to work but now connected the platform to speakers at home, during activities such as cooking and doing household tasks, physical exercise, family moments, or for consumption of information through podcasts and meditation content.

Before the pandemic, Spotify predicted that, given its rapidly growing user base, advertising revenue would be the pillar of the brand's strategy. Although the model in question already showed some signs of maturity, the pandemic drastically cut advertisers' budgets and affected Spotify's revenue.

In response, the platform began to develop original content in the form of podcasts, signing exclusive agreements with public figures and opinion makers around the world. The change in strategy has positioned –and continues to position – Spotify as an information and opinion-forming platform as well. This change in positioning has evolved, with Spotify announcing in September 2022 the acquisition of more than 300,000 audiobooks.

Also thinking about the circumstances of users during the pandemic, Spotify has expanded the offer of its playlists based on moods, festive seasons, or activities. If the practice of sports at home became common during confinement and has remained until today, the musical options in Spotify playlists also, with options for relaxation exercises, running, and bodybuilding, among others.

Spotify's evolution continues with the development of its sophisticated personalization algorithms that make the service unique to each consumer. The Daily Mix and Discover Weekly lists are examples of this individualization on the service, as they offer a daily or weekly set of new music tracks that the platform knows the user will like, based on their streaming history. At the end of each year, the brand also launches Spotify Wrapped, which brings together the main artists, songs, and musical genres of the year preferred by each consumer and invites them to share the results with friends on social networks. Spotify also lets people listen live to what friends are listening and discover their songs and playlists.

During the pandemic, Spotify also presented a new feature to its premium users (paid subscribers) allowing them to, as a group, control a streaming session, by individually adding songs that would be played. This has made music consumption an increasingly collaborative and joint experience.

3.4 Airbnb: from hosting to hosts

Airbnb is an accommodation platform where owners open the doors of their homes to accommodate tourists who visit their areas of residence. Contrasting with hotels, Airbnb offers more real homes adjusted to local lifestyles so that the travel experience is more immersive but, at the same time, welcoming, personal, and with a "homey" touch. This proposal is transversal to all Airbnb offers, even in properties made available under the Airbnb Plus or Airbnb Luxury umbrella.

Airbnb thus mediates the relationship between two types of customers: on the one hand, owners who register as hosts on its platform and, on the other hand, travelers who use the platform to find short-term accommodation that meets their needs on each trip.

Airbnb had presented in 2018 the Airbnb Experiences in which local hosts offered different adventures – from cooking to guided tours – in their cities to the tourists who visited them.

With the restrictions imposed on mobility and the widespread fear of the dangers of Covid-19, Airbnb saw its hosts having mass cancellations in the reservations of their properties and their experiences. Tourism, as the word meaning explicit, is an industry that lives from travel and, therefore, from moving to physical locations. Without the possibility of traveling, the platform would lack critical mass and would leave hosts without their businesses, whether it is main or side business.

To prepare for an imminent future and work on awareness, the platform launched several online initiatives such as providing accommodation for displaced doctors and nurses and long-term stays. Having also organized online training in the form of tutorials on how hosts could perform contactless check-ins improve check-in security or the cleanliness of their spaces without harming their health and that of their guests. For guests, it provided the same kind of reassuring and informed communication about safety and cleanliness procedures adopted by hosts around the world.

Airbnb has also always maintained direct and close contact with its community through communications from the CEO with its concerns, changes, and recommendations. Situations of conflicts of interest with cancellations between the host community and guests were also managed, always making it clear that each case had its own specificities and that the unpredictability and fragility resulting from the pandemic meant that the policies and forms could not be approached from traditional resolutions. In this way, they had the difficult job of having to deal with the expectations of two audiences with different interests and improving the communication, response, and interaction channels for each of them on their platform.

The Airbnb brand saw experiences to counter pandemic restrictions by creating more than 700 online experiences offered by hosts around the world. Through these types of services, the tourist, confined to his home, was taken on a digital journey to the culture of another country through experiences. Hosts now, instead of on-site guided tours and accommodation, offered online events focused on cooking, meditations, virtual tours, storytelling, and crafts, among others.

If, in the first phase of mobility restrictions, no one thought about taking trips, Airbnb took advantage of online experiences to put guests and hosts back in touch. With this digital rapprochement, the brand went from being just a platform that people use to plan trips to being a platform that also develops a global mindset for exchanging know-how about other cultures and world diversity. This mindset, in addition to bringing both communities closer together, enhances information, access, hope, and desire to visit places in loco – especially for those who were confined and had a lot of free time at home. But it is also a way to generate up and cross-selling by creating one-to-one bridges between locals and future tourists, who can stay in touch for the near future of travel and online or offline experiences.

Finally, to continue this path of bringing the two audiences closer together, creating value for each of them, Airbnb worked on emotions by inviting former guests to share the most important moments they had experienced with hosts over the years 2019 and 2020. These testimonials generated very personal emails that Airbnb sent to hosts, with words about their past guests. At the same time, the platform launched new and more flexible cancellation policies and invited hosts to join it with the promise of getting more bookings in the near future.

All these brands demonstrate that humanizing brands and their sociability through communication and individualized experiences promotes close and empathetic relationships with the consumer. These characteristics, in a world where the consumer is more empowered than ever, with constant access to information about products and services and, therefore, more aware and demanding, becomes distinctive values for brands and a way of turning a customer into a fan or evangelizer. It is no longer enough to offer a quality product or service with a good perception of the quality-price ratio. Today it is necessary to add value and create meaning by building close and trusting relationships. Being a social brand is a sine qua non to exist in the post-pandemic world.

References

Barrero, J. M., Bloom, N., & Davis, S. J. (2021). 'Why working from home will stick', Working Paper 28731, National Bureau of Economic Research.

Brown, R., & Kulik, J. (1977). Flashbulb memories. *Cognition*, *5*(1), 73–99.

Dias, P., & Teixeira-Botelho, I. (2020). Smarketing: How mobile marketing is changing Portugal. Edições Sílabo.

eMarketeer. (2021). *Global Ecommerce Update 2021*. Insider Intelligence eMarketeer.

EY Future Consumer Index. (2022). Ernst & Young Global Limited.

Food Delivery App Report – Research, Insights and Statistics https://www.businesofapps.com/data/food-delivery-app-report/?utm_source=food+delivery&utm_medium=click&utm_campaign=hyperlink+report

Forbes (2020). *5 Big numbers that show Netflix's massive growth continues during the coronavirus pandemic*. https://www.forbes.com/sites/jonathanponciano/2020/10/19/netflix-earnings-5-numbers-growth-continues-during-the-coronavirus-pandemic/?sh=69bae88a225e

Katz, J. E., & Aakhus, M. (2002). Perpetual contact: Mobile communication, private talk, public performance. Cambridge: Cambridge University Press.

Marcus, J. S. (2022). *COVID-19 and the shift to remote work*. Bruegel Policy Contribution Issue nº 09/2022, Bruegel. (Policy Paper).

OECD. (2020). *Productivity gains from teleworking in the post COVID-19 era: How can public policies make it happen*? Policy Brief, Organisation for Economic Co-operation and Development.

Pariser, E. (2012). *The filter bubble: What the Internet is hiding from You*. New York: Penguin Books.

Sandvine. (2020). The global internet phenomena report: COVID-19.Waterloo, ON: Sandvine.

United Nations Environment Programme. (2019). *Emissions gap report 2019*. Nairobi: UNEP.

4 Challenges and opportunities of branding after Covid-19

Ricardo Miranda
Partner Wonder\Why

This chapter begins with an overview of how the digital realm is both triggered by human imagination and ambition and how it is shaping human evolution.

Afterwards it presents an overview of how brands were affected by Covid-19 several lockdowns, what was at stake, and how they managed to adapt and take advantage of the stay-at-home exceptional circumstances.

Finally, the last section outlines a series of critical themes that will provide crucial answers for better understanding the future of brands in this post-Covid 19 period, based on the concept of social brand as a model for building strong relationships with consumers.

4.1 The digital transformation revolution

Branding is an evolutionary process intertwined with human beings own evolutionary process. The first is a consequence of the ladder.

Both are being profoundly marked by digital transformation.

Digitalization is the great leap forward. Not just for corporations, economic processes, products, services, but above and beyond for people. From research projects to having meetings, meeting people, finding jobs, looking for romantic partners, booking hotels, renting houses, working from home, reading newspapers, playing games, buying stuff, hanging out with friends… all these used to be activities that required some kind of physical motion in the physical world. Not anymore. Now it's all available in an alternate reality we call digital, prompted mostly by the internet, social media, cloud platforms, and a latest addition called the multiverse, all of them made possible by a paraphernalia of tech devices such as smartphones, laptops, tablets, smartwatches, headsets, and more.

But it doesn't end here.

Creation itself, probably the last residue of human endeavor, is being outsourced to the digital realm via generative artificial intelligence (AI) programs such as OpenAi or MidJourney.

This evolving process doesn't happen by chance.

DOI: 10.4324/9781003382331-5

Human imagination has always pushed the boundaries of the natural world. Gun powder, skyscrapers, water dams, syringes, popcorn, toothpaste… they all came flushing down the human imagination fountain. Prior to being assembled with natural materials, all of these material objects we take for granted were first conceived in the human mind. When we look around, most of what we see today was dreamt, planned, and built by people. We have always looked at nature, saw things that weren't there, and transformed it to make our imagination constructs real and visible.

Human beings have thus used their imagination to transform nature since time immemorial.

However, this paradigm changed over the last century. Somehow just transforming the landscape wasn't enough. Human beings now feel limited by nature. The ultimate frontier is to bypass it completely. The next evolutionary test is to replace nature with an alt-nature where everything is conceived by humans. A human-created alternative natural world that isn't limited by the laws of physics. A world which can mirror imagination itself.

We started calling it virtual reality – a term coined by Jaron Lanier, Founder of VPL Research, in 1987 – but this concept now feels outdated. Evolution is showing us that the digital space is replacing the physical space as the preferred time-spending place. Digitality – the condition of living in a digital culture which comes from Nicholas Negroponte (1995) book, *Being Digital* – is quickly becoming the "real reality". The default.

Digitalization is the process by which human beings slowly but firmly outsource most of the human experiences to an artificial interactive realm. We go shopping at Amazon, Alibaba, Ebay, Farfetch, plus countless sites and apps, and soon the Metaverse. We listen to music at Spotify, Apple Music, and YouTube. We have meetings and give classes via Teams, Zoom, Webex, etc. We brief ChatGPT (Generative Pre-trained Transformer) to write school essays, short stories, brand content, and newspaper articles. We ask Google for information, but soon we will ask AI platforms, which won't just provide us with beats of information so we can figure out the answer, but which will be able to truly answer our question, like a person.

4.2 Covid-19 lockdown brand change accelerator

This process was accelerated by the ultimate stay home experience, the Covid-19 lockdowns.

People were confined, but all the digital appliances available made people feel connected. Work didn't stop. Meetings didn't stop. Projects didn't stop. Classes didn't stop. Brands didn't stop.

It's not so much that everything changed – which it did for a while – but that everyone understood at the same time, everywhere, that this was a life-changing event, and that everything was going to change. A "new normal" – a term used since First World War to describe a state by which an economy, society, and community settle following a crisis, when this differs from the situation that previously existed and first announced this century by technology investor Roger McNamee following the dot-com crash (Fast Company magazine, April 2003) – was going to set in. Work-from-home became the norm. Home-schooling children became an extra job for parents. Being with friends and relatives became a challenge. Buying all the things we need to simply live our lives became a test to supply-chain networks.

What changes people will change brands.

Covid-19 did not transform brands *per se* but it changed the social landscape where they operate. People wanted continuity – the products and services they physically bought delivered to their homes: shopping, networking, education, health, and sports exercise. Suddenly all these subjects seemed unavailable, unless they could be reached online and brought to their homes. In a flash, every living person realized he or she needed Zoom, Uber Eats, Glovo, Netflix, LinkedIn, Spotify, and League of Legends, as never before.

For the first time in Human History people were locked up in their homes on a massive scale, aware of this fact and connected. Unable to interact with their daily products in the physical world, consumers lost the ability not only to simply go out and buy stuff but also to base their buying decisions on sight, touch, smell, and taste. So, to acquire the products they needed, they had no option but to introduce new digital brands in their lives. These brands became the bridge between traditional product brands and consumer' homes. They provided continuity to their lives. A sign that some sort of normalcy was achievable. Most brands already existed, but the confinement gave them a premium boost and allowed them to become mainstream. UberEATS, Glovo, WhatsApp, Zoom, Teams, Netflix, Disney+, Spotify, Apple Music, and countless others now seemed key features essential not only to basic living but also to enjoy life, in a time when social physical interaction was impossible.

Trust became the name of the game. In a way it always was, but with the pandemic spreading brand trustworthiness reached new heights because it fed on primal fear.

Covid-19 was the big scare. People were deeply frightened. Brands needed to provide them with safety, security, reliability, dependability, and reassurance. Some digital brands were already known to most people so there were no qualms about activating them in their computers and mobile phones. Others required a leap of faith.

Branding was put to the test. Through trial-and-error, consumers tested new brand apps, new products, new services, and new social media platforms. Some brands performed and gained customers. Most didn't.

This was a golden opportunity for brands to deepen and expand their digital bodies.

Established brands accelerated their digital transformation. A process that should take years made digital brand systems, platforms, and applications available in a matter of months, even weeks. If the digital universe was the only universe where customers could interact with their brands, UIX design (User Interface XML) became a life-and-death issue for brands. New tutorials to tell stakeholders how to conduct their brand interactions were created and launched. New digital platforms were introduced. Podcasts followed a path from marginal media channel to mainstream channel. Brand content was now a key feature to entertain people.

Digital brands became the entry point into the digital realm for millions of people who would otherwise remain in traditional channels to acquire products and services: more people, more targets, more types of stakeholders, more financial investment, more experimentation, and more change.

The creation of brand online communities became an absolute staple for modern brands. Globalization had already made the emergence of these virtual groups possible. Since the dawn of the 21st century that the internet allowed brand users to access digital places where they could do a plethora of things: discuss the brand among themselves, provide feedback, interact with the brand, exert pressure and influence in order to condition the brand's future outline, create consumer-generated content, have access to previews and exclusive information, know first-hand about new products and perform an evaluation, get free samples, become beta testers, sponsor the brand project (via Patreon), and more. In short, users became members.

As previously stated in this book, brands are not just product differentiators, a system of values or an image: they are a manifestation of the social human condition. If people change, brands change.

4.3 What now?

There's no denying that brands are everywhere, as stated in Chapter 1. Covid-19 lockdowns established its relevance with absolute clarity. There is now a common perception that just having a business is not enough. Businesses become better equipped to deal with structural change when they have their own brands. Brands become the intangible "glue" that connects businesses and users, thus protecting its sustainability.

Branding used to be a key component of big business. Now it's a key component of business – period.

It allows for a business to differentiate itself from the competition and stand out.

It allows for stakeholders to interact with the product, the service, and the company and develop a relationship, thus ensuring brand loyalty and advocacy.

It provides reassurance for customers and prospective partners.

It creates vibrant communities of customers, members, fans, and brand ambassadors.

It is a highly effective tool for employing and retaining talent.

It is a valuable financial asset which can be bought and sold.

It is a vital sign of a healthy economic market.

However, in this massive landscape overpopulated by brands, new brands should wonder if there is even a place for them in the world. The question should be: what does the new brand bring to the world that the world needs?

The post-Covid-19 economic landscape is quickly becoming a thriving and thrilling era for brands and branding, filled not only with opportunities and challenges but also with threats and warnings.

4.3.1 The rise of brand entrepreneurs

The technology that prompted globalization is also promoting entrepreneurship. New generations are launching their own businesses more and more, whether they're just hobbyish pet projects or their main financial sustenance.

For several reasons.

It's now much easier to launch brands than ever before. All it takes is a technical device and an idea. Everyone now feels that, as Snap's 2003 prophetic song goes, "I've got the power". The emergence of social media influencers' personal branding projects provides well-known case studies of business sustainability being driven by the number of followers and interactions they achieve on Instagram, YouTube, or TikTok.

Also, although not everything in the digital economy is for free (computers, smartphones, brand registration…), the most important key factors are, like the access to a potential customer base via social media channels, Instagram, TikTok, YouTube, LinkedIn, Facebook, and Twitter, all of them free of charge.

Time and minimum financial investment are all it takes for an entrepreneur to make brands happen.

4.3.2 Authenticity zeitgeist

"Authenticity is not a movement (…) it's a manifestation of the spirit of our times" (Olins, 2014, p.22). In an era where digitality prevails and everything seems possible, the criteria for deciding which brands to choose rests on critical themes such as authenticity.

Brand authenticity can be defined as "the extent to which consumers perceive a brand to be faithful and true toward itself and its consumers, and to support consumers being true to themselves" (Morhart et al., 2015, p.202).

Wally Olins (2014, p.14) also states that "authentic brands come from everywhere, but they have to be based around provenance". Their legitimacy comes from the place of origin, geographical or symbolical. This infuses them with a panache of purity, honesty, and sustainability. To be authentic a brand has to have an origin story, a legacy, and a mission and be true to them. Consumers will be predisposed to choose the brands if they perceive this authenticity as something real. It's not just what the brand says. It's what it says and what it does, avoiding what social psychologist Leon Festinger, at the beginning of the 1950s, called cognitive dissonance, where "inconsistencies among cognitions (i.e., knowledge, opinion, or belief about the environment, oneself, or one's behavior) generate an uncomfortable motivating feeling" (David C. Vaidis, Alexandre Bran, Oxford Bibliographies, 2017).

In a world populated by an ever-growing number of millions of brands, where any brand seems able to popup from thin air, authenticity works as a lighthouse, helping consumers decide which brands to choose, thus mitigating "the paradox of choice" (Schwartz, 2004) – the more the choices, the less able human beings are to choose.

Also, consumers seem prone to reject brands which are not perceived as authentic, feeling that by choosing unworthy brands they will become themselves devaluated as human beings – the same way, in a press conference for Euro 2020, football player Cristiano Ronaldo rejected a Coke can labeling it as a harmful product for health.

Gilles Lipovetsky (2021, p.264). calls this authenticity trend a "fever". He states that

> Our time is witnessing an increase, particularly among young people, of a new consumer spirit marked by a mandate to give meaning to the acts of shopping. For the committed consumer, purchasing a product is much more than just a behavior focused on seeking private pleasure, it is adopting and defending a set of values, affirming a vision of the world, contributing to improving the present, preparing a better future. At a time when the "ecological crisis" raises fears and concerns about the future of the planet, the act of consuming becomes (…) a way of acting for the common good.

Authenticity thus derives from sustainability. People feel that by choosing authentic brands they will be helping the planet and feel good about themselves.

4.3.3 Driven by purpose

Brands don't just produce goods so consumers can buy them. Not anymore. Modern brands have to care for something other than their sale sheets. As stated in Chapter 1, in an era of stakeholder-focused brands (Merz et al., 2009), brands' job number 1 is to find out what their key stakeholders care for… and care for it.

Dove beauty products brand makes for an iconic best practice.

The brand believes society's perception of female beauty is distorted. As a direct result, millions of women begin having low self-esteem from a young age, which becomes a mental health issue throughout their lives. So, Dove helps women everywhere to develop a positive relationship with the way they

FIGURE 4.26 RuF4 dimer and RuF4-Li complex

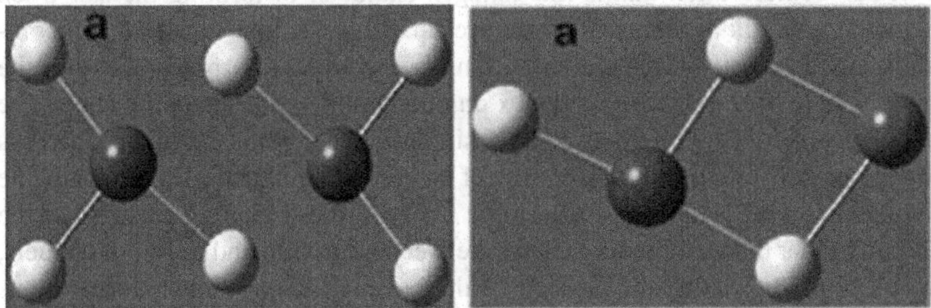

FIGURE 4.27 PdF3 dimer and PdF3-Na complex

FIGURE 4.28 MnF4 dimer and MnF4-Na complex

checked by their frequency calculation, which clearly shows that these dimers are stable as all the calculated frequencies are real. Now we discuss the interaction of superhalogen clusters with alkali atoms like Na and Li atom. Our reported results suggest that all the shown TMFs after interacting with the alkali atoms forms the

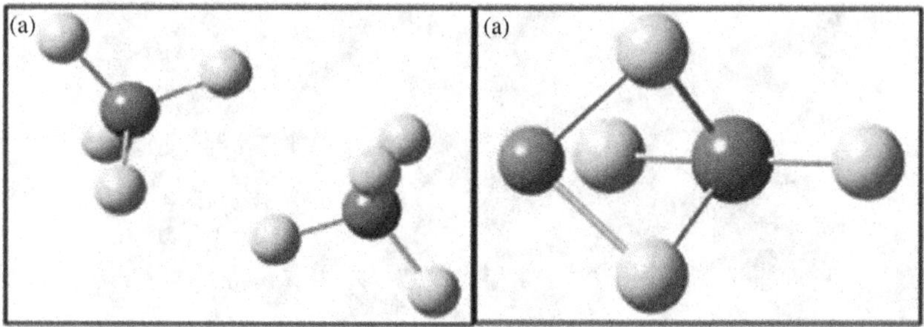

FIGURE 4.29 FeF4 dimer and FeF4-Li complex

stable salt configuration that is like halogen character of forming the salt when interacted with alkali atoms. Reported results suggests that superhalogen complexes interacts well with the alkali atom Na and Li that is shown above in the Figure 4.23 to 4.29. For RhF_4 complex on interacting with Na atom, it shows the binding energy of $NaRhF_4$ to be 5.16 eV, similarly the binding energy of $NaIrF_4$ is found to be 4.69 eV, the binding energy of $NaPtF_4$ is found to be 17.30 eV, and the binding energy of $NaPdF_3$ is found to be 5.31 eV. These binding energies of the salts are higher than the binding energy between a Na atom and a F atom, namely, 4.31 eV. This shows that the predicted salt formation is more stable in nature as compared to the Na-F complex. For $NaMnF_4$ complex the binding energy is found to be 3.95 eV that is little bit lower than the binding energy between a Na atom and a F atom, namely, 4.31 eV. And for $LiRuF_4$ the binding energy is found to be 4·94 eV that is also little bit lower than the binding energy between Li and F atoms, *viz*., 5·45 eV. By and large the salts formed by TMFs on interacting with alkali atoms increases the HOMO–LUMO gap as compared to neutral TMFs that shows that stability of theses TMFs when interacted with alkali atoms increases. This confirms that TMFs mimics the halogen nature, hence, we can conclude that superhalogen interaction with alkali atom form the salt complex that is like the halogen character.

4.5 CONCLUSION

Exhibiting high EA, most TMFs have superior superhalogen characteristics that can be exploited in materials science applications. The Review focused on various TMFs *viz*., RhF_n (n = 1–7), IrF_n (n = 1–7), PtF_n (n = 1–6), RuF_n (n = 1–7), PdF_n (n = 1–7), MnF_n (n = 1–6), FeF_n (n = 1–6) have been carried out in detailed manner. Transition metal atoms of subject TMFs have been found to be capable of binding with up to six or seven F atoms. These complexes exhibit EA values higher to the EA values of halogens. All TMFs have been studied by theoretical calculations in both neutral and anionic forms. It was observed during analysis, that anionic forms usually have weaker bond strengths as compared to bond strengths of corresponding neutral forms. The same is attributed to the extra electron acquired by the anionic form, which delocalizes over the whole cluster resulting in weakening of bond strength. Further, these TMFs have been checked for their stability by analysing their dissociation channels to F atom and F_2 molecule. It was found that, most TMFs are stable against fragmentation into F atom and F_2 molecule. The most important parameter to predict the superhalogen nature of these complexes is EA. The calculated EA values of most subject TMFs suggest that the clusters with $n \geq 4$ are superhalogens because their EAs are higher than the chlorine EA value of 3.62 eV. Due to this reason, most subject TMFs readily behave as superhalogens. Particularly, high n value ($n > 4$) TMFs provide superior superhalogens character by having very high EA values. Further the superhalogen forms the dimer complexes and when interacted with alkali atoms they produce the stable salt like configuration that is like the halogen character.

ACKNOWLEDGEMENT

Authors would like to thank editor and co-editor of the book on "*Superhalogens and Superalkalis: Bonding, Reactivity, Dynamics and Applications*" to give us the opportunity to contribute as a chapter on "Transition metal fluorides as superhalogen."

REFERENCES

1. Gutsev, G. L.; Bartlett, R. J.; Boldyrev, A. I.; Simons, J. Adiabatic Electron Affinities of Small Superhalogens: LiF2, LiCl2, NaF2, and NaCl2. *J Chem Phys*, 1997, *107* (10), 3867–3875. https://doi.org/10.1063/1.474764.
2. Hotop, H.; Lineberger, W. C. Binding Energies in Atomic Negative Ions: II. *J Phys Chem Ref Data*, 1985, *14* (3), 731–750. https://doi.org/10.1063/1.555735.
3. Berzinsh, U.; Gustafsson, M.; Hanstorp, D.; Klinkmüller, A.; Ljungblad, U.; Mårtensson-Pendrill, A.-M. Isotope Shift in the Electron Affinity of Chlorine. *Phys Rev A (Coll Park)*, 1995, *51* (1), 231–238. https://doi.org/10.1103/PhysRevA.51.231.
4. Gutsev, G. L.; Boldyrev, A. I. DVM-Xα Calculations on the Ionization Potentials of MX_{k+1}^- Complex Anions and the Electron Affinities of MX_{k+1} "Superhalogens". *Chem Phys*, 1981, *56* (3), 277–283. https://doi.org/10.1016/0301-0104(81)80150-4.
5. Bartlett, N. Xenon Hexafluoroplatinate (V) Xe+ [PtF6]-. *Proc Chem Soc London*. 1962, p 218.
6. Graudejus, O.; Elder, S. H.; Lucier, G. M.; Shen, C.; Bartlett, N. Room Temperature Syntheses of AuF_6^- and PtF_6^- Salts, $Ag^+AuF_6^-$, $Ag^{2+}PtF_6^{2-}$, and $Ag^{2+}PdF_6^{2-}$, and an Estimate for $E(MF_6^-)$ [M = Pt, Pd]. *Inorg Chem*, 1999, *38* (10), 2503–2509. https://doi.org/10.1021/ic981397z.
7. Srivastava, A. K.; Kumar, A.; Tiwari, S. N.; Misra, N. Application of Superhalogens in the Design of Organic Superconductors. *New J Chem*, 2017, *41* (24), 14847–14850. https://doi.org/10.1039/C7NJ02868G.
8. Czapla, M.; Anusiewicz, I.; Skurski, P. Does the Protonation of Superhalogen Anions Always Lead to Superacids? *Chem Phys*, 2016, *465–466*, 46–51. https://doi.org/10.1016/j.chemphys.2015.12.005.
9. Srivastava, A. K.; Kumar, A.; Misra, N. Superhalogens as Building Blocks of a New Series of Superacids. *New J Chem*, 2017, *41* (13), 5445–5449. https://doi.org/10.1039/c7nj00129k.

10. Rasheed, T.; Siddiqui, S. A.; Kargeti, A.; Shukla, D. V.; Singh, V.; Pandey, A. K. Exploration of Superhalogen Nature of $Pt(CN)_n$ Complexes (n = 1–6) and Their Abilities to Form Supersalts and Superacids: A DFT–D3 Study. *Struct Chem*, 2021, *32* (6), 2209–2221. https://doi.org/10.1007/s11224-021-01786-y.
11. Kargeti, A.; Sharma, G.; Rasheed, T. Exploration of Superacidic Properties of $HMnF_n$ (n = 1–6) Using Density Functional Theory. *Macromol Symp*, 2019, *388* (1), 1900040. https://doi.org/10.1002/masy.201900040.
12. Czapla, M.; Ciepła, O.; Brzeski, J.; Skurski, P. Formation of Enormously Strongly Bound Anionic Clusters Predicted in Binary Superacids. *J Phys Chem A*, 2018, *122* (43), 8539–8548. https://doi.org/10.1021/acs.jpca.8b07514.
13. Freza, S.; Skurski, P. Enormously Large (Approaching 14 EV!) Electron Binding Energies of $[H_nF_{n+1}]$– (n = 1–5, 7, 9, 12) Anions. *Chem Phys Lett*, 2010, *487* (1–3), 19–23. https://doi.org/10.1016/J.CPLETT.2010.01.022.
14. Giri, S.; Behera, S.; Jena, P. Superhalogens as Building Blocks of Halogen-Free Electrolytes in Lithium-Ion Batteries. *Angew. Chem. Int. Ed.*, 2014, *53* (50), 13916–13919. https://doi.org/10.1002/anie.201408648.
15. Golub, I. E.; Filippov, O. A.; Kulikova, V. A.; Belkova, N. V.; Epstein, L. M.; Shubina, E. S. Thermodynamic Hydricity of Small Borane Clusters and Polyhedral Closo-Boranes. *Molecules*, 2020, *25* (12), 2920. https://doi.org/10.3390/molecules25122920.
16. Elliott, B. M.; Koyle, E.; Boldyrev, A. I.; Wang, X.-B.; Wang, L.-S. MX_3–Superhalogens (M = Be, Mg, Ca; X = Cl, Br): A Photoelectron Spectroscopic and ab Initio Theoretical Study. *J Phys Chem A*, 2005, *109* (50), 11560–11567. https://doi.org/10.1021/jp054036v.
17. Anusiewicz, I.; Sobczyk, M.; Dąbkowska, I.; Skurski, P. An ab Initio Study on MgX_3^- and CaX_3^- Superhalogen Anions (X=F, Cl, Br). *Chem Phys*, 2003, *291* (2), 171–180. https://doi.org/10.1016/S0301-0104(03)00208-8.
18. Anusiewicz, I.; Skurski, P. An ab Initio Study on BeX_3^- Superhalogen Anions (X = F, Cl, Br). *Chem Phys Lett*, 2002, *358* (5–6), 426–434. https://doi.org/10.1016/S0009-2614(02)00666-8.
19. Sikorska, C.; Smuczyńska, S.; Skurski, P.; Anusiewicz, I. BX_4^- and AlX_4^- Superhalogen Anions (X = F, Cl, Br): An ab Initio Study. *Inorg Chem*, 2008, *47* (16), 7348–7354. https://doi.org/10.1021/ic800863z.
20. Gutsev, G. L.; Jena, P.; Bartlett, R. J. Structure and Stability of BF3∗F and AlF3∗F Superhalogens. *Chem Phys Lett*, 1998, *292* (3), 289–294. https://doi.org/10.1016/S0009-2614(98)00716-7.
21. Pradhan, K.; Gutsev, G. L.; Jena, P. Negative Ions of Transition Metal-Halogen Clusters. *J Chem Phys*, 2010, *133* (14), 144301. https://doi.org/10.1063/1.3489117.
22. Koirala, P.; Willis, M.; Kiran, B.; Kandalam, A. K.; Jena, P. Superhalogen Properties of Fluorinated Coinage Metal Clusters. *J Phys Chem C*, 2010, *114* (38), 16018–16024. https://doi.org/10.1021/jp101807s.
23. Wang, Q.; Sun, Q.; Jena, P. Superhalogen Properties of CuF_n Clusters. *J Chem Phys*, 2009, *131* (12), 124301. https://doi.org/10.1063/1.3236576.
24. Yang, J.; Wang, X.-B.; Xing, X.-P.; Wang, L.-S. Photoelectron Spectroscopy of Anions at 118.2nm: Observation of High Electron Binding Energies in Superhalogens MCl_4^- (M=Sc, Y, La). *J Chem Phys*, 2008, *128* (20), 201102. https://doi.org/10.1063/1.2938390.
25. Yang, X.; Wang, X.-B.; Wang, L.-S.; Niu, S.; Ichiye, T. On the Electronic Structures of Gaseous Transition Metal Halide Complexes, FeX_4^- and MX_3^- (M=Mn, Fe, Co, Ni, X=Cl, Br), Using Photoelectron Spectroscopy and Density Functional Calculations. *J Chem Phys*, 2003, *119* (16), 8311–8320. https://doi.org/10.1063/1.1610431.

26. Scheller, M. K.; Compton, R. N.; Cederbaum, L. S. Gas-Phase Multiply Charged Anions. *Science (1979)*, 1995, *270* (5239), 1160–1166. https://doi.org/10.1126/science.270.5239.1160.
27. Siddiqui, S. A.; Pandey, A. K.; Rasheed, T.; Mishra, M. Investigation of Superhalogen Properties of RhF_n (*n*=1–7) Clusters Using Quantum Chemical Method. *J Fluor Chem*, 2012, *135*, 285–291. https://doi.org/10.1016/j.jfluchem.2011.12.009.
28. Siddiqui, S. A.; Rasheed, T. Quantum Chemical Study of IrF_n (*n* = 1–7) Clusters: An Investigation of Superhalogen Properties. *Int J Quantum Chem*, 2013, *113* (7), 959–965. https://doi.org/10.1002/qua.24053.
29. Siddiqui, S. A.; Rasheed, T.; Pandey, A. K. Quantum Chemical Study of PtF_n and $PtCl_n$ (*n* = 1–6) Complexes: An Investigation of Superhalogen Properties. *Comput Theor Chem*, 2012, *979*, 119–127. https://doi.org/10.1016/j.comptc.2011.10.023.
30. Siddiqui, S. A.; Rasheed, T.; Bouarissa, N. Investigation of Superhalogen Behaviour of RuF_n (*n* = 1–7) Clusters: Density Functional Theory (DFT) Study. *Bull Mater Sci*, 2013, *36* (4), 743–749. https://doi.org/10.1007/s12034-013-0514-8.
31. Siddiqui, S. A.; Bouarissa, N. Superhalogen Properties of PdF_n (N=1–7) Clusters Using Quantum Chemical Method. *Solid State Sci*, 2013, *15*, 60–65. https://doi.org/10.1016/j.solidstatesciences.2012.08.024.
32. Rasheed, T.; Siddiqui, S. A.; Bouarissa, N. Quantum Chemical Investigations on Superhalogen Properties of MnF_n (*n* = 1.6) Nano-Complexes and the Consequential Possibility of Formation of New MnF_{n-} Na Salt Species. *J Fluor Chem*, 2013, *146*, 59–65. https://doi.org/10.1016/j.jfluchem.2013.01.010.
33. Rasheed, T.; Siddiqui, S. A.; Pandey, A. K.; Bouarissa, N.; Al-Hajry, A. Investigations on the Frontier Orbitals of FeFn (n=1–6) Superhalogen Complexes and Prediction of Novel Salt Series Li-(FeFn). *J Fluor Chem*, 2017, *195*, 85–92. https://doi.org/10.1016/j.jfluchem.2017.01.014.
34. Hohenberg, P.; Kohn, W. Inhomogeneous Electron Gas. *Phys Rev*, 1964, *136*, B864.
35. Kohn, W.; Sham, L. J. Self-Consistent Equations Including Exchange and Correlation Effects. *Phys Rev*, 1965. https://doi.org/10.1103/PhysRev.140.A1133.
36. Becke, A. D. A New Mixing of Hartree-Fock and Local Density-Functional Theories. *J Chem Phys*, 1993, *98* (2), 1372–1377. https://doi.org/10.1063/1.464304.
37. Frisch, M. J., et al., *Gaussian 09, Revision A.1*, Gaussian, Inc., Wallingford, CT; 2009.
38. Dennington, R.; Keith, T. A.; Millam, J. M. *GaussView 6, Gaussian*, Semichem Inc., Shawnee Mission, KS; 2016.

5 Metalloboranes as Unusual Superhalogens

Ruby Srivastava

5.1 INTRODUCTION

The concept of "superatoms" was introduced in cluster chemistry for the atomic clusters of given size and composition, which mimic the properties of elements in the periodic table. Superatoms are nanoscale building blocks, which are used for designing materials with tailored functionalities. A vast library of chemically tuneable superatoms has been created in past four decades that display unique physical and chemical properties.

Superhalogens are highly electronegative atomic clusters whose electron affinities exceed those of halogens. Electron affinity being the most important parameter affects the chemical reactivity. Halogens have the highest electron affinities among all elements. Chlorine has highest electron affinity (3.62) among the halogens. Superhalogens are a known family from the past three decades and all these complexes follow either the 8-electron rule or 18-electron rule. Superhalogens have potential to promote unusual reactions and they act as weakly coordinating anions. These clusters are used for design and synthesis of bulk materials. Superhalogens are metal-halogen complexes that require one electron to close their electronic structure. Superhalogens are composed of a metal atom surrounded by halogen atoms [1]. Here the central metal atom can either be a main group element [2–9] or a transition metal atom [10–12] and the halogen ligand can be substituted for electronegative element such as oxygen, hydroxyl radical *etc.* [13–24]. Superhalogens are also used to stabilise the unusually high oxidation states of metals. The name "superhalogens" was given by Gutsev and Boldyrev in 1981, when they first observed a central metal atom decorated with halogen ligands complexes with electron affinity exceeding Cl [25,26]. Usually superhalogens are binary complexes, *i.e.*, made from two elements. A lot of unexplored work lies in replacing the halogen atoms by pseudohalogens (CN, SCN, NCO) or by using different central metal atoms such as Li, Na, Be, Mg, Ca, B and Al [27]. A class of polynuclear magnetic superhalogens Mn_nCl_{2n+1} (n = 1, 2, 3 . . .) are also used to build salts [28]. For example, $KMnCl_3$ exhibit antiferromagnetic properties. Superhalogens can oxidize neutral species (SiO_2, NH_3, *etc.*) with moderately high ionization potentials to form stable ionic salts [29]. As superhalogens increase the work function of graphene [30], it is

DOI: 10.1201/9781003384205-5

used to design graphene-based electrodes. Another class of extraordinary compounds is "superalkalis" that have lower IEs than those of alkali metal atoms. Superhalogens and superalkalis can be used to form new supersalts with unique properties. These interesting species are used to design novel oxidants [31–33] or nonlinear optical (NLO) molecules [34–36] by formation of novel salts-super/hyper salts [37]. Superhalogens [38] can serve as the building blocks for hydrogen storage materials [39], halogen-free electrolytes in metal-ion batteries, lithium ion batteries [40], hybrid perovskite solar cells [41], and ion conductors with increased ionic conductivities [42]. Superhalogens can be used to design multiferroic materials [43]. As molecules with high electron affinity (EA) form very stable negative ions, superhalogens are used to purify air, lift mood [44] and act as strong oxidizing agents [45]. They can oxidize species with high vertical ionization energies (VIEs); thereby, forming novel and unusual salts.

The unconventional superhalogens are designed as borane-based superhalogens and pseudohalogen-based superhalogens. Boranes and their derivatives are created by replacing B with C (carboranes), B with metal atoms (metalloboranes) and H with halogens (F and Cl). Carboranes belong to a class of heteroboranes, and composed of B, C and H atoms. These electron-delocalized clusters are polyhedra or fragments of polyhedral with wide applications [46–48]. In this class, it is anticipated that as $CB_{11}H_{12}$ is isoelectronic with $B_{12}H_{13}$, it may be a superhalogen. Similarly, $M(B_{12}H_{12})$ (M = Li, Na, K, Rb, Cs) is isoelectronic with $CB_{11}H_{12}$, it could be a superhalogen. If we look into the structural geometries, C replaces a B atom in the $B_{12}H_{12}$, while in $M(B_{12}H_{12})$ alkali metal donate an electron to the $B_{12}H_{12}$ moiety. These are unconventional superhalogens, as in conventional superhalogens, an alkali metal needs minimum two halogens and here it is sufficient to use only $B_{12}H_{12}$ [49]. Like halogens, pseudohalogens need one electron to close their electronic shell, and form very stable single charge negative ions. Pseudohalides are composed of two or more atoms. For example: CN, NCO, SCN, N3, *etc.* Core atoms as Li, Na, Be, Mg, Ca, B and Al are used to form superhalogens [27]. The nature of bonding between these atoms does not affect the chemical reactions.

In this chapter, we will discuss the unusual behaviour of metalloboranes, a class of superhalogen clusters, using properties as electronic structures, geometry, reactivity, bond strengths and quantum theory of atoms in molecule (QTAIM) analysis.

5.2 METALLOBORANES

A class of compounds in which metal atom is bound to boron hydride $[BH_4]$-group is known as metalloboranes. As inclusion of these metal fragments does not disrupt the large borane cage, these metalloboranes are considered as clusters. According to the bonding rules, metalloboranes involving boron with more than four bonds are considered as clusters [50]. Metallaboranes provide a crucial link between boron atom to metal clusters and metal-hydrocarbon π complexes. With the discovery of boranes and their metal derivatives, it has been observed that boron and metals have the similar kind of bonding as the boron atoms of boranes. Also, the bonding between metal atoms and borane ligands is similar to that of bonding between metal atoms and unsaturated hydrocarbons in organometallic compounds. Further, it is observed that boron and metal clusters, metal-hydrocarbon π-complexes, aromatic ring systems, many small-ring hydrocarbons and a few carbo-cations; all follow same structural and bonding patterns [51]. Coupling of borane cages involve more than one metal atom, which may or may not bond together. Mostly, the interaction of metal with borane cage either expands the cage or substitute metal formally. The comprehensive articles [52–53] are very informative regarding bonding characteristics of metalloboranes for the interested readers.

Metalloboranes can be classified as follows:

1. Ionic hydroborates, ex: $NaBH_4$, $Ba(BH_4)_2$.
2. Metal hydroborates: containing hydrogen bridge bonds, M–H–B. ex: $Al(BH_4)_3$, $(Ph_3P)_2CuBH_4$.
3. Metal carbaboranes: include π-bonded "sandwich compounds."
4. Compounds containing metal-boron bonds except those in the above category (3).

Metalloboranes are also classified as *closo*, *nido*, *arachno*, *crypto*, and *commo*. Few of the metalloborane structures are given in Figure 5.1. The reactivity of metalloboranes can be divided in two parts: (1) boron-rich metalloboranes and (2) metal-rich metalloboranes. In boron-rich metalloboranes, cluster arrangement plays a vital role; while in metal-rich metalloboranes, the chemistry is governed by the reactivity of transition metals and boron atoms.

A new class of magnetic superhalogens were also found where the electron affinities are higher than those of the constituent superhalogens. For example: $Fe(BO_2)_4$, $Fe(NO_2)_4$ and $Fe(NO_3)_4$ are the magnetic superhalogens that have higher electron affinities than BO_2, NO_2 and NO_3 and the extra electron is localized over the ligands in anions [29]. The extra stability of clusters is justified by electron counting rules such as the 8-electron rule [54], the 18-electron rule [55], the aromaticity rule [56] and the Wade-Mingos rule [57], which are discussed in the next section.

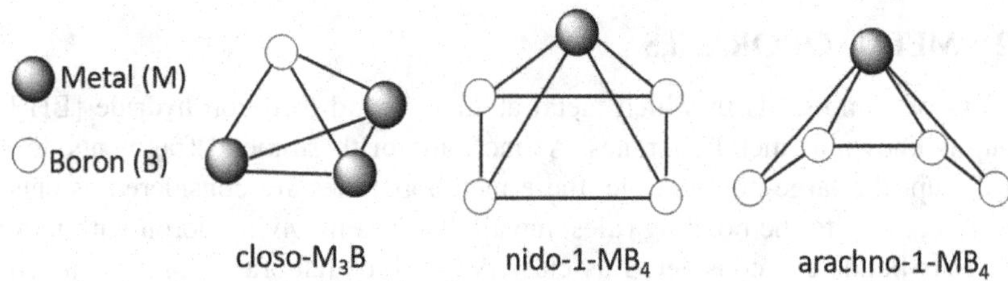

FIGURE 5.1 The structures of *closo*, *nido*, *arachno* metalloboranes.

5.3 ELECTRON COUNTING RULES

The stability and chemistry of clusters are associated with the electron counting rules. For example: the inertness of noble gases and reactivity of alkali metals and halogens are due to the octet rule [54,58]. The 8-electron rule and the 18-electron rule are responsible for stable negative ions.

5.3.1 8-Electron Rule

In octet rule, eight electrons are required to complete the valency of a simple element and attain a noble gas configuration. Elements are combined in such a manner that each element has a full octet. Halogens with ns^2 np^5 configuration need one electron to satisfy the octet rule. Since only one electron is needed to satisfy the 8-electron rule, the EAs of halogens atoms are higher (3.0–3.6 eV). Most of the main group elements follow octet rule, with only few exceptions. Octet rule has been used to design and synthesize superhalogens (low atomic number species) from the past few decades [20–24].

5.3.2 18-Electron Rule

The stability of many transition metal complexes follows the 18-electron rule. The rule indicates that the central atom needs (2 *s*-electrons, 6 *p*-electrons, 10 *d*-electrons) to complete the electronic shell. The rule is applicable for superhalogens composed of only metal atom clusters $M@Au_{12}$ (M = V, Nb, Ta) [28]. The 18-electron rule [59] is responsible for transition metal compounds. For example: $Cr(C_6H_6)_2$ and $V(CO)_6^-$. In nutshell the superhalogen need an optimal size and one electron less than that required to complete the electronic shell.

5.3.3 Aromaticity Rule

Aromaticity rule follows four conditions, such as: it must be cyclic; every atom in the ring must be conjugated; the molecule must have [4n+2] π electrons; and molecules have flat structure. Some rare exceptions follow only first three conditions. Aromaticity is associated with planar conjugated cyclic systems (benzene) and utilize free delocalized π electrons to improve its stability and unusual reactivity patterns. The molecules are designed either by altering cyclopentadienyl ligands or by substituting CH groups with isoelectronic N atoms in multiple benzo-annulations of cyclopentadienyl. These aromatic superhalogens has higher electron affinities (5.59 eV). The aromaticity rule has given a new class of superhalogens that consists of neither a metal nor a halogen atom [59].

5.3.4 Wade-Mingos Rule

The Wade-Mingo rule is based on polyhedral skeleton electron pair theory. Skeleton electron counting scheme is developed to justify the structure of carboranes. According to the Wade's rule, a closed deltahedral cluster (n vertices) requires (n + 1) electrons to completely occupy the cluster bonding molecular orbitals (MOs). It is assumed that each boron atom in a closed cage possess one terminal hydrogen atom. Borane clusters are divided into five categories according to the degree of openness.

1. *closo*-closed deltahedral cage $[B_6H_6]^{2-}$
2. *nido*-open cage related to closed deltahedral cage by removing one vertex, as B_5H_9
3. *arachno*-open cage related to closed deltahedral cage by removing two vertices, as B_4H_{10}
4. *hypho*-open cage related to closed deltahedral cage by removing three vertices, as $B_5H_9(PMe_3)_2$
5. *conjuncto*-open or closed fused cages, as 2,2'-$\{B_{10}H_{13}\}_2$.

It is rationalized by following rules:

Step 1: The number of skeletal electrons is calculated as sum of the following contributions. (1) 2 electrons are contributed for each B-H bond. (2) 3 electrons are contributed for each C-H bond. (3) Each additional hydrogen contributes 1 electron. (4) Lastly the anionic charge on the cluster is to be added to above contributions:
Skeletal electron count = 2(B-H) + 3(C-H) + 1(additional H) + anionic charge on cluster.

Step-2: Calculate the number of pairs of electrons

$$\text{Number of electron pairs} = \frac{\text{skeleton electron counts}}{2}$$

Step-3: Let the number of vertices for B and C atoms is given by "n." Comparing the number of electron pairs with number of vertices (n), the cluster's structure can be identified as follows:

A borane cage also follow Wade-Mingos rule, in which (n + 1) pairs of electrons are required to stabilize the boron cage, where n is the number of vertices in

TABLE 5.1
The number of electron pairs for different types of boranes.

No of electron pairs	Type of boranes
n +1	*Closo*
n + 2	*Nido*
n + 3	*Arachno*
n + 4	*Hypo*

the boron polyhedron. $B_{12}H_{12}^{2-}$ has an icosahedral symmetry, where n = 12. Total 48 electrons are available for bonding in which 24 electrons occupy 12 B-H covalent bonds, leaving the remaining 24 electrons for cage bonding. The stability of $B_{12}H_{12}$ is according to the Wade-Mingos rule [56].

$B_{12}(CN)_{12}$ (2-) is a *closo*-borane (*e.g.*, $B_{12}H_{12}$(2-)) derivative, obtained by replacing H with CN, having EA (5.28 eV) of its second electron, which is experimentally verified as stable dianion [28]. $B_{12}H_{12}$ (2-) is thermodynamically stable dianion and the stability of dianion is increased when H atoms are replaced by a more electronegative CN moiety. In a recent study it was seen that $B_{12}(CN)_{12}$(2-) is more stable than its mono anion. The unusual stability of B_{12} $(CN)_{12}$(2-) is according to Wade-Mingos rule of the boron polyhedron and the octet rule of the CN moiety simultaneously. Results indicated that $B_{12}(CN)_{12}$(3-) is not stable against electron emission [59].

5.4 DISCUSSION

Let's discuss the two unusual types of metalloboranes: (1) closed-shell metalloboranes ($Zn(B_{12}H_{12})$ and $Al(BeB_{11}H_{12})$) and (2) metalloboranes ($Zn(B_{12}H_{11})$, $Zn(CB_{11}H_{12})$ and $Al(B_{12}H_{12})$), computationally studied by Prof. Jena et al. [60] in detail. The optimise structures of neutral closed-shell metalloboranes (($Zn(B_{12}H_{12})$ and $Al(BeB_{11}H_{12})$) are given in Figure 5.2 and the optimized structures of neutral metalloboranes ($Zn(B_{12}H_{11})$, $Zn(CB_{11}H_{12})$, and $Al(B_{12}H_{12})$) are given in Figure 5.3. The second group of metalloboranes has one electron more than needed for electron shell closure. If we look into the structural stability of *closo* B_nH_n, it is governed by Wade–Mingos rule, which is explained in the previous section. For electron shell closure structure, (4n + 2) electrons are required. Since B_nH_n clusters have 4n electrons, the stability for the structure need two more electrons. $B_nH_n^{2-}$ is a stable cluster.

According to Wade–Mingos rule, since Zn is divalent, $Zn(B_{12}H_{12})$ (EA= 4.30 eV) should also form a closed shell cluster. Similarly, $Al(BeB_{11}H_{12})$ (EA=3.61 eV) should also form electron closure structure, as Al is trivalent atom. Accordingly,

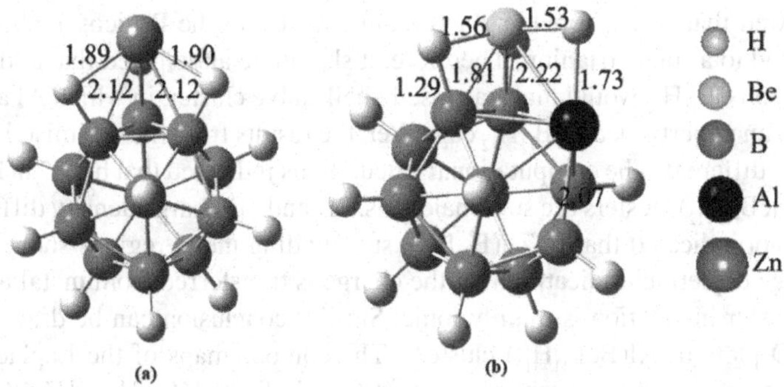

FIGURE 5.2 Optimised structures of closed-shell metalloboranes (a) ($Zn(B_{12}H_{12})$ and (b) $Al(BeB_{11}H_{12})$). Few bondlengths (Å) are given with the structures. Figures are adopted from Banjade, H. et al. **Nanoscale**, 2022, **14**, 1767–1778. Ref. [60].

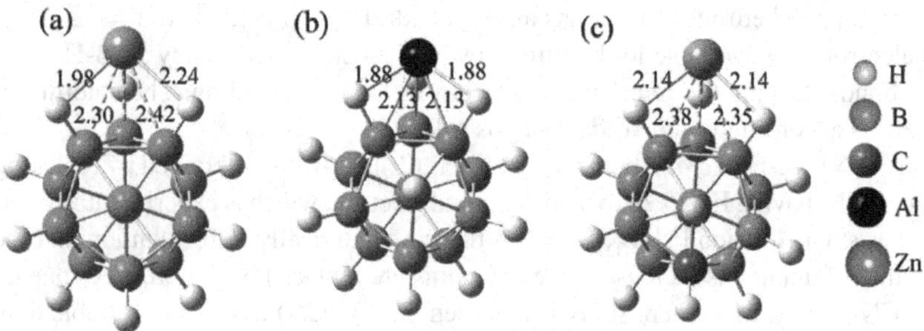

FIGURE 5.3 Optimised structures of metalloboranes (a) $Zn(B_{12}H_{11})$, (b) $Al(B_{12}H_{12})$, (c) $Zn(CB_{11}H_{12})$. Few bondlengths (Å) are given with the structures. Figures are adopted from Banjade, H. et al. **Nanoscale**, 2022, **14**, 1767–1778. Ref. [60].

these clusters are expected to have small EA (< EA of halogens). But the DFT calculations carried out on these clusters showed the opposite trend. The two group clusters emerge out to be superhalogens regardless of Wade-Minogs rule. The conventional rules for the two groups will state that the first group should have lower EAs and the second group will be of superalkalis.

In this study, the concept "unconventional superhalogen" means that the added electron will occupy the doped metal atom, which has a positively charged state due to the transfer of charge to the cluster. The two *closo*-dodecaborates and their derivatives, *e.g.*, $(B_{12}H_{12})^{2-}$ and $(BeB_{11}H_{12})^{3-}$ are very stable. Now in another two clusters $Zn(B_{12}H_{12})$ and $Al(BeB_{11}H_{12})$, Zn, being divalent will donate two electrons to stabilize $B_{12}H_{12}$; while Al, being trivalent, can donate three electrons to stabilize $Be(B_{11}H_{12})$. Therefore, it is expected that these closed shells clusters should possess small EAs. The optimized geometries of $Zn(B_{12}H_{12})$ indicated that Zn comes out of the cage after optimization while in $Al(BeB_{11}H_{12})$, Al is in the cage (with some distortion) and Be is out of the cage. See Figure 5.2.

In the same manner the equilibrium geometries of neutral Al $(BeB_{11}H_{12})$ clusters showed that before replacing a B atom with Be in the B_{12} icosahedral cage would lead to a stable trianion. Therefore, it should be anticipated that adding an Al atom to $BeB_{11}H_{12}$ would make a closed-shell stable cluster, in which Al atom is bound to an unperturbed $BeB_{11}H_{12}$ cage. Yet, the results from the optimised geometries are different. The computational calculations indicated that both $Zn(B_{12}H_{12})$ and $Al(BeB_{11}H_{12})$ clusters are superhalogens. 2D and 3D charge density difference (CDD) plot indicated that in $Zn(B_{12}H_{12})$, surrounding metal regions show apparent charge depletion, indicating that the charge is transferred from metal and the metal-cluster interaction is mainly ionic. Similar conclusion can be drawn from the CDD plots of $Al(BeB_{11}H_{12})$ clusters. The contour maps of the Laplacian of electron density with bond critical points (BCPs) indicated Zn–H and Zn–B interaction in $Zn(B_{12}H_{12})$ and the interaction of Be and Al with $Al(BeB_{11}H_{12})$ clusters. The positive values of the Laplacian at BCP from the CDD plot indicated ionic nature of bonding in $Al(BeB_{11}H_{12})$. See Figure 5.4.

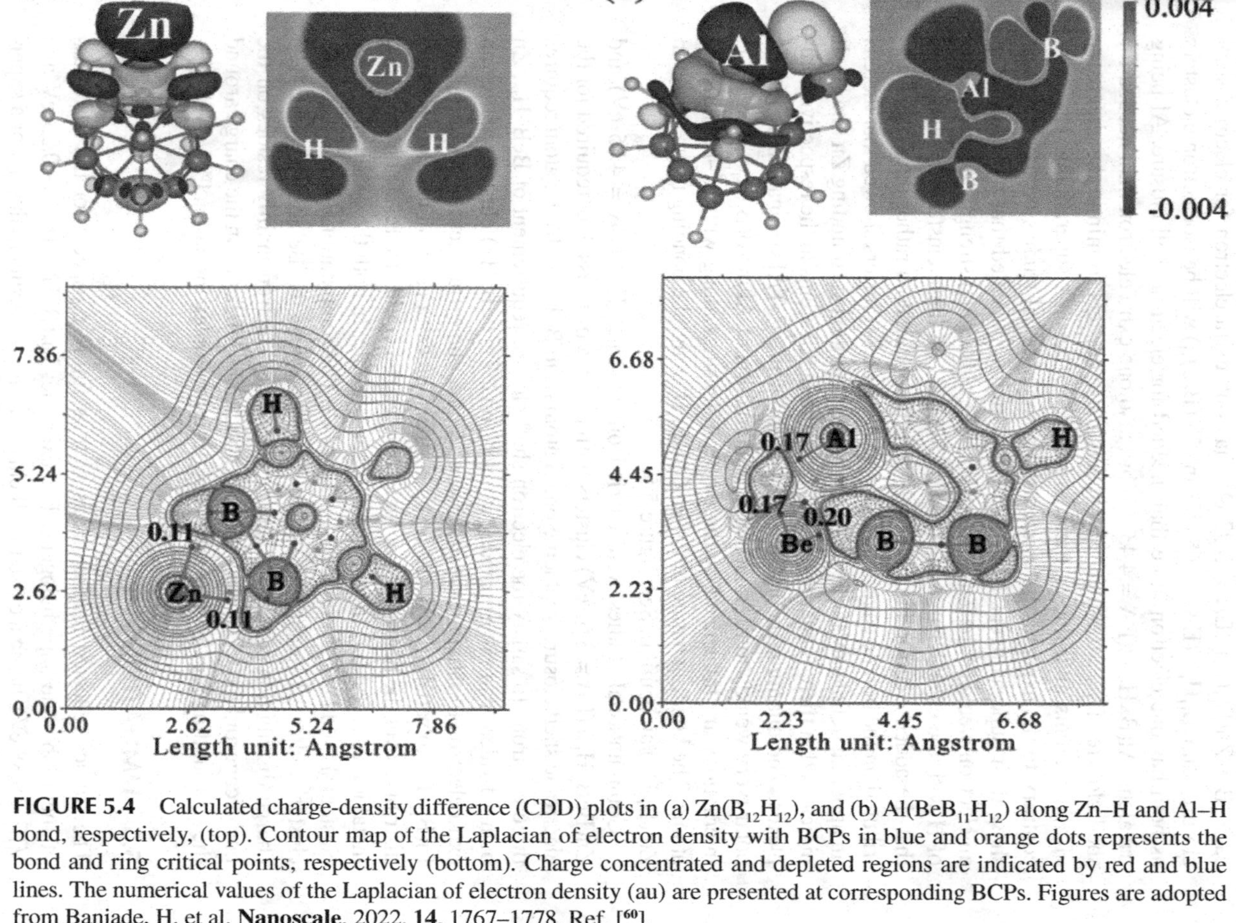

FIGURE 5.4 Calculated charge-density difference (CDD) plots in (a) $Zn(B_{12}H_{12})$, and (b) $Al(BeB_{11}H_{12})$ along Zn–H and Al–H bond, respectively, (top). Contour map of the Laplacian of electron density with BCPs in blue and orange dots represents the bond and ring critical points, respectively (bottom). Charge concentrated and depleted regions are indicated by red and blue lines. The numerical values of the Laplacian of electron density (au) are presented at corresponding BCPs. Figures are adopted from Banjade, H. et al. **Nanoscale**, 2022, **14**, 1767–1778. Ref. [60].

If we look into $B_{12}H_{12}$ clusters, it is seen that if one H is removed from $B_{12}H_{12}$, $B_{12}H_{11}$ cluster will require only one extra electron to satisfy the electron shell closure rule. With Zn being divalent, it can contribute two electrons to the boron cage. But $Zn(B_{12}H_{11})$ (EA = 3.93 eV) has one extra electron for electron shell closure. $Zn(CB_{11}H_{12})$ (EA = 3.68 eV) and $Zn(B_{12}H_{11})$ will be isoelectronic clusters as both has one electron more than needed for electron shell closure. Al being trivalent, $Al(B_{12}H_{12})$ ((EA = 4.43 eV) has also one extra electron than needed for shell closure. This extra electron expects these clusters to mimic the chemistry of alkali atoms and, due to their large size, behave as superalkalis. However, contrary to the expected behaviour, the computational studies on these clusters show their superhalogens behaviour. The studies indicated that the addition of an electron weakens the Zn bonding with the cluster and strengthen the bonding of H with the cage. Charge distribution analysis on these clusters also indicated that the added electron mainly resides on the metal atom rather than being distributed on the borane cage. The CDD plots for these clusters showed that there is an apparent deficiency in the charge around the region surrounding Zn and Al. It indicates that there is charge transfer from the metal atoms to the respective complexes. The quantum theory of atoms in molecule (QTAIM) analysis indicated polar (very weakly ionic, noncovalent) interaction of Al in $Al(B_{12}H_{12})$. The Al–H interactions are influenced by the electron densities along with Al–H and B–H atoms. The bonding nature of Zn in $Zn(CB_{11}H_{12})$ and the bonding nature of Zn in Zn $(B_{12}H_{11})$ are similar. See Figure 5.5.

Computational studies were carried out on $Li(B_{12}H_{12})$ (EA = 4.85 eV) and $Zn(BeB_{11}H_{12})$ (EA = 3.67 eV) clusters with one electron less than required for the electronic shell closure, replacing one B atom from $B_{12}H_{12}$ with a Be atom requires three electrons to satisfy the electron shell closure requirement of $BeB_{11}H_{12}$. Zn being divalent can only provide two electrons. $Zn(BeB_{11}H_{12})$ cluster requires an extra electron to satisfy its shell closure. Again, one extra electron is required for $Li(B_{12}H_{12})$ to satisfy the Wade–Mingos rule. The extra electron has the option to distribute over the boron cage or the extra electron could also reside on the doped metal ion as it remains in a positively charged state. In both the conditions, these clusters should have EA larger than that of Cl. The computational studies on these clusters indicated weakening of the B–H bond near the Be adsorbed site. The charge distribution indicated that the additional electron mainly resides on the boron cage and the CDD plot reflects a noticeable deficiency in the charge around the surrounding region of metal atoms during complex formation [[60]].

5.5 LIMITATIONS

The studies reflected limitations as well as future prospects for the new class of metalloborane superhalogens. For example: $Li_2(B_{12}H_{12})$ (EA = 0.25 eV and VIE = 9.38 eV), being a closed shell cluster is neither a superhalogen nor a superalkali. $Li_3(B_{12}H_{12})$ has one electron more than needed for shell closure and it is a superalkali (EA = 0.22 eV and VIE = 3.24 eV). The studies showed that even if the alkali metal atoms donate electrons to the boron cage, resulting species are

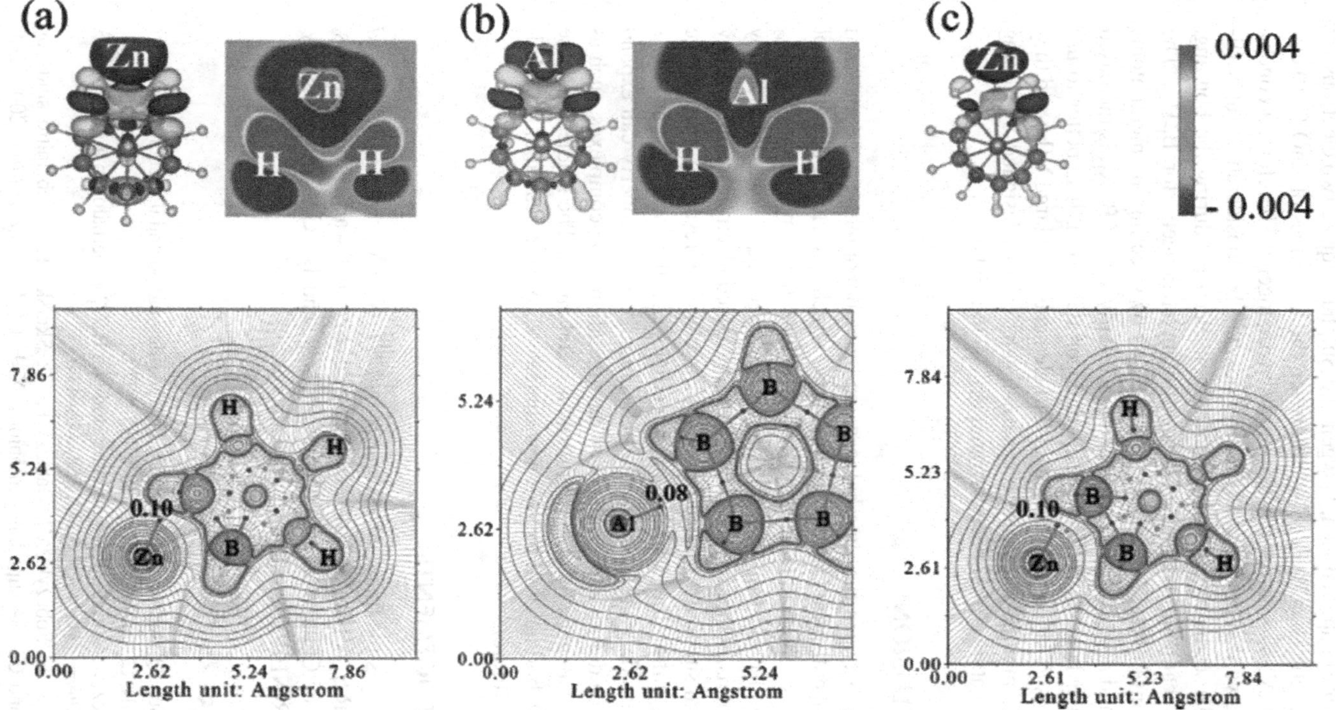

FIGURE 5.5 Calculated charge-density difference (CDD) plots in (a) $Zn(B_{12}H_{11})$, (b) $Al(B_{12}H_{12})$, (c) $Zn(CB_{11}H_{12})$ along Zn–H and Al–H bond and Zn-H, respectively (top). Contour map of the Laplacian of electron density with BCPs in blue and orange dots represents the bond and ring critical points, respectively (bottom). Charge (concentrated and depleted) regions are indicated by red and blue lines. The numerical values of the Laplacian of electron density (au) are presented at corresponding BCPs. Figures are adopted from Banjade, H. et al. **Nanoscale**, 2022, **14**, 1767–1778. Ref. [60].

not superhalogens. This can be explained from the significance of *closo*-boranes in the behaviour of superhalogen. Oxygen and Sulphur require two electrons for closed shell structures as $B_{12}H_{12}$ does. However, the EA values of ZnO (2.30 eV) and ZnS (2.39 eV) shows that these are not superhalogens. Therefore, a conclusion can be drawn that not all metal compounds will be considered as superhalogens just because the metal atom donates their electrons during the bonding. Superhalogens cannot be formed by substituting chalcogens for $B_{12}H_{12}$. The design of unconventional superhalogens is formed by selecting the metal atoms of high VIEs and negative ions components. Similarly, EA of $B_{12}X_{12}$ is increased when X is either halogen (F, Cl) or superhalogen (CN, BO). It is better to use transition metals (TMs) other than Zn, as TM elements have large VIEs and they carry magnetic moments. Magnetic superhalogens can be designed by doping transition metal atoms to produce ferromagnetic semiconductors.

5.6 CONCLUSIONS

The difference between the superhalogens and metalloboranes can be analysed by the fundamental principal of superhalogens, which govern the behaviour of these complexes. The traditional superhalogens have the composition of MX_{k+1} in which the added electron is distributed over a large phase space provided by the halogen atoms, while in metalloboranes, the added electron mainly occupy the doped metal site to counterbalance its positive charge, resulting from charge transfer. Also, unlike the conventional superhalogens, the metalloborane class of superhalogens defy the electron shell closure rule. This property can expand the scope for the design and synthesis of new metalloborane complexes, such as magnetic superhalogens by doping transition metal atoms. The studies on these unusual metalloboranes will bring a new dimension to the research of superhalogen family in future.

ACKNOWLEDGEMENTS

R.S. acknowledges the financial assistance by DST WOSA project (SR/WOS-A/CS-69/2018). R.S. is thankful to her Mentor Dr. Shrish Tiwari, Bioinformatics, CSIR-Centre for Cellular and Molecular Biology, and Dr. G. Narahari Sastry, Director, CSIR-NEIST for the technical support.

REFERENCES

1. Jena, P.; Sun, Q. Super atomic clusters: Design rules and potential for building blocks of materials, *Chem. Rev.* 2018, 118, 5755–5870.
2. Anusiewicz, I.; Skurski, P. An ab initio study on BeX_3^- superhalogen anions (X = F, Cl, Br). *Chem. Phys. Lett.*, 2002, 358, 426–434.
3. Anusiewicz, I.; Sobczyk, M.; Dabkowska, I.; Skurski, P. An ab initio study on MgX3– and CaX3– superhalogen anions (X=F, Cl, Br). *Chem. Phys.* 2003, 291, 171–180.

4. Gutsev, G. L.; Bartlett, R. J.; Boldyrev, A. I.; Simons, J. Adiabatic electron affinities of small superhalogens: LiF2,LiF2, LiCl2,LiCl2, NaF2,NaF2, and NaCl2. *J. Chem. Phys.* 1997, 107, 3867–3875.
5. Gutsev, G. L.; Jena, P.; Bartlett, R. J. Structure and stability of BF3∗F and AlF3∗F superhalogens. *Chem. Phys. Lett.* 1998, 292, 289–294.
6. Wang, X. B.; Ding, C. F.; Wang, L. S.; Boldyrev, A. I.; Simons, J. First experimental photoelectron spectra of superhalogens and their theoretical interpretations. *J. Chem. Phys.* 1999, 110, 4763–4771.
7. Sikorska, C.; Smuczynska, S.; Skurski, P.; Anusiewicz, I. BX_4^- and AlX_4^- superhalogen anions (X = F, Cl, Br): An ab initio study. *Inorg. Chem.* 2008, 47, 7348–7354.
8. Elliott, B. M.; Koyle, E.; Boldyrev, A. I.; Wang, X. B.; Wang, L. S. Compounds of superatom clusters: Preferred structures and significant nonlinear optical properties of the BLi 6 -X (X = F, LiF 2, BeF 3, BF 4) motifs. *J. Phys. Chem. A* 2005, 109, 11560–11567.
9. Marchaj, M.; Freza, S.; Skurski, P. Why are SiX5–and GeX5–(X= F, Cl) stable but not CF5–and CCl5–? *J. Phys. Chem. A* 2012, 116, 1966–1973.
10. Yang, X.; Wang, X. B.; Wang, L. S.; Niu, S. Q.; Ichiye, T. On the electronic structures of gaseous transition metal halide complexes, FeX4- and MX3- (M=Mn, Fe, Co, Ni, X=Cl, Br), using photoelectron spectroscopy and density functional calculations. *J. Chem. Phys.* 2003, 119, 8311–8320.
11. Yang, J.; Wang, X.; Xing, X.; Wang, L. Photoelectron spectroscopy of anions observation of high electron binding energies in superhalogens MCl4–MCl4– (M=Sc-M=Sc, Y, La). *J. Chem. Phys.* 2008, 128, 201102.
12. Wang, Q.; Sun, Q.; Jena, P. Superhalogen properties of CuF(n) clusters. *J. Chem. Phys.* 2009, 131, 124301.
13. Siddiqui, S. A.; Pandey, A. K.; Rasheed, T.; Mishra, M. Investigation of superhalogen properties of RhFn (n=1–7) clusters using quantum chemical method. *J. Fluorine Chem.* 2012, 135, 285–291.
14. Anusiewicz, I. Electrophilic substituents as ligands in superhalogen anions. *J. Phys. Chem. A* 2009, 113, 6511–6516.
15. Boldyrev, A. I.; Simons, J. Vertical and adiabatical ionization potentials of MH^-_{K+1} anions, *ab initio* study of the structure and stability of hypervalent MH_{k+1} molecules. *J. Chem. Phys.* 1993, 99, 4628–4637.
16. Boldyrev, A. I.; Vonniessen, W. The first ionization potentials of some MH_{k+1}^- and $M_2H_{2k+1}^-$ anions calculated by a Green's function method. *Chem. Phys.* 1991, 155, 71–78.
17. Enlow, M.; Ortiz, J. V. Aromatic carboxylate superhalogens and multiply charged anions. *J. Phys. Chem. A* 2002, 106, 5373–5379.
18. Gutsev, G. L.; Jena, P.; Zhai, H. J.; Wang, L. S. Electronic structure of chromium oxides, CrOn– and CrOn (n = 1–5) from photoelectron spectroscopy and density functional theory calculations. *J. Chem. Phys.* 2001, 115, 7935–7944.
19. Gutsev, G. L.; Khanna, S. N.; Rao, B. K.; Jena, P. FeO4: A unique example of a closed-shell cluster mimicking a superhalogen. *Phys. Rev. A* 1999, 59, 3681–3684.
20. Gutsev, G. L.; Rao, B. K.; Jena, P.; Wang, X. B.; Wang, L. S. Origin of the unusual stability of MnO4-. *Chem. Phys. Lett.* 1999, 312, 598–605.
21. Gutsev, G. L.; Weatherford, C. A.; Pradhan, K.; Jena, P. Density functional study of neutral and anionic AlO_n and ScO_n with high oxygen content. *J. Comput. Chem.* 2011, 32, 2974–2982.
22. Pradhan, K.; Gutsev, G. L.; Weatherford, C. A.; Jena, P. A systematic study of neutral and charged 3d-metal trioxides and tetraoxides. *J. Chem. Phys.* 2011, 134, 144305.

23. Swierszcz, I.; Anusiewicz, I. Neutral and anionic superhalogen hydroxides. *Chem. Phys*. 2011, 383, 93–100.
24. Swierszcz, I.; Skurski, P. Stable anions formed by organic molecules substituted with superhalogen functional groups. *Chem. Phys. Lett*. 2012, 537, 27–32.
25. Gutsev, G. L.; Boldyrev, A. I. DVM-Xα calculations on the ionization potentials of MXk+1− complex anions and the electron affinities of MXk+1 "superhalogens." *Chem. Phys*. 1981, 56, 277–283.
26. Gutsev, G. L.; Bartlett, R. J.; Boldyrev, A. I.; Simons, J. Adiabatic electron affinities of small superhalogens: LiF_2, $LiCl_2$, NaF_2, and $NaCl_2$. *J. Chem. Phys*. 1997, 107, 3867–3875.
27. Smuczynska, S.; Skurski, P. Halogenoids as ligands in superhalogen anions. *Inorg. Chem*. 2009, 48, 10231–10238.
28. Ding, L. P.; Shao, P.; Lu, C.; Zhang, F. H.; Liu, Y.; Mu, Q. Prediction of the iron-based polynuclear magnetic superhalogens with pseudohalogen CN as ligands. *Inorg. Chem*. 2017, 56(14), 7928–7935.
29. Sikorska, C.; Skurski, P. Moderately reactive molecules forming stable ionic compounds with superhalogens. *Inorg. Chem*. 2011, 50, 6384–6391.
30. Bae, G.; Cha, J.; Lee, H.; Park, W.; Park, N. Effects of defects *and* non-coordinating molecular. *Carbon* 2012, 50, 851–856.
31. Marchaj, M.; Freza, S.; Rybacka, O.; Skurski, P. Why are SiX5− and GeX5− (X = F, Cl) stable but not CF5. *Chem. Phys. Lett*. 2013, 574, 13.
32. Czapla, M.; Freza, S.; Skurski, P. Ionizing benzene with superhalogens. *Chem. Phys. Lett*. 2015, 619, 32.
33. Sikorska, C. When a nanoparticle meets a superhalogen: A case study with C_{60} fullerene. *Phys. Chem. Chem. Phys*. 2016, 18, 18739.
34. Wang, S.-J. Li, Y.; Wang, Y.-F.; Wu, D.; Li, Z.-R. Structures and nonlinear optical properties of the endohedral metallofullerene-superhalogen compounds Li@C_{60}–BX_4 (X = F, Cl, Br). *Phys. Chem. Chem. Phys*. 2013, 15, 12903.
35. Sun, W.-M.; Zhang, X.-L.; Pan, K.-Y.; Chen, J.-H.; Wu, D.; Li, C.-Y.; et al. On the possibility of using the Jellium model as a guide to design bimetallic superalkali cations. *Chem. Eur. J*. 2019, 25 (17), 4358–4366.
36. Sun, W. M.; Wu, D. Recent progress on the design, characterization, and application of superalkalis. *Chem. Eur. J*. 2019, 25 (41), 9568–9579.
37. Hou, G.-L.; Feng, G.; Zhao, L.-J.; Xu, H.-G.; Zheng, W.-J. Structures and electronic properties of $(KI)_n^{-/0}$ (n = 1–4) and $K(KI)_n^{-/0}$ (n = 1–3) clusters: Photoelectron spectroscopy, isomer-depletion, and *ab initio* calculations. *J. Phys. Chem. A* 2015, 119, 11154.
38. Willis, M.; Götz, M.; Kandalam, A. K.; Ganteför, G. F.; Jena, P. Hyperhalogens: Discovery of a new class of highly electronegative species. *Angew. Chem., Int. Ed*. 2010, 49, 8966–8970.
39. Jena, P. Materials for hydrogen storage: Past, present, and future. *J. Phys. Chem. Lett*. 2011, 2, 206–211.
40. Giri, S.; Behera, S.; Jena, P. Superhalogens as building blocks of halogen-free electrolytes in lithium-ion batteries. *Angew. Chem., Int. Ed*. 2014, 53, 13916–13919.
41. Fang, H.; Jena, P. Super-ion inspired colorful hybrid perovskite solar cells. *J. Mater. Chem. A* 2016, 4, 4728–4737.
42. Fang, H.; Jena, P. Sodium superionic conductors based on clusters. *ACS Appl. Mater. Interfaces* 2019, 11, 963–972.
43. Gao, Y.; Wu, M.; Jena, P. A family of ionic supersalts with covalent-like directionality and unconventional multiferroicity. *Nat. Commun*. 2021, 12, 1331.

44. Goel, N.; Etwaroo, G. R. Bright light, negative air ions and auditory stimuli produce rapid mood changes in a student population: A placebo-controlled study. *Psychol. Med.* 2006, 36, 1253–1263.
45. Marshakov, A. I.; Chebotareva, N. P.; Lukina, N. B. Influence of the anionic composition of the supporting electrolyte on the kinetics of the dissolution of iron in acid oxidizing media. *Prot. Met.* 1992, 28, 301–307.
46. Znao, R. F.; Zhou, F. Q.; Xu, W. H.; Li, J. F.; Li, C. C.; Li, J. L.; Yin, B. Superhalogen-based composite with strong acidity-a crossing point between two topics. *Inorg. Chem. Front.*, 2018, 5, 2934–2947.
47. Yang, H.; He, H. M.; Li, N.; Jiang, S.; Pang, M. J., Li, Y.; Zhao, J. G. Design of a novel series of hetero-binuclear superhalogen anions $MM'X_4^-$ (M = Li, Na; M' = Be, Mg, Ca; X = Cl, Br). *Front Chem.* 2022, 10, 936936.
48. Chen, Y.; Du, F.; Tang, L.; Xu, J.; Zhao, Y.; Wu, X.; Li, M.; Shen, J.; Wen, Q.; Cho, C. H.; Xiao, Z. Carboranes as unique pharmacophores in antitumor medicinal chemistry. *Mol. Ther. Oncolytics*. 2022, 24, 400–416.
49. Pathak, B.; Samanta, D.; Ahuja, R.; Jena, P. Borane derivatives: A new class of super- and hyperhalogens. *ChemPhysChem*. 2011, 12(13), 2423–2428.
50. Housecrof, C. E. *Boranesand Metalloboranes*. John Wiley & Sons, New York, 1990.
51. Grimes, R. N. Boron clusters come of age. *J. Chem. Educ.* 2004, 81(5), 657–673. DOI: 10.1021/ed081p657.
52. Wegner, P. A.; *Boron Hydride Chemistry*, E.L. Muetterties, Ed., Academic Press, New York, 1975, Chapter 12, 431–480.
53. Greenwood, N. N. The synthesis, structure, and chemical reactions of metalloboranes. *Pure Appl. Chem.* 1977, 49, 791–801.
54. Grimes, R. N. Role of metals in borane clusters. *Acc. Chem. Res.* 1983, 16, 22–26.
55. Langmuir, I. The arrangement of electrons in atoms and molecules. *J. Am. Chem. Soc.* 1919, 41, 868–934.
56. Langmuir, I. Types of valence. *Science*. 1921, 54, 59–67.
57. Schleyer, P. V. R. Introduction: Aromaticity. *Chem. Rev.* 2001, 101, 1115–1118.
58. Wade, K. The structural significance of the number of skeletal bonding electron-pairs in carboranes, the higher boranes and borane anions, and various transition-metal carbonyl cluster compounds. *J. Chem. Soc. D*, 1971, 792–793. DOI: 10.1039/C29710000792.
59. Child, B. Z., Giri, S.; Gronert, S.; Jena, P. Aromatic superhalogens. *Chemistry*. 2014, 20(16), 4736–4745.
60. Banjade, H.; Fang, H.; Jena, P. Metallo-boranes: A class of unconventional superhalogens defying electron counting rules. *Nanoscale*, 2022, 14, 1767–1778.

6 Design and Potential Applications of Superalkalis

Wei-Ming Sun

6.1 INTRODUCTION

Reducing matters with low ionization energies (IEs) continues to act as a crucial role in chemical industry and material science. The alkali metals exhibit the lowest ionization energies (5.39–3.89 eV) [1] among all the chemical elements in the periodic table. However, it was found that a class of polyatomic molecules or clusters possess even lower IEs than that (3.89 eV) of cesium atom, and were termed "superalkalis" by Gutsev and Boldyrev in 1982 [2]. As a result, such extraordinary species possess very strong reducing ability and, therefore, can be utilized to synthesize unusual charge-transfer salts [3–6], reduce stable gas molecules [7–17] and catalyse specific reactions [18–21]. Moreover, as a typical kind of superatoms [22–25], superalkalis can behave like "man-made" alkali metal atoms when they are assembled into extended nanostructures with tailored properties, such as nonlinear optical materials [26–32], supersalts [33–37], superbases [38–40], hydrogen storage materials [41–47], nanocrystals [48,49] and perovskites [50–54]. Hence, they have a great potential to be used as exciting building blocks of nanoscale materials with highly tunable properties useful for a great variety of potential technologies.

Owing to the above-mentioned intriguing features, superalkalis have received increasing attention in past two decades, and a large amount of efforts have been devoted to designing and characterizing more and more new superalkalis [55]. Up to now, the scope of superalkalis has been extended from conventional mononuclear superalkalis [2,56] to binuclear superalkali species [57,58], and then to polynuclear superalkalis [59–64], nonmetallic superalkalis [65–69], aromatic superalkalis [70–73], magnetic superalkalis [74–76] and organic superalkalis [77–80]. This chapter will highlight the theoretical design and potential applications of superalkalis to present a broad overview of the currently used strategies in the superalkali design and the promising directions for the further application of these unique species.

6.2 DESIGN STRATEGIES OF SUPERALKALIS

6.2.1 Octet Rule

The octet rule [81–83] is a simple electron counting rule that was developed in early 1900s to show that atoms of low atomic number (< 20) tend to combine

 DOI: 10.1201/9781003384205-6

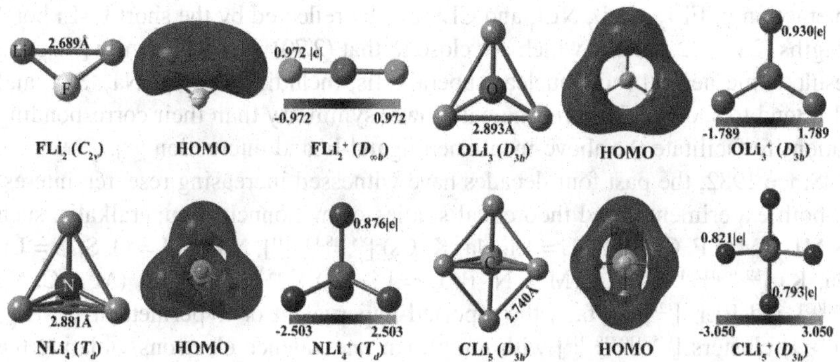

FIGURE 6.1 The optimised structures and HOMOs of FLi_2, OLi_3, NLi_4 and CLi_5, and the natural population analysis (NPA) charges on Li ligands of the corresponding cations. Reproduced with permission from Ref. [55]. Copyright 2019 Wiley.

in such a way that each of them possesses eight electrons in their own valence shells, endowing them the same electronic configurations as noble gas atoms. Just based on this simple rule, the classical mononuclear superalkalis with a simple formula of ML_{k+1} (M is an electronegative atom with the maximal formal valence k and L is an alkali metal atom) was firstly proposed by Gutsev and Boldyrev in 1982 [2]. Such ML_{k+1} superalkalis were designed by decorating an electronegative central atom with one more alkali-metal ligand than the central M atom needs. Typical examples are FLi_2 [84–88], OLi_3 [89–96], NLi_4 [2,56] and CLi_5 [99,100] as shown in Figure 6.1.

In these ML_{k+1} superalkalis, k is the number of valence electrons needed to close the electronic shell of M atom and one more alkali metal atom offers an extra electron for closing the electronic shell of M. So, these ML_{k+1} clusters prefer to lose one valence electron to generate $[ML_{k+1}]^+$ cations, which are very stable with the positive charges separately distributed on all of the k+1 alkali metal atoms (see Figure 6.1). The electrostatic attraction between the cationic $L^{\delta+}$ ligands and anionic $M^{\delta-}$ core yields the strong ionic M-L bonds in the resulting $[ML_{k+1}]^+$ cations. Usually, the $[ML_{k+1}]^+$ cations have high-symmetric structures with alkali metal ligands being far away from each other because of the electrostatic repulsion of the cationic ligands. From Figure 6.1, it is noted that the excess electron in each neutral superalkali always accommodates in the highest occupied molecular orbitals (HOMO) composed of 2s atomic orbital (AO) of Li ligands [55]. As these HOMOs are antibonding with respect to the M-L interactions [2], the energy levels of HOMOs of ML_{k+1} are much higher than that of the AO for Li atom, which leads to much lower ionization energies (IEs) of these ML_{k+1} superalkalis than that (5.39 eV) [1] of Li. Though the stoichiometries of mononuclear ML_{k+1} superalkalis violate the octet rule, they still possess certain thermodynamic stability due to the attractive electrostatic interaction between L and M atoms as well as the covalent interaction between L ligands [56]. As shown in Figure 6.1, the ligand-ligand interaction in FLi_2, OLi_3, NLi_4 and CLi_5 can be reflected by the short Li-Li bond lengths (2.689–2.893 Å), which are close to that (2.701 Å) of Li_2 dimer [55]. As a result, some neutral mononuclear superalkalis, including FLi_2, $ClNa_2$, SLi_3 and PLi_4 tend to show global minima with lower symmetry than their corresponding cations to facilitate the above-mentioned ligand-ligand interaction [56].

Since 1982, the past four decades have witnessed increasing research interest in both experimental and theoretical studies of mononuclear superalkalis, such as ML_2 (M = F, Cl, Br, I; L = Li, Na, K, Cs) [84–88,101–111], ML_3 (M = O, S; L = Li, Na, K) [89–98,112–118], ML_4 (M = N, P; L = Li, Na) [2,56,119,120], MLi_5 (M = C, Si) [99,100] and BLi_6 [121]. In fact, the reported hypervalent or hypermetalated ML_{k+n} ($n > 1$) clusters [105–110,112–114] with more than one valence electrons violating the octet rule can also be regarded as superalkali species considering their lower IEs than those of alkali metal atoms. Up to now, most of such superalkalis have been detected by various spectrometric experiments, including the laser ablation, electron impact ionization, surface ionization, thermal ionization, Knudsen effusion cell, plasma desorption and neutralization-reionization [84–89,102–111]. For example, a typical mass spectrum of Li_nF_{n-1} (n = 2–4) was detected in a pulsed supersonic cluster beam expanded from a laser-ablation source (see Figure 6.2) [85]. Peaks at m/z = 33, 59 and 85 can be assigned to Li_2F, Li_3F_2 and Li_4F_3, respectively. By such an experimental investigation, the IEs of resulting species can be measured by photoionization to verify their superalkali identities.

In the initial studies on the design of mononuclear superalkalis, researchers mainly focus on using the group 14–17 elements as the central cores until 2007, when Wu and his coworkers [121] reported the superalkali BLi_6 taking boron atom as the centre. This mononuclear superalkali has a regular octahedron structure

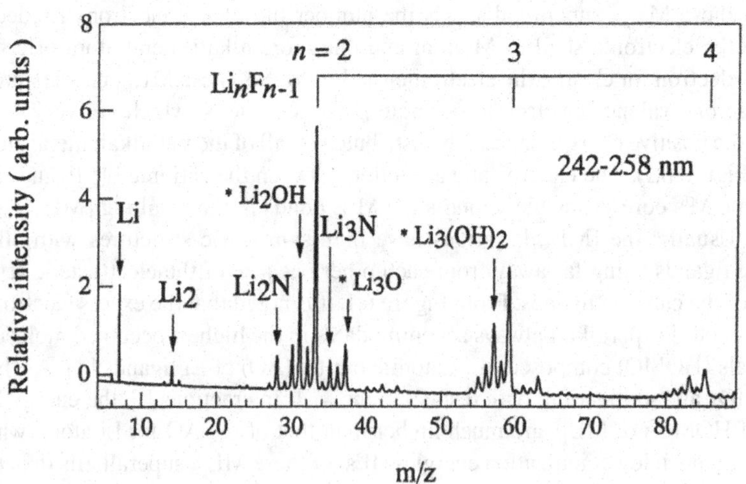

FIGURE 6.2 Typical mass spectrum of the cluster beam accumulated over a spectral range 242–258 *nm*. Asterisks indicate impurity signals arising from water vapor in air [85]. Copyright 2019. Elsevier Science B.V.

with an electronic shell structure of $1S^21P^62S^1$. Due to the extra electron than the closed shell, this BLi_6 cluster exhibits a low IE value of 3.82 eV. To obtain novel superalkalis with higher reducibility, the mononuclear BLi_6 superalkali has been enlarged to a binuclear B_2Li_{11} cluster by Wu and coworkers in 2009 [57]. This is the first example of binuclear superalkalis with an extended formula of M_2L_{2k+1} (L is alkali metal, k is the maximal formal valence of nonmetal M atom), where the $2k+1$ alkali metal ligands can provide one more electron than the atomic shells of two nonmetal M atoms atomic orbitals need. Hence, one excess electron will accommodate the lithium cage or network, leading to the low IEs of M_2L_{2k+1} superalkalis. By using the Saunders "kick" method [122], both of the global minima of B_2Li_{11} and $B_2Li_{11}^+$ were found to be a capsule-like geometry, where two encapsulated boron atoms directly link each other and the Li atoms distribute around the formed B–B bond (see Figure 6.3a). The increase in the number of ligands without increase in inter-ligand repulsion from mononuclear BLi_6 to binuclear B_2Li_{11} benefits the delocalization of positive charge on $B_2Li_{11}^+$, resulting in a lower vertical electron affinity of 3.49 eV than the IE (3.82 eV) of BLi_6.

Afterwards, Tong et al. further design another series of binuclear superalkali cations $M_2Li_{2k+1}^+$ (k =1, 2, 3, 4 for M = F, O, N, C, respectively) by utilizing the above M_2L_{2k+1} formula [58]. As shown in Figure 6.3b, the geometric structures of the most stable $M_2Li_{2k+1}^+$ cations are closely related to the electronegativity of central atoms. In the global minima of $F_2Li_3^+$, $O_2Li_5^+$ and $N_2Li_7^+$, two central negative atoms are bridged by lithium atoms, while two central carbon atoms directly link each other in the lowest-energy structure of $C_2Li_9^+$. The lowest-energy isomers of these binuclear superalkali cations show very low vertical electron affinities (VEAs) of 2.94–3.40 eV, which reflects the superalkali identities of the corresponding neutral M_2Li_{2k+1} clusters.

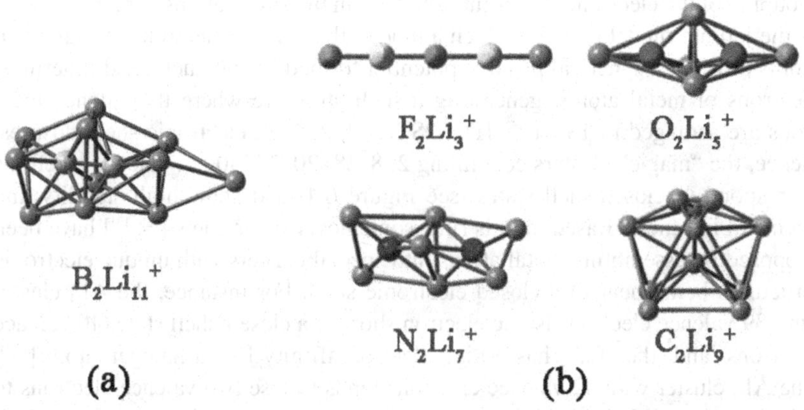

FIGURE 6.3 The global minima of binuclear (a) $B_2Li_{11}^+$, (b) $F_2Li_3^+$, $O_2Li_5^+$, $N_2Li_7^+$, $C_2Li_9^+$ superalkali cations reproduced from references 57 (Copyright 2009 American Institute of Physics) and 58 (Copyright 2011 American Chemical Society).

In 2012, Xi et al. [123] reported that the local minima of $COLi_7^+$ have low VEAs ranging from 3.090 to 3.581 eV and, thus, can be regarded as heterobinuclear superalkali cations follow the formula of $(XY)L_{x+y+1}$ (L is alkali metal, x and y are the maximal formal valence of nonmetal X and Y atoms, respectively). Later, Hou et al. [124] explored a series of $CNLi_n$ (n = 2–10) clusters with low VIE values of 3.780–5.674 eV. Amongst, the $CNLi_8$ that conforms to the XYL_{x+y+1} formula also has the lowest VIE of 3.780 eV, which verifies the heterobinuclear superalkali characteristics of $CNLi_8$. Consequently, this $(XY)L_{x+y+1}$ formula may be used for designing more heterobinuclear superalkalis in the future.

Moreover, according to the octet rule, Wu and coworkers [125] have made a new attempt to design superalkali species by combining the alkaline-earth-metal atoms with fluorine atoms to construct a class of $M_kF_{2k-1}^+$ (M = Mg, Ca; k = 2, 3) cations. Different from the above superalkalis, no alkali metal atoms was used in these newly proposed superalkalis. All the atoms meet with the octet rule in these alkali-metal-free $M_kF_{2k-1}^+$ superalkali cations. As compared with diatomic $(M\text{-}F)^+$ (M = alkaline-earth atom), these $M_kF_{2k-1}^+$ cations possess much larger size and more F^- ligands. Thereby, the corresponding neutral M_kF_{2k-1} molecules have the HOMOs dominated by loosely bound *s* electron from M atom(s), where the electron cloud protrudes out of the molecule due to the repulsion of F^- ligands [125]. Resultantly, the M_kF_{2k-1} molecules tend to lose the extra electron, which results in the low VEAs of the corresponding cations. This study proposed that the alkaline-earth-metal atoms could partner with halogens to construct superalkali cations by following the $M_kX_{2k-1}^+$ (M is alkaline-earth-metal and X is halogen element; k is the number of the involved M atoms).

6.2.2 Jellium Model

In the metallic clusters, the quantum confinement of electron gas results in a grouping of the electronic states into shells as in the single atoms, which is known as the jellium model [126,127]. In such a model, the valence electrons of individual atoms move in a spherical positive potential formed by the nuclei and innermost electrons of metal atoms, generating a shell structure where the valence electrons are arranged in $1S^2$, $1P^6$, $1D^{10}$, $2S^2$, $1F^{14}$, $2P^6$ . . . electronic shell closures. Hence, the "magic" clusters containing 2, 8, 18, 20, 34, 40 . . . valence electrons correspond to closed-shell states (see Figure 6.4) and show high stability and chemical inertness. Based on this rule, some novel superatoms [128–131] have been proposed by assembling metal atoms into special clusters with unique electronic structures being near to a closed electronic shell. For instance, the Al_{13} cluster with 39 valence electrons is one electron short of a closed-shell state (40 valence electrons) and, therefore, has a high electron affinity like a halogen atom [128]. The Al_{14} cluster with 42 valence electrons tends to lose two valence electrons to achieve a closed shell, and has been proposed as a superalkaline earth atom [129]. When two electrons are required to close the electronic shells of $Al_{12}Be$, this cluster exhibits similar characteristics to the chalcogen elements, and has been named as quasi-chalcogen superatom [131].

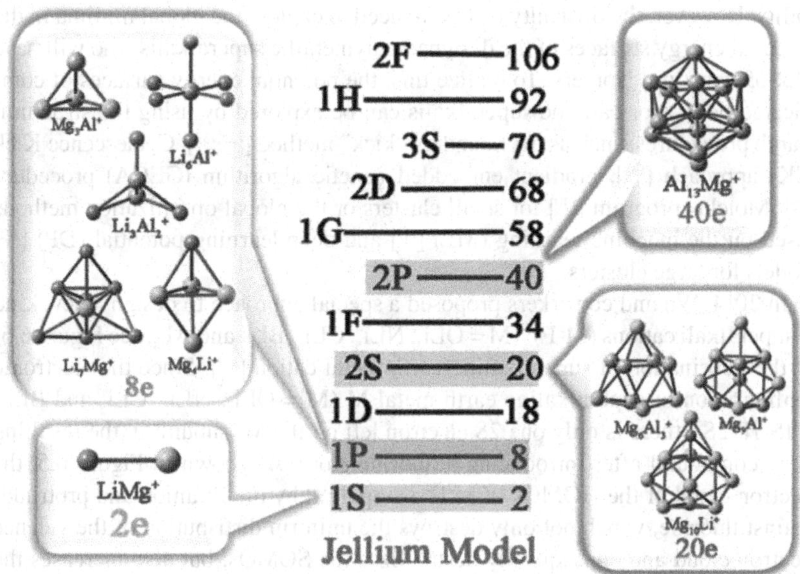

FIGURE 6.4 The jellium model and global minima of the bimetallic superalkali cations $LiMg^+$, Li_6Al^+, Li_7Mg^+, $Li_3Al_2^+$, Mg_3Al^+, Mg_4Li^+, $Mg_6Al_3^+$, Mg_9Al^+, $Mg_{10}Li^+$ and $Al_{13}Mg^+$ cations (136). Copyright 2019 Wiley.

In the same vein, the jellium model has been also used as a meaningful guidance to design metallic superalkali clusters [20,132–136]. According to this simple rule, the metal clusters containing one more valence electron than the $1S^2$, $1P^6$, $1D^{10}$, $2S^2$, $1F^{14}$, $2P^6$ . . . electronic shell closures need will show a tendency to lose one electron to generate cluster cations with high stability and chemical inertness. Thus, the Li_3, Al_3 and $Al_{12}P$ clusters with 3, 9 and 41 valence electrons, respectively, contain one extra electron than needed for the electron shell closures, which results in their low IE of 4.08 ± 0.05 eV [132], 4.75 eV, [133] and 5.37 ± 0.04 eV [134], respectively. Therefore, these species can be classified as pseudoalkalis or alkali-like superatoms because their IEs are not below the threshold of the atomic IE of 3.89 eV for Cs atom in the periodic table.

To systemically prove the potential application of jellium model as a guidance to rationally design superalkalis, Sun et al. [136] proposed a series of bimetallic cationic clusters with 2, 8, 20 and 40 valence electrons by using Li, Mg and Al as atomic building blocks in 2019 (see Figure 6.4). As the corresponding neutral clusters prefer to lose one electron to achieve closed shell states, all these studied cations exhibit low VEAs in the range of 3.42–4.95 eV. The high stability of these cations is confirmed by their considerable HOMO-LUMO gaps and binding energies per atom. As there are endless ways to assemble various metal atoms with different valence electrons into polymetallic cationic clusters, this study provides a universal rule to design a great number of superalkali species with high reducing ability. However, the difficulty is that we need to explore the global minima in the potential energy surfaces of the designed polymetallic superalkalis who will have a lot of structural isomers. To realize this, the potential energy surfaces of complicated neutral or cationic superalkalis can be explored by using the structural search procedures, such as the Saunders "kick" method [122,137], Coalescence Kick (CK) approach [138], gradient embedded genetic algorithm (GEGA) procedure [139], Molclus program [140] for small clusters or the global optimization methods based on the machine learning (ML) [141] and deep learning potential (DP) [142] models for large clusters.

In 2014, Wu and coworkers proposed a special approach to design a new kind of superalkali cations $(M\text{-}F)^+$ (M = OLi_4, NLi_5, CLi_6, BLi_7 and Al_{14}, see Figure 6.5) by the fluorination of superalkaline-earth metal cation [143]. Since the electronic configuration of superalkaline earth metal M (M = OLi_4, NLi_5, CLi_6 and BLi_7) is $1S^2 1P^6 2S^2$, there is only one 2S electron left on the M^+ subunit in the resulting M^+F^- compound after introducing a fluorine atom. As shown in Figure 6.5, the electron cloud of the HOMOs of M^+F^- is repulsed by the F^- anion and protrudes against fluorine, which not only destroys the uniform distribution of the valence electron cloud and consequently destabilizes the SOMOs, but also increases the diffusion degree of electron cloud [143]. The diffuse SOMOs demonstrate that the 2S valence electrons of M^+F^- tend to escape from the M–F molecules, which brings forth their lower VEA values (3.42–3.69 eV) than that of Cs atom. As for Al_{14}^+, its HOMO orbital is found to vary from 2S to 1G state upon the attachment of a F^- anion along with its HOMO orbital energy being greatly raised from −7.84 to −4.48 eV. As a result, the $(Al_{14}\text{-}F)^+$ cation also possesses a much lower VEA value (4.87 eV) than that (5.42 eV) of Al_{14}^+.

6.2.3 Hückel's Rule

The stability of aromatic molecules is governed by the well-known Hückel's rule [144] that requires $(4n + 2)$ π electrons for a planar molecule to be aromatic. Classic examples of aromatic molecules are pyrrole (C_4H_5N) and benzene (C_6H_6) with 6 π electrons. If any tetravalent carbon atom in the pyrrole and benzene rings was replaced with a pentavalent nitrogen atom, there will be one more than needed to satisfy Hückel's rule in each ring of $C_3H_5N_2$ [77] and C_5NH_6 [133] (see Figure 6.6). Due to the extra electron, these derivatives tend to lose one π electron to regain aromaticity, resulting in the low IE values of 3.84 and 3.95 eV for $C_3H_5N_2$ [77] and C_5NH_6 [133], respectively. Therefore, they are named aromatic heterocyclic superalkalis. Moreover, Reddy et al. [77] found that the IE of such a heterocyclic $C_3H_5N_2$ molecule can be further lowered by substituting their peripheral hydrogen atoms by electron-donating groups, such as methyl or tertiary butyl groups. For instance, the IE values of $C_3N_2(CH_3)_5$ and $C_3H_3N_2[C(CH_3)_3]_2$ molecules have been reduced to 3.08 and 3.66 eV, which are less than the ionization energy of Cs (3.89 eV). However, the opposite effect is observed when the hydrogen atoms of $C_3H_5N_2$ are substituted by electron-withdrawing groups, including CN and F.

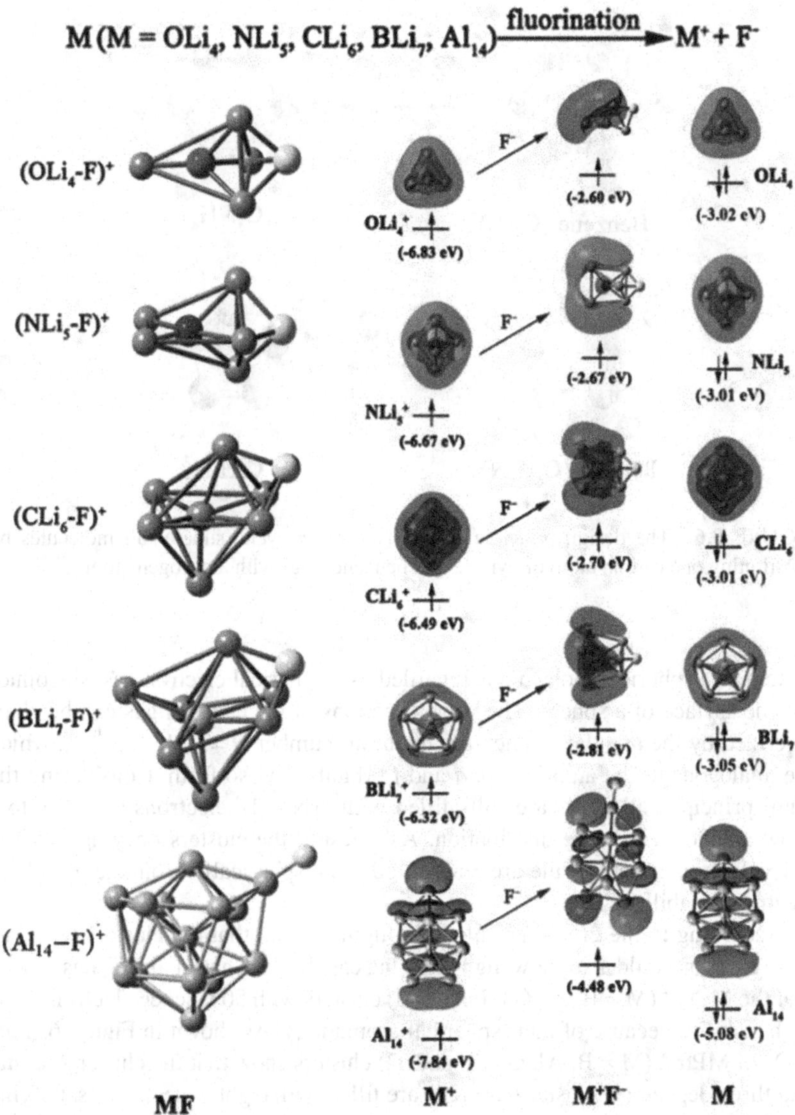

FIGURE 6.5 Global minima of the $(M–F)^+$ (M = OLi_4, NLi_5, CLi_6, BLi_7 and Al_{14}) cations and the HOMOs of M^+, M–F and M [143]. The corresponding orbital energies are shown in parentheses. Copyright 2014 American Chemical Society.

6.2.4 The $2(N + 1)^2$ Rule

As a spherical analogue to the Hückel's rule [144], a $2(N + 1)^2$ rule has been developed by Hirsch et al. [145] to describe the spherical aromaticity of I_h-symmetric fullerene or other spherical clusters. In this electron counting rule, the electron

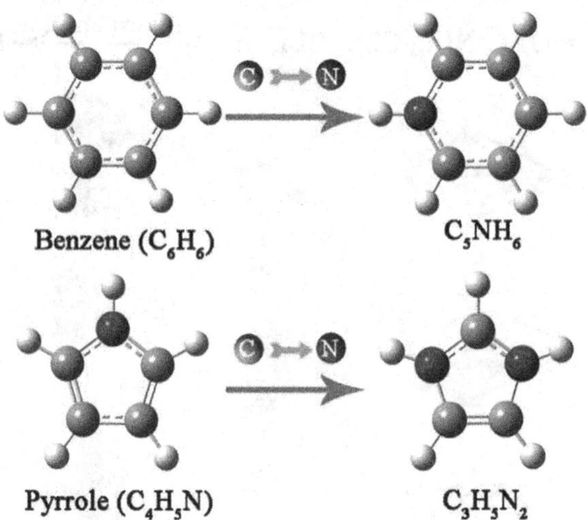

FIGURE 6.6 The design principle of aromatic heterocyclic superalkali molecules by substituting one carbon atom of pyrrole and benzene rings with a nitrogen atom.

system of a spherical molecule is regarded as a spherical electron gas, surrounding the surface of a sphere. The wave functions of the electron gas can be characterized by the angular momentum quantum numbers $l = 0, 1, 2, 3, \ldots$, which are analogous to the atomic *s*, *p*, *d* and *f* orbitals, and so forth. Considering the Pauli principle, all shells are fully filled with $2(N + 1)^2$ electrons and therefore show a spherical charge distribution. As a result, the clusters obeying the $2(N + 1)^2$ electron-counting rule are considered to be spherically aromatic with high electronic stability.

According to the $2(N + 1)^2$ rule, it is highly hoped that the clusters with one extra electron could also show high reducing capability. Chen et al. [146] has proved that the MPb_{12}^+ (M = B, Al, Ga, In and Tl) clusters with 50 valence electrons have high stability because of their spherical aromaticity. As shown in Figure 6.7, the MOs of MPb_{12}^+ (M = B, Al, Ga, In and Tl) clusters show that the cluster *s* orbital and three degenerate cluster *p* orbitals are filled with eight π electrons, satisfying the $2(N + 1)^2$ rule with N = 1. As a result, the calculated vertical ionization energies (VIEs) of the neutral MPb_{12} (M = B, Al, Ga, In and Tl) clusters are 5.03, 4.36, 4.43, 4.45 and 4.81 eV, respectively, which are all smaller than that of Li (5.39 eV) [146]. Such clusters with IEs in the range of 3.89 ~ 5.39 eV should be considered as pseudoalkalis or alkali-like superatoms, though they are also regarded as superalkalis in many previous reports. Similarly, $M@Si_{16}^+$ (M = V, Nb and Ta) has 18 electrons in π states and 50 electrons in σ states. Consequently, both π and σ states follow the $2(N + 1)^2$ rule of spherical aromaticity [146,147] with N = 2 for π states and N = 4 for σ states. Hence, the corresponding neutral $M@Si_{16}$ (M = V, Nb and Ta) can also be classified as alkali-like superatoms.

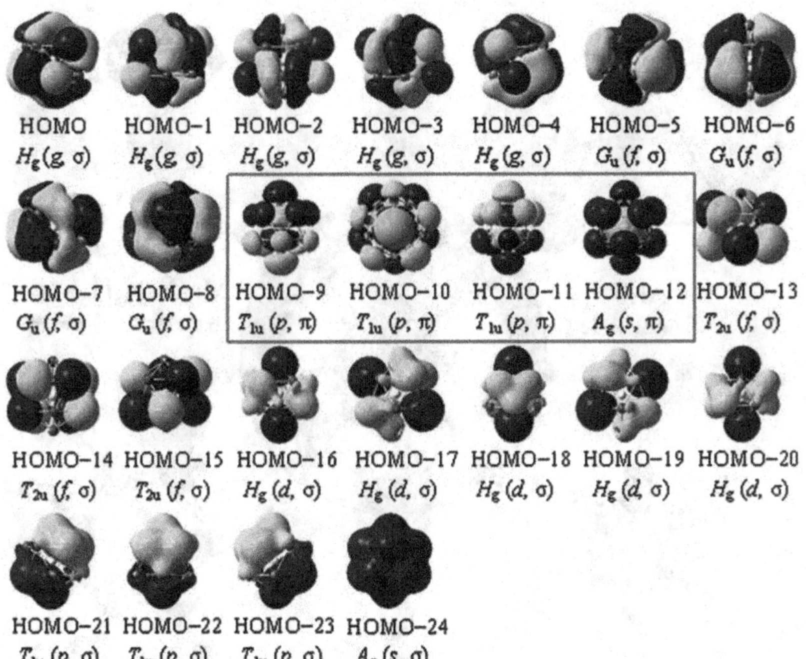

FIGURE 6.7 Representation of the 25 delocalized HOMOs of the most stable structure of the cationic MPb_{12}^+ (M =B, Al, Ga, In and Tl) clusters [146]. The $A_g(s, \pi)$ and T_{1u} (p, π) orbitals in the red rectangle are cluster π orbitals. Copyright 2006 American Institute of Physics.

6.2.5 The 18-Electron Rule

The 18-electron rule is generally applied to transition metal complexes, which require 18 electrons to fill the $s^2p^6d^{10}$ orbitals of the transition metal atoms. Classic examples obeying the 18-electron rule are ferrocene [$Fe(C_5H_5)_2$] and chromium bis-benzene [$Cr(C_6H_6)_2$]. With the $3d^64s^2$ ($3d^54s^1$) electronic configuration of Fe (Cr) atom and 5 (6) π electrons from each of C_5H_5 (C_6H_6) ligands, both of [$Fe(C_5H_5)_2$] and [$Cr(C_6H_6)_2$] have 18 valence electrons, which can account for their special stability. Based on such a rule, Wang and coworkers [133] proposed a stable $Mn(B_3N_3H_6)_2^+$ sandwich cation (see Figure 6.8), which is isoelectronic with $Cr(C_6H_6)_2$. The high chemical and thermal stability of this cation has been proved by the high HOMO–LUMO gap of 6.79 eV and *ab initio* molecular dynamics (AIMD) simulations at 800 K. With one electron more than needed to satisfy the 18-electron rule, the ionization potential of neutral $Mn(B_3N_3H_6)_2$ is calculated to be 4.35 eV. Besides, the magnetic moment of Mn in the neutral $Mn(B_3N_3H_6)_2$ cluster remains at 3 μ_B, indicating that this complex can be regarded as a magnetic superalkali.

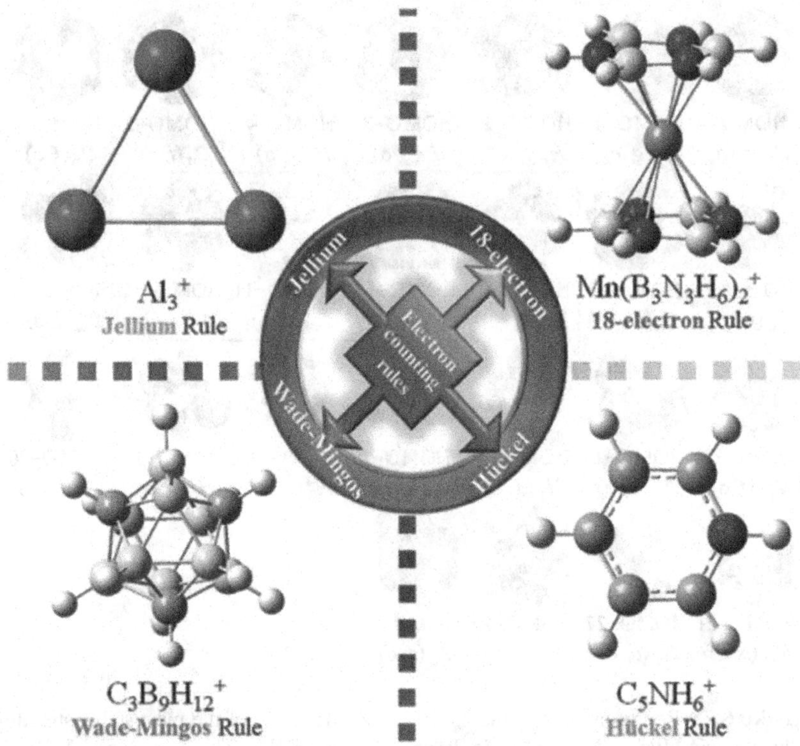

FIGURE 6.8 Optimised geometries of four different kinds of superalkalis designed by using several electron counting rules beyond the octet rule [133]. Copyright 2017 The Royal Society of Chemistry.

6.2.6 Wade–Mingos Rule

The poly skeletal electron pairing theory (PSEPT) was developed to explain the stability of boron-based electron deficient compounds (such as boranes) by Wade [148–150] and Mingos [151,152]. The Wade–Mingos rule, states that $(2n + 1)$ pairs of electrons are needed to stabilize closo-boranes ($B_nH_n^{2-}$) where n is the number of vertices in the boron polyhedron. For instance, the $B_{12}H_{12}^{2-}$ has an icosahedron structure with 12 vertices occupied by 12 B atoms and 12 H atoms radially bonded to these B atoms. With four electrons derived from each BH pair, there are 48 electrons in $B_{12}H_{12}$, wherein 24 electrons forms the 12 B-H covalent bonds and the rest 24 electrons contribute to cage bonding. However, according to the Wade–Mingos rule, 50–24 = 26 electrons (n = 6) are required for cage bonding [133]. Therefore, the $B_{12}H_{12}$ molecule is stable only as a dianion.

Accordingly, via replacing three boron atoms of $B_{12}H_{12}$ with carbon atoms, a neutral $B_9C_3H_{12}$ cage with 51 electrons $(9\times3+3\times4+12\times1)$ has been designed by Zhao et al. [133]. This molecule possesses one more electron than $(2n + 1)$ pairs of electrons needed for the Wade–Mingos shell closure rule and, thus, tends to for a stable C_{3v}-symmetric $B_9C_3H_{12}^+$ with a *closo*-dodecaborate structure, which is similar to the $B_{12}H_{12}^{2-}$ structure (see Figure 6.9). The calculated ionization potential of $B_9C_3H_{12}$ is as small as 3.64 eV, which verifies its superalkali identity [133].

6.2.7 Decorating the Polyatomic Y^{N-} Anion with Alkali Metal Atoms or Electron-Donating Group

As an extension of the ML_{k+1}^+ formula for traditional mononuclear superalkalis, a YL_{n+1}^+ (where L is alkali metal atom and *n* is the valence state of anionic Y^{n-} groups or clusters) formula has been extensively used to design a lot of superalkalis by decorating the polyatomic Y^{n-} anion with alkali metal atoms or electron-donating group. Such polyatomic Y^{n-} anions include various functional groups (such as CN^-, OH^-, SO_3^{2-}, O_4^{2-}, O_5^{2-}, CO_3^{2-}, $C_2O_4^{2-}$, SO_4^{2-}, PO_4^{3-}, AsO_4^{3-} and VO_4^{3-}), aromatic cluster anions (such as inorganic Be_3^{2-}, B_3^-, Al_3^-, Al_4^{2-}, Al_6^{2-}, Ge_4^{4-}, Si_5^{6-}, N_4^{2-}, N_5^- and $B_6H_6^{2-}$ as well as organic $C_4H_4^{2-}$, $C_5H_5^-$, $C_7H_7^{3-}$ and $C_8H_8^{2-}$), Zintl polyanions Y^{k-} (Si_5^{2-}, Ge_5^{2-}, Sn_5^{2-}, Pb_5^{2-}, Ge_9^{4-}, Ge_{10}^{2-}, As_4^{2-}, As_7^{3-} and As_{11}^{3-}), anionic clusters containing planar tetracoordinate carbon (ptC) or planar pentacoordinate carbon (ppC), and so on. Next, such superalkalis with different Y^{n-} cores will be introduced in the following sections.

6.2.7.1 Functional Groups

As early as in 1990s, the halogenoid (CN^-) has been used as core to design hypervalent M_2CN (M = Li and Na) molecules [153,154]. The joint experimental and *ab initio* study on the designed Li_2CN and Na_2CN molecules has proved their VIEs to be 5.39 ± 0.17 eV [153] and 4.92 ± 0.20 eV [154], respectively. Four years later, Tanaka et al. have studied the structures and ionization energies of $Li_n(OH)_{n-1}$ (n = 2–5) clusters by *ab initio* study [155] and photoionization efficiency curves [156]. Amongst, the IE of $Li_2(OH)$ is examined to be 4.49 eV [155] and 4.35±0.12 eV) [155] from the theoretical calculation and experimental detection. These molecules may be considered as the early examples of heterobinuclear pseudoalkali species.

In 2010, Anusiewicz and coworkers [157] designed a class of YNa_2 (X = SH, SCH_3, OCH_3, CN, N_3) superalkalis by using the *ab initio* methods. The calculated adiabatic ionization energies by CCSD(T) method are in the range of 3.711 ~ 4.053 eV. These Na_2X molecules can form stable $[Na_2X]^+[Y]^-$ salts with the species exhibiting various electron affinities (Y = $MgCl_3$, Cl, NO_2). Later, they proposed another series of YL_2 (Y = SCN, OCN, CN; L = Li, Na) superalkalis [158] by using the halogenoids Y– as the central cores. The vertical ionization potentials of these YL_2 molecules were calculated at the outer valence Green function level (OVGF) with the 6–311+G(3df) basis sets. Results show that all Na_2X molecules possess smaller VIEs than the IE of Na (5.14 eV) and are termed superalkali molecules, whereas the Li_2X radicals exhibit slightly larger IEs. In the same year, they also studied the hypermetalated Na_4OCN molecules taking the cyanate group (OCN^-) as the core [159]. By introducing more Na ligands, the VIE of 4.747 eV for the most stable Na_2OCN has been reduced to that of 4.478 eV for the global minimum of Na_4OCN.

To further decrease the IE values of such superalkalis, the anionic groups with higher valence were introduced by Wu's group [59–63]. In 2012, they proposed a series of YLi_3^+ (Y = CO_3, SO_3, SO_4, O_4 and O_5) cations by utilising divalent functional groups Y^{2-} as the central cores [59]. The structural integrities of Y^{2-} are preserved in the lowest-energy structures of $CO_3Li_3^+$, $SO_3Li_3^+$ and $SO_4Li_3^+$ cations, whereas the central cores are split into two monovalent ion units in the most stable $O_4Li_3^+$ and $O_5Li_3^+$ cations (see Figure 6.9a). They proposed that the structural integrity of Y^{2-} and the arrangement of the lithium ligands are two major influencing factors on the VEAs of these YLi_3^+ species [59]. Afterwards, they further designed another group of polynuclear YLi_3^+ (Y = O_2, CO_4, C_2O_4 and C_2O_6) whose structures and stabilities are closely related to the characters of central cores [60]. As shown in Figure 6.9b, the central group of $C_2O_4Li_3^+$ only features a slight distortion, whereas the central cores of $O_2Li_3^+$, $CO_4Li_3^+$ and $C_2O_6Li_3^+$ are divided into various units. With larger structures and more Li atoms than YLi_2, these YLi_3^+ species have much lower VEAs of 3.04–3.36 eV and, therefore, can be classified as novel polynuclear superalkali cations.

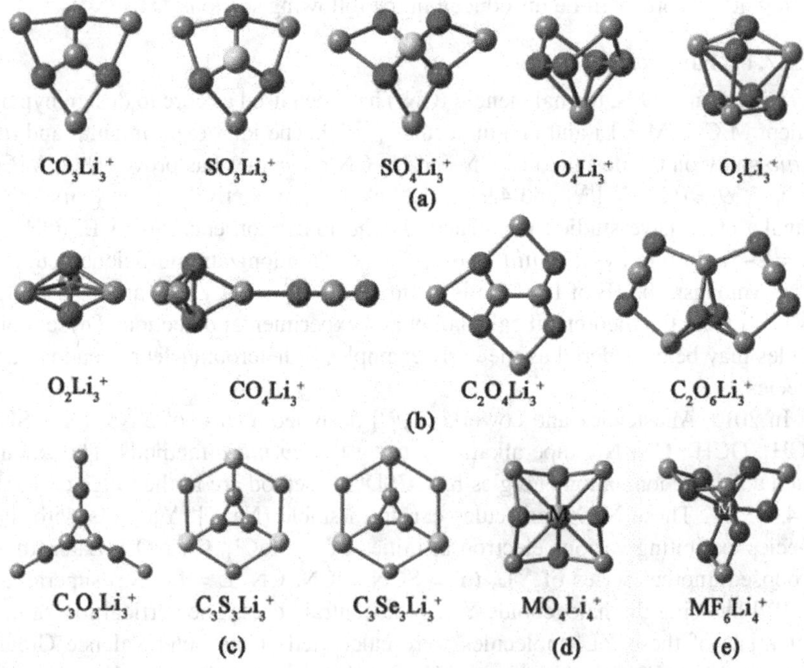

FIGURE 6.9 The global minima of polynuclear (a) YLi_3^+ (Y = CO_3, SO_3, SO_4, O_4 and O_5), (b) YLi_3^+ (Y = O_2, CO_4, C_2O_4 and C_2O_6), (c) $C_3X_3Li_3^+$ (X = O, S and Se), (d) YLi_4^+ (Y = PO_4, AsO_4 and VO_4) and (e) $MF_6Li_4^+$ (M = Al, Ga and Sc) superalkali cations [55]. Copyright 2019 Wiley.

Besides the well-known divalent functional groups, some complicated cyclic (pseudo)oxocarbons dianions have been also applied to design a class of polynuclear superalkali cations $C_3X_3Li_3^+$ (see Figure 6.9c), namely $C_3(NCN)_3Li_3^+$, $C_3X(NCN)_2Li_3^+$, $C_3[C(CN)_2]_3Li_3^+$, $C_3(NCN)[C(CN)_2]_2Li_3^+$, $C_3X[C(CN)_2]_2Li_3^+$ (X = O, S and Se) by Tong et al. [61] Their results reveal that the IEs of the corresponding neutral species range from 2.52 to 4.51 eV, which offers a new strategy to achieve superalkalis by using more complicated monocyclic or even polycyclic (pseudo)oxocarbons dianions as the central cores. Later, the trivalent acid radicals Y^{3-} (PO_4^{3-}, AsO_4^{3-} and VO_4^{3-}) [62] and regular octahedral groups MF_6^{3-} (M = Al, Ga and Sc) [63] have been also used to design a series of polynuclear YLi_4^+ cations by Liu et al. in 2014. Their calculations demonstrate that these YLi_4^+ cations with the same type Y^{3-} cores possess similar global minima, where the central cores always preserve their structural and electronic integrity in the resulting superalkali cations (see Figures 6.9d and 6.9e). Moreover, these YLi_4^+ cations not only show much larger binding energies and HOMO–LUMO gaps than those of mononuclear superalkali cation NLi_4^+, but also possess much lower VEAs (2.618–3.212 eV) than the threshold of 3.89 eV for Cs atom. This indicates that these YLi_4^+ cations are stable superalkali members, and the high valent functional groups are nice cores for the design of polynuclear superalkalis.

6.2.7.2 Aromatic Cluster Anions

Aromaticity is a concept initially invented to account for the unusual stability of an important type of organic molecules: the aromatic compounds [144]. However, the aromaticity concept has been extended to include the "all-metal aromaticity" for all-metal clusters (such as a square Be_3^{2-}, Al_4^{2-} and Al_6^{2-}) [160,161], "three-dimensional aromaticity" or "spherical aromaticity" for fullerenes, polyhedral boranes and related structures [162,163]. Hence, a lot of aromatic cluster anions, such as inorganic Be_3^{2-}, B_3^-, Al_3^-, Al_4^{2-}, Al_6^{2-}, Ge_4^{4-}, Si_5^{6-}, N_4^{2-}, N_5^- and $B_6H_6^{2-}$ as well as organic $C_4H_4^{2-}$, $C_5H_5^-$, $C_7H_7^{3-}$ and $C_8H_8^{2-}$ have been reported in the past decades, which provides an opportunity to introduce the concept of aromaticity to the territory of superalkali. To reveal this, Sun et al. [70] designed a series of aromatic superalkali cations YLi_{k+1}^+ by utilising the aromatic cluster anions Y^{k-} (*i.e.*, Be_3^{2-}, B_3^-, Al_3^-, Al_4^{2-}, Al_6^{2-}, Ge_4^{4-}, Si_5^{6-}, N_4^{2-}, N_5^-, $B_6H_6^{2-}$, $C_4H_4^{2-}$, $C_5H_5^-$, $C_7H_7^{3-}$ and $C_8H_8^{2-}$) as the central cores to combine with k+1 Li^+ ligands. Except for N_4^{2-} and $C_4H_4^{2-}$, all the aromatic anions retain their identities inside the resulting YLi_{k+1}^+ cations (see Figure 6.10). Thus, these cations not only feature aromaticity and high stability, but also show low VEAs of 2.88–4.78 eV. This work firstly introduces the aromatic members into the superalkali family and provides hints for further researches on the design and synthesis of such species with aromatic building blocks.

In 2018, Giri and coworkers [71] explore the plausible existence of organometallic superalkalis consisting of an all-metal aromatic trigonal Au_3 core and the pyridine (Py) and imidazole (IMD) ligands based on the first principles calculations. The calculated ionization energy (IE) values of the obtained complexes, $Au_3(Py)_3$ and $Au_3(IMD)_3$ are as low as 2.56 and 2.11 eV, respectively, reflecting

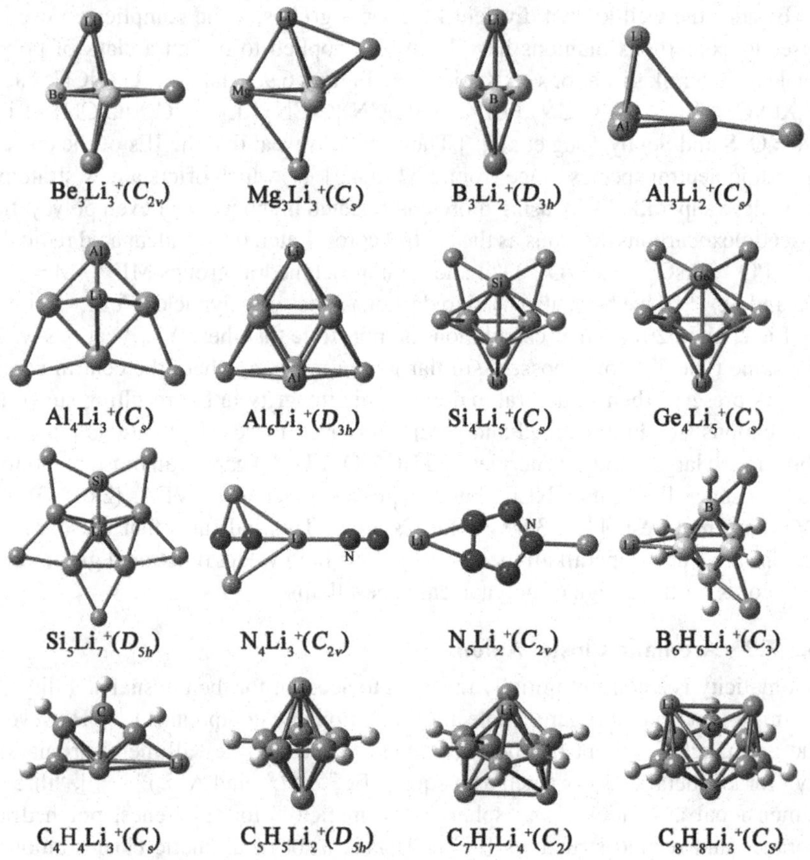

FIGURE 6.10 Global minima of the MLi_{n+1}^+ cations at the B3LYP/6–311++G(3df,3pd) level [70]. Copyright 2013 American Chemical Society.

their superalkali behaviours. In the same year, Zhai and coworkers [72] proposed a star-like $CBe_5Au_5^+$ cluster with planar pentacoordinate carbon (ppC) (see Figure 6.11a) as a superalkali cation with low VEA of 3.75 eV. Their bonding analysis reveals the three-fold (π and σ) aromaticity in this $CBe_5Au_5^+$ cation, indicating its aromatic superalkali identity. In 2020, Srivastava [73] designed a kind of star-like MC_6Li_6 (M = Li, Na and K) complexes by decorating hexalithiobenzene with an alkali atom (see Figure 6.11b). All the MC_6Li_6 complexes are aromatic as suggested by their nucleus independent chemical shifts (NICSs) [164]. The calculated adiabatic ionization energies of MC_6Li_6 are as low as 2.60–2.78 eV, which suggests they are also aromatic superalkalis. The *closo*-boranes B_nH_n (n = 6, 12) with spherical aromaticity have been also used to systematically study the role of size and composition on the superalkali properties of $Li_mB_nX_n$ (m = 1, 2, 3; n = 6, 12; X = H, F, CN) by Jena and coworkers [165].

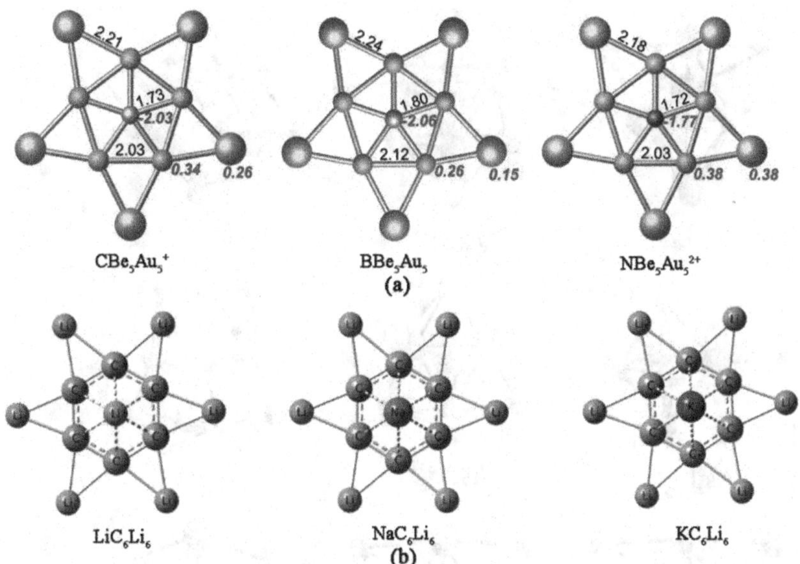

FIGURE 6.11 (a) Optimised global-minimum structures of $CBe_5Au_5^+$, BBe_5Au_5 and $NBe_5Au_5^{2+}$ reproduced from Ref. 72 (Copyright 2018 American Chemical Society) and (b) the equilibrium structures of neutral MC_6Li_6 (M = Li, Na and K) complexes from ref. 73 (Copyright 2020 Informa UK Limited, trading as Taylor & Francis Group).

6.2.7.3 Zintl Polyanions

Zintl ions were posthumously named after Edward Zintl, who investigated polyatomic anions of the posttransition metals and semimetals in liquid ammonia in the early 1930s [166]. Typical multiply charged Zintl polyanions are composed of group 13, 14 and 15 elements, such as E_9^{3-}, E_9^{4-} and E_5^{2-} (E = Ge, Sn, Pb), As_7^{3-}, Sb_7^{3-}, As_{11}^{3-}, Sb_{11}^{3-}, Sb_4^{2-} and Bi_4^{2-} [166]. Due to the versatile chemical reactivity of Zintl ions, they can not only form the building blocks of the Zintl phase but also serve as precursors for synthesizing various nanostructured materials [167]. Therefore, the possibility of designing polynuclear superalkali cations YLi_{k+1}^+ by decorating typical Zintl polyanions Y^{k-} (Si_5^{2-}, Ge_5^{2-}, Sn_5^{2-}, Pb_5^{2-}, Ge_9^{4-}, Ge_{10}^{2-}, As_4^{2-}, As_7^{3-} and As_{11}^{3-}) with lithium cations was examined by Sun and coworkers [64]. They found that all the Zintl cores preserve their geometric and electronic integrity in the global minima of these YLi_{k+1}^+ cations (see Figure 6.12). The stability of these cations is confirmed by the strong ionic bonds between the Zintl core and Li^+ ligands, as well as their large HOMO–LUMO gaps and positive dissociation energies. In particular, the separately distributed Li^+ ligands of YLi_{n+1}^+ are beneficial to effectively decentralize the excess positive charge, yielding the low VEAs (2.93–4.93 eV) of these cations. The Zintl anion Ge_9^{4-} can be transformed into a chalcogen-like superatom by doping a beryllium atom, forming a stable Zintl dianion Ge_9Be^{2-} [168]. Then, this Ge_9Be superatom can also combine with three Li atoms to generate a new superalkali Ge_9BeLi_3 [165].

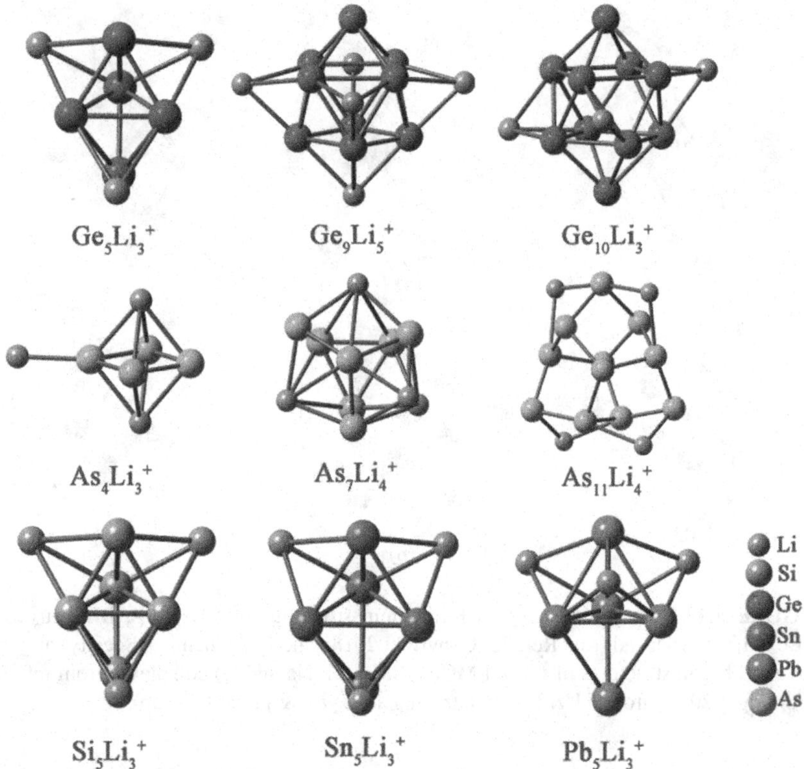

FIGURE 6.12 The most stable structures of YLi_{k+1}^+ (Y = Si_5, Ge_5, Sn_5, Pb_5, Ge_9, Ge_{10}, As_4, As_7 and As_{11}) cations [64]. Copyright 2017 Elsevier B.V.

Besides the alkali metal ligands, the organic ligands were also found to be capable of decorating Zintl anions to design superalkalis. Giri et al. [169] explored this possibility by performing a systematic study of a P_7^{3-} Zintl core decorated with organic ligands [R = Me, CH_2Me, $CH(Me)_2$ and $C(Me)_3$]. The calculated IEs of these P_7R_4 complexes are 4.27 ~ 4.79 eV, and hence can be classified as superalkalis. This is indeed surprising as the ligands of conventional superalkalis consists of alkali metal atoms that can easily release their electron, while in these proposed complexes the Zintl core is surrounded by organic molecules. They found that the electron releasing (ER) nature of organic ligands is responsible for creating a superalkali. This study opens the door to the design and synthesis of a new class of organo-Zintl superalkali moieties apart from the traditional ones composed of only inorganic elements.

6.2.7.4 The ptC and ppC-Based Anionic Clusters

The concept of ptC, a carbon atom bonded with four ligands in an in-plane fashion, was first introduced by Monkhorst in 1968 [170]. Up to now, several exciting ptC clusters including Al_4C^-, Al_4CNa^-, CAl_3Si^-, CAl_3Ge^- and Al_4CH^- were experimentally observed in the gas phase via photoelectron spectroscopy (PES) [171,172]. All these ptC clusters follow the 18-electron counting. As a crucial extension of ptC, molecules with ppC are also of interest for pursuit. The first ppC and relevant molecules, "hyparenes," were proposed by Wang and Schleyer in 2001 [173]. Stimulated by this pioneering work, a number of ppC clusters were explored computationally, including the anionic CBe_5^{4-}, $CAl_3Be_2^-$, $CAl_2Be_3^{2-}$ and $ECBe_5^-$ (E = Al, Ga) [72,174,175]. Such anionic clusters containing ptC and ppC can be also designed as planar superalkalis by decorating with monovalent ligands, such as hydrogen, alkali metal and halogen atoms [72,171,172,174,175].

In 2016, Guo et al. [171] designed a series of $X_3Li_3^+$ (X = C, Si and Ge) containing three planar tetracoordinate X atom, whose VEAs ranges from 2.88 to 3.02 eV. Thereby, they are first examples of superalkali cations with three ptX (X = C, Si and Ge) atoms. Further analyses proved that these $X_3Li_3^+$ cations are π aromatic with two delocalized π electrons, which endows them high chemical and thermodynamic stability. Three years later, they proposed another class of ternary 12-electron ptC clusters, $CBe_3X_3^+$ (X = H, Li, Na, Cu, Ag) based on a rhombic CBe_3^{2-} anion [172]. Bonding analyses show that the ptC core is governed by delocalized 2π/6σ bonding, that is, double π/σ aromaticity, which collectively conforms to the 8-electron counting. Predicted VEAs of these ptC clusters are 3.13 ~ 5.48 eV, indicative of superalkali or pseudoalkali cations.

In addition, Guo et al. also devoted themselves to exploring new superalkali cations by using ppC-based anions as the central cores [72,174,175]. Besides the aromatic $CBe_5Au_5^+$ superalkali cation with ppC mentioned above, they also constructed a series of star-like ppC or quasi-ppC $CBe_5X_5^+$ (X = F, Cl, Br, Li, Na, K) with double (π and σ) aromaticity [174]. The predicted VEAs for $CBe_5X_5^+$ range from 3.01 to 3.71 eV for X = F, Cl, Br and 2.12–2.51 eV for X = Li, Na, K, confirming their identities as superalkali cations. Using the strategy to remove the O ligands from the previously reported polynuclear superalkali cation $CO_3Li_3^+$ [59], they also designed a new planar pentacoordiante carbon (ppC) cluster $CO_2Li_3^+$ with 18 valence electrons [175]. Interestingly, this cation shows a low VEA of 3.08 eV, and can be regarded as a superalkali cation.

6.2.8 Substitution of the Ligands of Conventional Superalkalis

To evaluate the substituent effect of ligands on the geometric and electronic structures of traditional superalkalis, Tong et al. [117] have studied the geometries and energetic properties of neutral and cationic OL_3 (L = Li, Na, K) clusters. They found that the AIEs of OL_3 mainly depend on the peripheral L atoms, and can be efficiently lowered by replacing the L with the more electropositive one. In the same vein, the bimetallic mononuclear MFM′ and MOM'_2 clusters (M, M′ = Li, Na or K and M ≠ M′) with low IEs of 3.63–3.19 eV and 3.26–2.84 eV, respectively, have been designed by Srivastava [118], while the superalkalis NM_4 (M = Li, Na, K) with hetero-alkalis and low VIEs of 3.22 ~ 3.74eV have been designed by Zhang and Chen [119].

Considering the obvious substituent effect mentioned above, the superalkalis instead of alkali atoms have been also successfully utilized as ligands to achieve hyperalkali cations by Sun et al. [176] By taking FLi_2, OLi_3 and NLi_4 as ligands, they proposed a series of hyperalkali cations ML_2^+ [M = F, Cl, Br, LiF_2, BeF_3 and BF_4); L = FLi_2, OLi_3 and NLi_4] with even lower vertical electron affinities (2.36 ~ 3.56 eV) than those of their corresponding cationic superalkali ligands. Similarly, Paduani et al. [177] also proposed a similar recipe to design hyperalkali molecules by decorating an electronegative atom with superalkalis in number that exceeds its formal valence by one. Afterwards, they further designed a class of magnetic hyperalkalis by decorating Gd atom with superalkalis Li_3O and Li_4O [177]. Obviously, these studies offer a new method to further reduce the IEs of conventional superalkalis by utilizing superalkalis as ligands.

As hydrogen and alkali metal atoms belong to the same group in the period table, the hydrogen atoms have been also used as ligands to replace the Li ligands in the binuclear M_2L_{2k+1} superalkalis [58]. By doing this, Hou et al. [65] designed a series of binuclear $F_2H_3^+$, $O_2H_5^+$, $N_2H_7^+$ and $C_2H_9^+$ cations (see Figure 6.13). In these cations, the hydrogen bonds play a crucial role in stabilizing the $F_2H_3^+$, $O_2H_5^+$ and $N_2H_7^+$, while the two moieties of $C_2H_9^+$ are held together by weak Vander Waals interaction. The calculated VEAs of these nonmetallic cations are 3.55–4.48 eV, which indicates their identity as superalkali cations. This study not only verifies the feasibility of utilizing hydrogen atoms as ligands in designing superalkalis but also introduces nonmetallic members to the superalkali family for the first time.

Afterwards, Srivastava has designed the nonmetallic superalkalis BH_x^+ (x = 1–6) by using the hydrogen atoms as ligands [178], and investigated the superalkali behaviours of ammonium (NH_4^+) and hydronium (OH_3^+) cations [179]. He further designed a series of nonmetallic superakalis, including $C_nH_{4n+1}^+$ (n = 1–5) [79], $N_nH_{3n+1}^+$ (n = 1–5, 9) [66], $O_nH_{2n+1}^+$ (n = 1–5) [67] and $F_nH_{n+1}^+$ (n = 1–10) [68]. All these cations can be covered by a formula $X_nH_{nk+1}^+$ (X = C, N, O and F; k is the

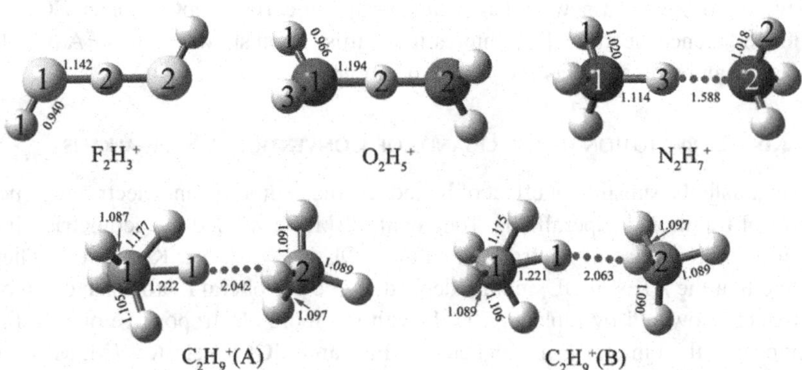

FIGURE 6.13 The equilibrium structures of the $M_2H_{2n+1}^+$ cations [65]. The bond lengths are in Å. Color legend: F, cyan; O, red; N, blue; C, gray; H, light gray. Copyright 2013 Elsevier B.V.

formal valence of X atom), in which one more hydrogen ligand than the X atoms need is contained. Therefore, the VEAs of $C_nH_{4n+1}^+$, $N_nH_{3n+1}^+$, $O_xH_{2x+1}^+$ and $F_nH_{n+1}^+$ can be dramatically decreased to 2.96 eV [79], 1.84 eV [66], 2.67 eV [67] and 1.60 eV [68], respectively. It is noted that a number of hydrogen bonds play an important role in stabilizing these large hydrogen-contained superalkali systems. Srivastava et al. also introduced a new type of superalkali cations, $X(CH_3)_{k+1}^+$ (X = F, O and N with valence *k*) by successively substituting the hydrogen atoms of XH_{k+1}^+ with methyl groups [180]. Their calculations reveal that the successive substitutions reduced their VEAs to 4.35 eV, 2.87 eV and 2.76 eV for $F(CH_3)_2^+$, $O(CH_3)_3^+$ and $N(CH_3)_4^+$, respectively. Similarly, the successive substitutions of methyl groups for the hydrogen atoms of XH_4 (X = N, P and As) result in novel superalkalis $NH_{4-x}(CH_3)_x$, $PH_{4-x}(CH_3)_x$ and $AsH_{4-x}(CH_3)_x$ (x = 0–4) with IEs of 3.97–2.77 eV, 4.10–2.76 eV and 4.40–2.78 eV, respectively [181].

Besides, Anusiewicz and coworkers have proposed a novel approach to reduce the IEs of metal oxide MO (M = Be, Mg, Ca, Fe, Co and Ni) by attaching an alkali metal atom to them [182,183]. In fact, the triatomic MOL (M = Be, Mg, Ca, Fe, Co and Ni; N = alkali metal atom) can also be regarded as the results of substituting two alkali metal ligands of traditional superalkalis OL_3 by a divalent metal atom. The calculated VIEs of the MOL (M = Be, Mg, Ca; L = Li, Na, K) are ca. 2 ~ 3 eV smaller than the IEs of the corresponding MO systems or that of the isolated M atom [182]. In particular, the VIE of CaOL (L = Li, Na, K) are 4.17 ~ 4.69 eV, which can be regarded as superalkalis. In the same vein, the ionization potentials (IPs) of MO (M = Fe, Co, Ni) are decreased by ca. 3 ~ 5 eV upon functionalization with alkali metal L (L = Li and Na) to give either MOL or MOL_2 [183]. Their results indicate that the VIEs of MOL_2 are in the range of 3.960 ~ 4.848 eV, which are much smaller than those of MOL. Therefore, the MOL_2 can be regarded as novel candidates of the mononuclear superalkalis.

6.2.9 Substitution of the Hydrogen Atoms of Organic Molecules

Srivastava et al. suggested a simple strategy to obtain closed-shell superalkalis by substituting alkali metal atoms for the hydrogen atoms of planar molecules, including borazine ($B_3N_3H_6$), [184] benzene (C_6H_6) and polycyclic hydrocarbons-benzene (C_6H_6) [78]. It is demonstrated that the VIE values of $B_3N_3H_{6-x}Li_x$ are decreased as the number of Li atoms increases from 1 to 6. For $x > 4$, the VIEs of $B_3N_3H_{6-x}Li_x$ are lower than that of Li atom, which indicates their superalkali nature [184]. By substituting all the hydrogen atoms of benzene (C_6H_6) and naphthalene ($C_{10}H_8$), anthracene ($C_{14}H_{10}$) and coronene ($C_{24}H_{12}$) with Li atoms, a series of lithiated species, *i.e.*, C_6Li_6, $C_{10}Li_8$, $C_{14}Li_{10}$ and $C_{24}Li_{12}$ with VIEs of 4.24–4.50 eV are obtained [78]. Thus, these species may behave as superalkalis, due to their lower IE than alkali metal. Moreover, all these Li-substituted species are aromatic, and possess planar and closed-shell structures. Hence, they are another kind of organic superalkalis with aromaticity. Besides, by doping one more alkali atom to the C_6Li_6, its IE can be sharply reduced from 4.48 eV [78] to 2.60 ~ 2.78 eV for MC_6Li_6 (M = Li, Na and K) [73], which demonstrates that the IEs of these organic superalkalis closed-shell structures can be further decreased by introducing one more excess electron via alkali-metal doping.

When the hydrogen atoms of the cyclopentadienyl rings in manganocene are periodically substituted with electron-donating groups, a new magnetic superalkali has been obtained by Parida et al. in 2019 [75]. They found that the IE (6.50 eV) of $Mn(C_5H_5)_2$ is reduced to that of 4.65 eV for $Mn(C_5(NH_2)_5)_2$ by substituting all the hydrogen atoms of cyclopentadienyl with the electron-donating NH_2 group, whereas the magnetic nature of Mn keeps intact.

6.2.10 Embedding Dopant Atom into Molecular Cages or Cryptands

Embedding dopant atoms into molecular cages is another special strategy to achieve superalkali species [185,186]. In an early work, the AIEs of endohedral complexes $M@C_{20}H_{20}$ (M = Li, Na, Be, Mg) were calculated to be 2.66–3.66 eV and, thus, were proposed to be superalkalis by Schleyer and coworkers [185]. Similarly, the $M@(HBNH)_{12}$ (M=Li, Na, K) species were also classified as special superalkalis with very small IEs of 2.42–2.93 eV [186]. The low IEs of such metal-encapsulated cages may be attributed to the strong metal-cage interactions under the encapsulation. Differently, as mentioned above, the low IEs of metal-encapsulated $V@Si_{16}$ and $Ta@Si_{16}$ because they possess one more valence electron needed for filling the electronic shell closure (68 electrons) according to the $2(N+1)^2$ rule [185,186].

In 2019, Boldyrev and his coworkers [187] reported the pretty low IEs (1.52 ~ 2.15 eV) of the most popular examples of alkali metal macrocyclic complexes, including four crown ether complexes and three [2.2.2] Cryptand complexes (see Figure 6.14) via DFT hybrid functionals. Such low IEs are significantly lower than those of alkali metal atoms, indicating that the investigated complexes can be defined as superalkalis. Interestingly, they found that the IEs of these complexes highly depend on the coordination sphere, and the destabilization of neutral complexes can be one of the main factors of such low IEs [187]. In fact, such complexes are known to be electrides [188], which have been usually used to stabilize multiple charged Zintl ions. In previous theoretical studies, the low IEs of electrides have been already observed [189,190]. The high reducing ability of such complexes derives from the diffuse excess electrons in them and, thus, the electrides belong to a kind of excess electron compounds [189]. As a result, another type of excess electron compounds, namely alkalides [191,192] also show low IE values. This Boldyrev's work [187] firstly builds a bridge between the electrides and superalkalis and, thus, opens up new directions in the design of novel superakali species by using the strategies to construct electrides and alkalides.

Inspired by the above finding, 3^6adamanzane (3^6adz), a complexant commonly used to synthesize electrides and alkalides, has been chosen as a representative to design a series of superalkalis by encapsulating different atoms into this aza-cryptand complexant by Sun's group [69,76,193]. A series of nonmetallic superalkalis $X@3^6$adz (X = H, B, C, N, O, F and Si) have been constructed and investigated by embedding nonmetallic atoms into 3^6adz (see Figure 6.15) [69].

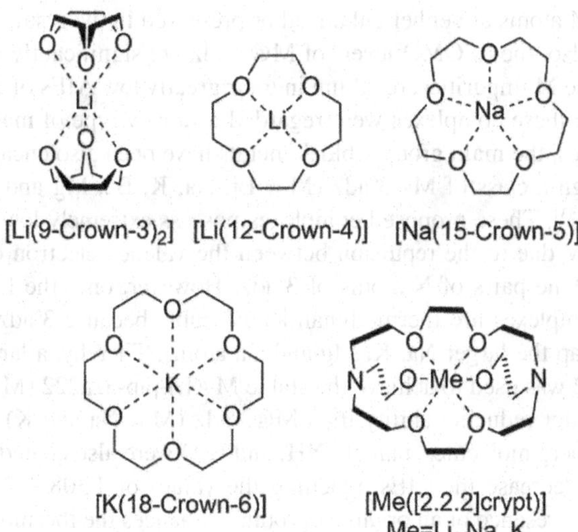

FIGURE 6.14 The structures of the investigated alkali metal complexes by Boldyrev and his coworkers [187]. Copyright 2019 Wiley-VCH Verlag GmbH & Co. KGaA, Weinheim.

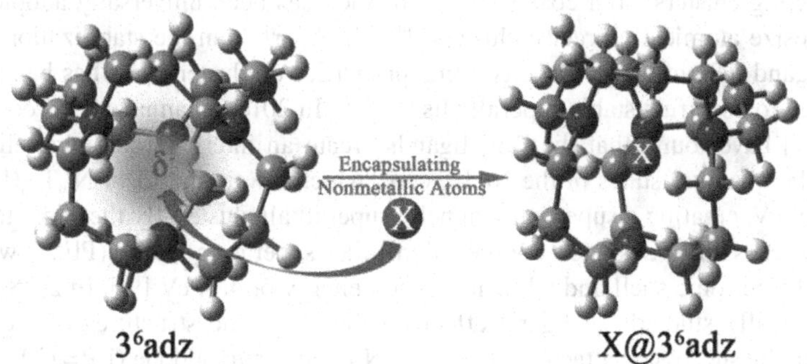

FIGURE 6.15 The schematic design strategy of X@3^6adz (X = H, B, C, N, O, F and Si) based on the cage-like 3^6adz complexant [69].

Under the repulsion of lone pairs of N atoms, the outmost valence electrons of X are destabilized to different degrees, leading to the obviously raise of HOMO level of X@3^6adz as compared with the isolated 3^6adz complexant. As a result, these proposed complexes exhibit are extraordinary low AIE values of 0.78 eV ~ 5.28 eV though X atoms possess very high ionization energies (IEs) of 8.15 eV ~ 17.42 eV [1]. In particular, as far as we know, the AIE of 0.78 eV for H@3^6adz is the current record for the lowest IE for superalkali species known up to now.

In 2022, a series of organometallic M@3^6adz (M = Sc ~ Zn) complexes have been obtained by embedding the 3*d* transition-metal atoms into the 3^6adz complexant. Under the intramolecular interaction between M and 3^6adz, the magnetic moments of M atoms are either enhanced or preserved in the resulting M@3^6adz complexes. Also, the αSOMO levels of M@3^6adz are significantly raised by the presence of the M impurities, resulting in their greatly low AIEs of 1.78–2.56 eV. Consequently, these complexes were regarded as a new type of magnetic superalkalis. Besides, the main group s-block metals have been also encapsulated into 3^6adz to design a class of M@3^6adz (M = Li, Na, K, Be, Mg and Ca) superalkali species [193]. These proposed complexes possess extremely low IE values of 1.94 ~ 2.34 eV due to the repulsion between the valence electron of embedded M atom and lone pairs of N atoms of 3^6adz. However, only the Li@3^6adz and Be@3^6adz complexes are thermodynamically stable because 3^6adz is not large enough to wrap the larger Na, K, Mg and Ca atoms. Thereby, a larger cryptand tri-pip-aza222 was used to achieve the stable M@tri-pip-aza222 (M = Na and K) complexes better reducing ability than M@3^6adz (M = Na and K). In the same year, the Rydberg molecules, namely NH_4 and H_3O were also coated with similar cryptands to decrease their IEs, reaching the values of 1.308 ~ 2.501 eV [194]. Moreover, the presence of an organic cryptand enhances the thermodynamic stability of Rydberg molecules making them stable toward the proton detachment.

6.2.11 Ligation Strategy

Protecting clusters from coalescing by ligands has been universally adopted to synthesize atomically precise clusters [195–197]. Apart from the stabilization role, the ligand-field effect on the electronic properties of cluster cores has been also applied to construct stable superalkalis [198–205]. In 2016, Khanna and his coworkers [198] have found that the PEt_3 ligands create an internal coulomb well that lifts the quantum states of the Ni_9Te_6 core, which lowers the IE of $Ni_9Te_6(PEt_3)_8$ to 3.39 eV, creating a superparamagnetic superalkali cluster. By the same token, they successfully designed another alkali-like superatom $Co_6Te_8(PEt_3)_6$ with a closed electronic shell and a low ionization energy of 4.74 eV [199]. In 2018, they systemically studied the ligand effect on the electronic structures of metallic clusters by successive attachment of three N-ethyl-2-pyrrolidone (EP= $C_6H_{11}NO$) ligands to a bare Al_{13} and doped MAl_{12} (M=B, C, Si and P) clusters containing 39, 40 and 41 valence electrons (see Figure 6.16) [200]. It is demonstrated that IEs of these metallic clusters with both filled and unfilled electronic shells can be substantially lowered by attaching ligands independent of the shell occupancy of the clusters. By attaching three EP ligands, all the AIE values of $Al_{13}(EP)_n$ and $MAl_{12}(EP)_n$ (M=B, C, Si and P) are lower than the IE (5.39 eV) of Li, because these weakly bound ligands obviously lift the electronic spectrum of the obtained charge-transfer complexes via crystal field like effect [200].

Reber et al. [201] reported that attaching phosphine ligands (PMe_3) to simple metal, noble metal, semiconducting, metal-oxide and metal-chalcogen clusters can sharply reduce their ionization energies. Several of the simple and noble metal ligated clusters are transformed into superalkalis with IEs nearly half that of cesium atom. They found that the reduction in IE can be split into initial and final state effects. The initial state effect derives in part from the surface

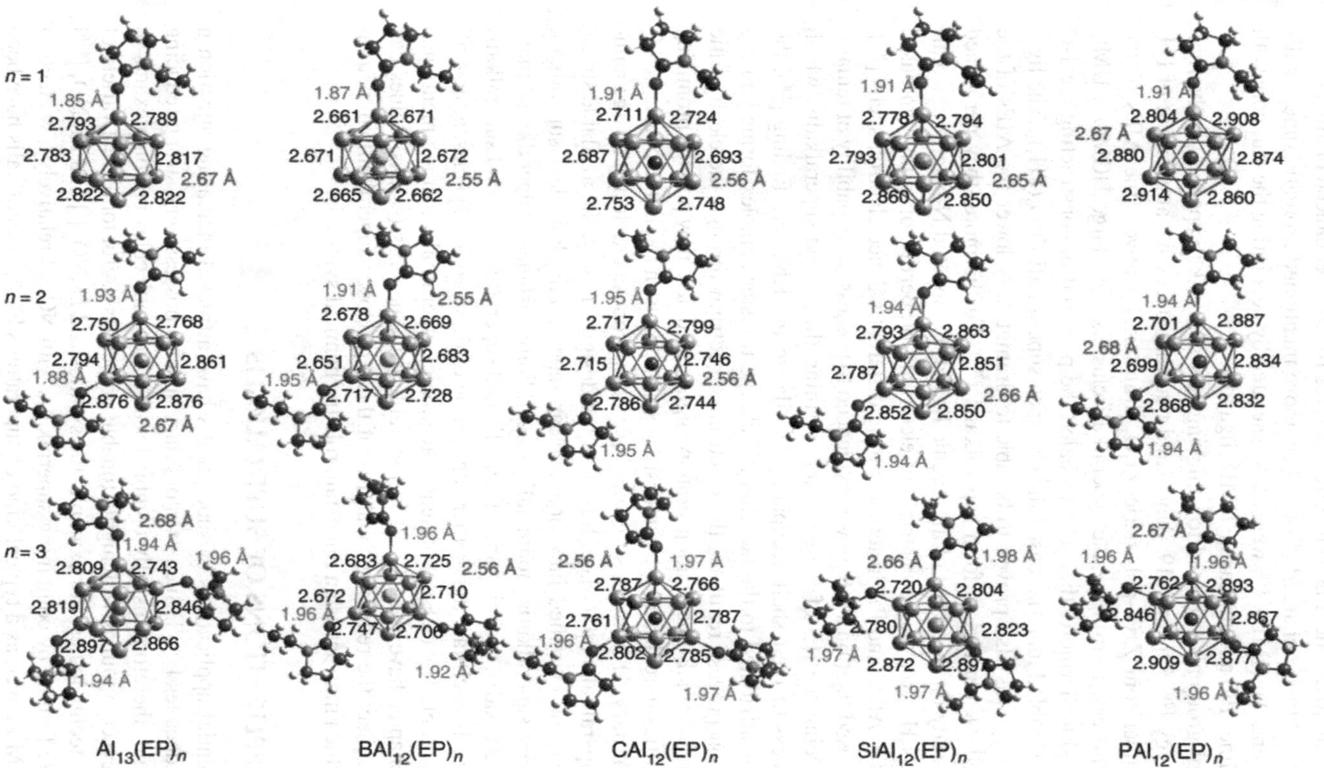

FIGURE 6.16 Ground state structures of neutral $Al_{13}(EP)_n$ and $MAl_{12}(EP)_n$ (M=B, C, Si and P) clusters where n=1–3. The Al–Al bond lengths are also shown and given in Å. The average bond lengths of the interior central atom to surface Al atoms are given in magenta text, while the Al–O bond lengths are given in light blue text [200].

dipole, but primarily through the formation of bonding/antibonding orbitals that shifts the HOMO. The final state effect derives from the enhanced binding of the donor ligand to the charged cluster. The above mentioned aromatic superalkalis $Au_3(Py)_3$ and $Au_3(IMD)_3$ [71] were also generated by decorating the Au_3 core with the pyridine (Py) and imidazole (IMD) ligands.

Cheng' group also devoted to achieving new superalkali species via the ligation strategy [202–204]. In 2019, they found that the successive attachment of PH_3 on the superatom ZrO–polymeric Zr_3O_3 dramatically lowers the AIEs of the ligated clusters, forming stable superalkali clusters with large HOMO–LUMO gaps [202]. Furthermore, they also revealed the potential of constructing the 1-D cluster-assembled material based the obtained superalkali $Zr_3O_3(PH_3)_5$. The ligation strategy was also proved to be able to dramatically lower the AIEs of the ligated Nb_2N_2 clusters to form superalkalis [203]. Furthermore, they performed another study on the effect of an organic ligand, methylated N-heterocyclic carbene ($C_5N_2H_8$), on the geometric and electronic properties of aluminum-based XAl_{12} (X = Al, C and P) clusters featuring different valence electron shells [204]. The proposed ligation strategy was evidenced to possess the capability of remarkably reducing the IEs of these clusters forming the ligated superalkalis, which is regardless of their shell occupancy. Similar to the Khanna's finding [200], the IE drop is attributed to the fact that the charge transfer complex formed during the ligation process regulates the electronic spectrum through the electrostatic Coulomb potential. The ligation strategy highlighted here may provide promising opportunities in realizing the synthesis of superalkalis in the liquid phase.

In summary, 11 kinds of strategies have been proposed to design a large number of superalkalis up to date. Based on the design strategy, the available superalkalis can be classified into mononuclear superalkalis, binuclear superalkali, polynuclear superalkalis, nonmetallic superalkalis, aromatic superalkalis, magnetic superalkalis, organic superalkalis, ligated superalkalis, special superalkalis and so on. To facilitate the further application of these unique species, almost all of the available IEs (VEAs) of neutral (cationic) superalkalis that are mentioned in this chapter have been listed in the Table 6.1. Such IEs (VEAs) values can directly reflect the reducing capability of these proposed superalkalis in neutral states, which may act as an important role in chemical synthesis.

6.3 APPLICATIONS OF SUPERALKALIS

The potential applications of superalkalis have attracted increasing attention n the past decades [26–54]. Such extraordinary species possess very strong reducing ability and, therefore, usually combine with other molecules or complexants to create new compounds with highly tunable properties useful for a great variety of potential technologies. For example, the synthesized Li_3NO_3 [3] and Na_3NO_3 [4–6] salts were formed by combining superalkalis with NO_2 of relatively low electron affinity. Moreover, as a typical kind of superatoms [22–25], superalkalis have been widely applied as substitutions for alkali metals in the design and synthesis of novel materials with tunable properties, especially nonlinear optical materials

TABLE 6.1
The IEs (VEAs) of selected neutral (cationic) superalkalis mentioned in this chapter.

Type	Species	IE(VEA)/eV	Ref
Mononuclear	ML_2 (M = F and Cl; L = Li, Na, Cs)	2.30–3.72	2
Superalkali	ML_2 (M = F and Cl; L = Li, Na)	3.75–3.90	56
	FLi_2	3.78±0.2	85,86
		3.90	84
	$ClLi_2$	4.93±0.2	86
		3.8±0.1	87
	$BrLi_2$	3.9±0.1	87
	ILi_2	4.0±0.1	87
	$ClNa_2$	4.21±0.2	86
	FLi_n (n = 2–4)	3.8–4.0	88
	K_2X (X=F, Cl, Br, I)	3.68–3.95	104
	K_nF (n = 2–6)	3.99–4.31	105
	Li_nBr (n = 2–7)	3.92–4.19	106
	K_nCl (n = 2–6)	3.57–3.69	107
	Li_3I	5.14 ± 0.25	108
	Li_5I	4.62 ± 0.25	108
	K_nI (n = 2–6)	3.46–3.98	109
	K_nBr (n = 3–6)	3.08–3.81	110
	K_2I	3.84	111
	OLi_3	3.59±0.02	95
		3.55±0.02	98
		3.59	92
	Li_nO (2< n < 8)	3.54–4.66	93
	ML_3 (M = O and S; L = Li and Na)	3.16–3.48	2
	OL_3 (L = Li and Na)	3.33–3.56	56
	ONa_3	3.69±0.15	115
	OM_3 (M = Li, Na, K)	2.842–3.558	117
	MOM'_2 (M, M′ = Li, Na, K, M≠M′)	3.26–2.84	118
	SLi_3	3.56	56
	ML_4 (M = N and P; L = Li, Na and H)	2.62–3.67	2
	MLi_4 (M = N and P)	3.40–3.56	56
	NLi_4	3.65	84
	NM_4 (M = Li, Na, K)	3.22–3.74	119
	$SiLi_5$	3.73	100
	BLi_6	3.75	121
	ML_2^+[M = (super)halogen; L = superalkali]	2.36–3.56	176

TABLE 6.1 *(Continued)*
The IEs (VEAs) of selected neutral (cationic) superalkalis mentioned in this chapter.

Type	Species	IE(VEA)/eV	Ref
	MON^+ (M = Be, Mg, Ca; N = Li, Na, K)	4.03–6.59	182
	MON_2 (M = Fe, Co, Ni; N = Li, Na)	3.960–4.848	183
Binuclear	$B_2Li_{11}^+$	3.49	57
Superalkali	$M_2Li_{2k+1}^+$ (M = F,O,N,C)	2.94–3.40	58
	F_2Li_3	4.32±0.2	85
	K_3Cl_2	3.71 ± 0.20	107
	K_3Br_2	4.07 ± 0.20	110
	K_3I_2	3.78	111
	$COLi_7^+$	3.090–3.581	123
	$CNLi_8$	3.78	124
	Li_2OH	4.49	155
		4.35± 0.12	156
Polynuclear	YLi_3^+(Y = CO_3,SO_3,SO_4)	3.03–3.34	59
Superalkali	YLi_3^+ (Y = O_2,CO_4,C_2O_4,C_2O_6)	3.04–3.36	60
	YLi_3^+ (Y = monocyclic (pseudo) oxocarbon)	2.52–4.51	61
	YLi_4^+ (Y = PO_4, AsO_4, VO_4)	2.44–4.67	62
	$MF_6Li_4^+$ (M = Al, Ga, Sc)	2.618–3.212	63
	YLi_{k+1}^+ (Y = Zintl polyanions)	2.93–4.93	64
	F_3Li_4	4.30±0.2	85
	K_4Cl_3	3.72±0.20	107
	K_4Br_3	4.08±0.20	110
	K_4I_3	3.83	111
	Na_2X (X = SH, SCH_3, OCH_3, CN, N_3)	3.720–4.053	157
	Li_2X and Na_2X (X=SCN, OCN, CN)	3.488–5.676	158
	Na_4OCN	4.117–4.640	159
	$Li_3B_nX_n$ (***n*** = 6, 12; X = H, F, CN)	2.84–5.49	165
	P_7R_4[R = Me, CH_2Me, $CH(Me)_2$, $C(Me)_3$]	4.27–4.79	169
	$CO_2Li_3^+$	3.08	175
Special superalkali	N_4Mg_6M (M = Li, Na, K)	4.414–4.710	13
	Li_9O_4	3.33	177
	$B_3N_3H_{6-x}Li_x$ (***x*** = 4–6)	4.18–4.98	184
	MPb_{12} (M = B, Al, Ga, In, Tl)	4.36–5.03	146
	$M_kF_{2k-1}^+$ (M = Mg, Ca; ***k*** = 2, 3)	3.30–5.37	125
	$B_9C_3H_{12}$	3.64	133
Superalkalis based on	Ca_3B	4.29	20
Jellium model	Al_3	4.75	133

TABLE 6.1 ***(Continued)***
The IEs (VEAs) of selected neutral (cationic) superalkalis mentioned in this chapter.

Type	Species	IE(VEA)/eV	Ref
	$Al_{12}P$	5.37±0.04	134
	$(MF)^+$ (M = superalkaline earth)	3.42–3.85	143
	Li_3	4.08±0.05	132
	Bimetallic superalkali	3.42–4.95	136
M-doped complexes	X@3^{6}adz (X = H, B, C, N, O, F, Si)	0.78–5.28	69
	M@$C_{20}H_{20}$ (M = Li, Na, Be, Mg)	2.66–3.66	185
	X@$(HBNH)_{12}$ (X=Li, Na, K)	2.42–2.93	186
	[M([2.2.2]crypt)] (M=Li, Na, K)	1.70–1.52	187
	M@3^{6}adz (M = Li, Na, K, Be, Mg, Ca)	1.94 – 2.34	193
	M@[bpy.bpy.bpy]cryptand	1.308–2.501	194
	M@[2.2.2]cryptand (M = Na, K, NH_4, H_3O)		
Aromatic superalkalis	YLi_{k+1}^+ (Y = aromatic anions)	2.88–4.78	70
	$CBe_5M_5^+$(M = Li,K,Cu,Au, H, F,Cl)	2.11–4.07	72
	$X_3Li_3^+$ (X = C, Si, Ge)	2.88–3.02	171
	CBe_3X_3 (X = Li, Na, Cu, Ag)	3.13–4.74	172
	$CBe_5X_5^+$(X = F, Cl, Br, Li, Na, K)	2.12–3.71	173
	MC_6Li_6 (M = Li, Na, K)	2.60–2.78	73
	$Au_3(Py)_3$; $Au_3(IMD)_3$	2.56; 2.11	71
Nonmetallic superalkalis	$M_2H_{2k+1}^+$ (M = F,O,N,C)	3.55–4.48	65
	$N_nH_{3n+1}^+$ (n = 1–5,9)	1.84–4.39	66
	$O_xH_{2x+1}^+$ (x = 1–5)	2.67–5.16	67
	$F_nH_{n+1}^+$ (n = 2–10)	1.60–4.48	68
	BH_6^+	2.94	178
	NH_4^+	4.50	179
	OH_3^+	5.25	179
	$X(CH_3)_{k+1}^+$ (X = F, O, N)	2.76–4.35	180
	$XH_{4-x}(CH_3)_x$ (X = N, P, As; x = 0=4)	2.76–4.40	181
Magnetic superalkalis	$Gd(Li_3O)_4$	3.75	74
	$Mn(C_5(NH_2)_5)_2$	4.46	75
	M@3^{6}adz (M = Sc~Zn)	1.78–2.56	76
	$Mn(B_3N_3H_6)_2$	4.35	133
Organic superalkalis	$C_3H_{5-x}N_2(CH_3)_x$(x = 0, 2, 5)	3.08–3.84	77
	C_6Li_6, $C_{10}Li_8$, $C_{14}Li_{10}$, $C_{24}Li_{12}$	4.24–4.50	78
	$C_xH_{4x+1}^+$ (x = 1–5)	2.96–4.35	79

TABLE 6.1 *(Continued)*
The IEs (VEAs) of selected neutral (cationic) superalkalis mentioned in this chapter.

Type	Species	IE(VEA)/eV	Ref
	1-alkyl-3-methylimidazolium (*n*MIM)	3.85–3.99	80
	C_5NH_6	3.95	133
Ligated superalkalis	$Ni_9Te_6(PEt_3)_8$	3.39	198
	$Co_6Te_8(PEt_3)_6$	4.47	199
	$MAl_{12}(EP)_n$ (M = B, C, Si, P; n = 0–3)	3.25–4.41	200
	$Au_{11}(PMe_3)_{10}$	2.23	201
	$Al_7(PMe_3)_7$	2.83	201
	$Au_{13}Cl_2(PMe_3)_{10}$	2.82	201
	$Zr_3O_3(PH_3)_5$	3.74	202
	$Nb_2N_2(C_5H_8N_2)_4$	3.00	203
	$XAl_{12}(C_5N_2H_8)_3$ (X = Al, C, P; n = 1–3)	3.44–4.11	204

[26–32], supersalts [33–37], superbases [38–40], hydrogen storage materials [41–47], nanocrystals [48,49] and perovskites [50–54]. More recently, the superalkalis were used to reduce stable CO_2, NO_x (x = 1, 2) and N_2 gas molecules with extremely high stability [7–17], and then catalyse the transformation of these active molecules via specific reactions [18–21] as new catalysts. These extensive applications of superalkalis in different fields will be summarized in following subsections.

6.3.1 Nonlinear Optical Materials

The potential application of superalkalis as building blocks to design promising nonlinear optical (NLO) materials has been extensively explored in the past two decades [26–32,55,205–262]. In 2008, Li et al. [26] proposed a new type of superalkali-(super)halogen compounds $(BLi_6)^+(X)^-$ (X = F, LiF_2, BeF_3, BF_4) with large first hyperpolarizabilities (β_0) ranging from 5167 to 17791 au, which opens a way to use superalkalis as building blocks to synthesize novel meaningful materials with unusual properties such as NLO properties. The interaction between superalkali BLi_6 and different shaped superhalogen X is found to be strong and ionic in nature. Besides, the examination of the variation of NLO properties with the size of $(BLi_6\text{-}BeF_3)_n$ assemblies shows the dependence of NLO properties on the chain length of $(BLi_6\text{-}BeF_3)_n$. Afterwards, another series of $(M)^+(BF_4)^-$ (M = FLi_2, OLi_3, NLi_4) compounds were also designed by combining superhalogen BF_4 with different mononuclear superakalis [27]. In this work, the different effects of superalkali and superhalogen subunits on the NLO properties of such superatom compounds are also revealed.

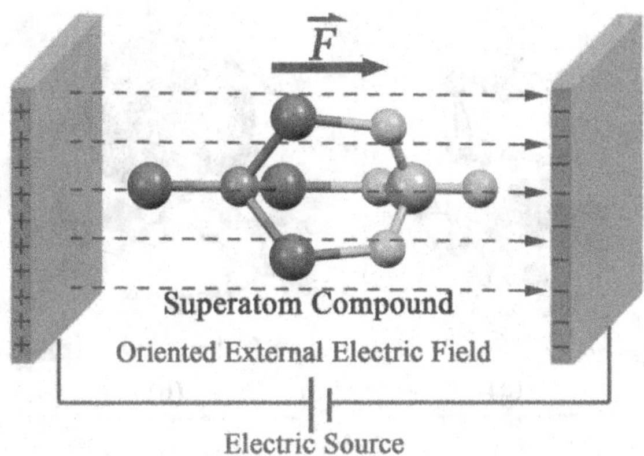

FIGURE 6.17 Schematic diagram of superatom compounds, *i.e.*, $(NLi_4)(BF_4)$ under oriented external electric field (OEEF) [206]. Copyright 2018 American Chemical Society.

In 2018, the effect of oriented external electric fields (OEEFs) on the geometric structures, electronic properties, bonding properties and NLO responses of three typical superatom compounds, *i.e.*, $(NLi_4)(BF_4)$ and $(BLi_6)X$ (X = BeF_3 and BF_4) has been investigated in detail by Sun et al. [206]. They found that the stability and NLO response of $(NLi_4)(BF_4)$ can be gradually enhanced by increasing the imposed OEEF from zero to the critical external electric field (F_c) along with the charge transfer direction ($NLi_4 \rightarrow BF_4$) (see Figure 6.17). In particular, the β_0 of $(NLi_4)(BF_4)$ is greatly enlarged from 2.84×10^3 au to 1.36×10^7 au by 4,772 times by increasing OEEF from 0 to 121×10^{-4} au. Similar cases are also true for (BLi_6) X. Consequently, this study offers an effective method to simultaneously enhance the bond energies and NLO responses of superatom compounds by imposing OEEF along with the charge transfer direction.

Inspired by a fascinating design of superatom compounds by combining superhalogen (Al_{13}) with superalkalis (M_3O, M = Na and K) [207], a class of donor–acceptor frameworks with considerable nonlinear optical responses were constructed by bonding superalkalis M_3O (Li_3O, Na_3O, K_3O, Li_2NaO, Li_2KO, Na_2LiO, Na_2KO, K_2LiO, K_2NaO and LiNaKO) with Al_{13} [208]. Their results revealed that the bonding superalkalis efficiently narrow the wide HOMO–LUMO gap and significantly enhance first hyperpolarizability of Al_{13}, due to the electron transfer in these superatom compounds. In this work, the effect of OEEFs on the NLO responses of M_3O–Al_{13} was also explored, which indicated that the first hyperpolarizabilities of these superatom compounds can be gradually enhanced by increasing the imposed OEEFs from zero to the critical external electric field along the charge transfer direction from M_3O to Al_{13}.

With higher tendency to lose valence electron than alkali metal atoms, superalkalis can serve as excellent source of excess electrons [189] instead of alkali metals to design novel excess electron compounds, such as electrides [209] and alkalides [210,211] with high NLO responses. In 2012, Wang et al. [28] reported the

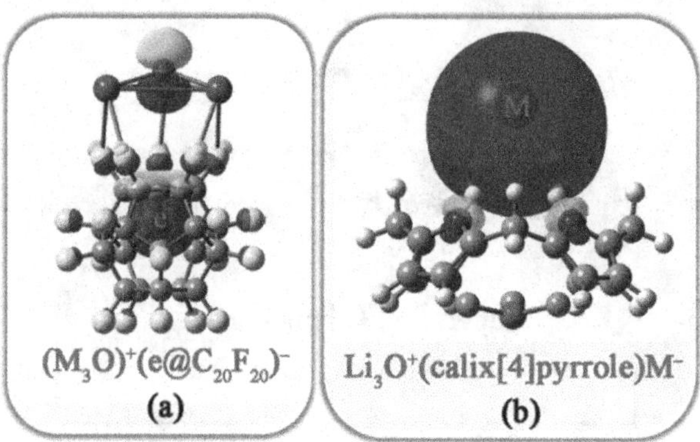

FIGURE 6.18 HOMOs of (a) electride $(M_3O)^+(e@C_{20}F_{20})^-$ (M = Na, K) with the excess electron mainly confined in the $C_{20}F_{20}$ cage reproduced from Ref. 28 (Copyright 2012 The Royal Society of Chemistry) and (b) alkalide Li_3O^+(calix[4]pyrrole)M– (M = Li, Na, K)[8h] with an alkali metal anion M– reproduced from Ref. 29 (Copyright 2014 American Chemical Society).

first evidence for superalkali-based electrides $(M_3O)^+(e@C_{20}F_{20})$– (M = Na, K), where the excess electron of superalkali is encapsulated in the $C_{20}F_{20}$ cage with a sufficient interior electron attractive potential (see Figure 6.18a). Later, they also proposed another acceptor-bridge-donor strategy for enhancing NLO response by using superalkalis as excess electron source of superalkali-based electrides [212]. Afterwards, Chen's group also proposed the $Li_3O@P_4$ and $Li_3O@C_3H_6$ [213] as well as a series of superalkali-boron-heterofullerene dyads M_3O-BC_{59} (M = Li, Na, K) [214] with considerable β_0 values.

Sun et al. [31] reported a successful attempt to design stable inorganic electrides by doping superalkali Li_3O onto the $Al_{12}N_{12}$ nanocage. Their results reveal that these designed $Li_3O@Al_{12}N_{12}$ species possess considerable first hyperpolarizabilities (β_0) up to 1.86×10^7 au. Moreover, the effects of superalkali and nanocage subunits on the NLO responses of $M_3O@Al_{12}N_{12}$ (M = Li, Na, K) and $Li_3O@X_{12}Y_{12}$ (X = B, Al; Y = N, P) are systemically investigated. Results show that respective substitution of Na_3O and $B_{12}P_{12}$ for Li_3O and $Al_{12}N_{12}$ can bring a larger β_0 for such electrides. Evidently, doping suitable nanocages with superalkalis is an effective strategy to construct thermally stable compounds as potential NLO molecules of high-performance. Thereafter, more and more endeavours have been devoted to designing stable NLO molecules by doping mononuclear superalkalis on different nanocages, including B_{40} [31], $Si_{12}C_{12}$ [216,217], $C_{24}N_{24}$ [218], $Zn_{12}O_{12}$ [219], $B_{12}P_{12}$ [220], $Al_{12}P_{12}$ [221], C_{20} [222], C_{24} [223], $Si_nAl_{12-n}N_{12}$ (n = 1, 2) [224], $B_{12}N_{12}$ [225] and $Al_{12}N_{12}$ [226]. In 2022, Gilani and coworkers [227] found that the mixed superalkalis (Li_2NaO, Li_2KO, Na_2LiO, Na_2KO, K_2LiO, K_2NaO and LiNaKO) are better choice than pure superalkalis for $B_{12}N_{12}$ nanocage to design high performance NLO materials. Other electrides have been also theoretically designed by doping superalkalis onto the hexamethylenetetramine (HMT) [228], hexalithiobenzene

(C_6Li_6) [229], $C_6S_6Li_6$ [230], all-cis-1,2,3,4,5,6-hexafluorocyclohexane ($C_6H_6F_6$) [231] and phenalenyl radical [232].

Superalkalis can also be utilized as excess electron source of alkalides [29], which was first revealed by Sun et al. in 2014. They proposed a series of superalkali-based alkalides, such as Li_3O^+(calix[4]pyrrole)M^-, where the diffuse electron cloud enwraps the M atom and creates an M^- anion (see Figure 6.18b). Compared with traditional alkali-metal-based alkalides, these novel alkalides exhibit the structural diversity as well as larger stabilities and NLO responses. Besides, the dependence of the NLO response of such alkalides on the species of involved superalkalis is also investigated. Afterwards, other several kinds of superalkali-based alkalides with different complexants, including NH_3 [30], aza222 [32] and facially polarized $C_6H_6F_6$ molecule [192] by them. This paves a new way to obtain novel unconventional alkalides by using superalkalis as building blocks. More recently, this strategy has been used to design a series of superalkali-based alkalides Li_3O@[12-crown-4]M (M = Li, Na and K) [233] with remarkable static and dynamic NLO properties.

Furthermore, superalkalis can be also applied as electron acceptors in designing a new kind of excess electron compounds termed superalkalides, where superalkali occupies the anionic site [234,235]. This new kind of excess electron compounds was firstly proposed by Mai et al. [234] via studying a series of $M^+(en)_3M'_3O^-$ (M, M' = Li, Na and K; en = Ethylenediamine) superalkalides. Their calculations show that these superalkalides have significantly large first hyperpolarizabilities (β_0) of 7.80×10^3 ~ 9.16×10^4 au. Also, the stabilities of $M^+(en)_3M'_3O^-$ are higher than the corresponding $M^+(en)_3M'^-$ alkalides because of the existences of the hydrogen bonds in them. Therefore, the designed superalkalides with excellent nonlinear optical properties and high stabilities are greatly promising candidates for NLO materials. In another work, the superalkalis M'_3O (M' = Li and Na) were combined with $C_6H_6F_6$ (L) to design novel superalkalides L-M-L-M'_3O with high thermodynamic stability and NLO responses [235].

Besides, a lot of NLO materials can be obtained by doping superalkalis onto the surface of various materials or complexants, including graphdiyne [236–240], borophene [241], phenalenyl [242–244], *g*-C_3N_4 [245,246], pyridinic vacancy graphene [247], phosphorene [248,249], teetotum boron clusters [250], borazine [251], graphene quantum dot [252,253], macrocyclic [hexa-]thiophene [254], biphenylene based sheets [255], $C_6H_6F_6$ [256], Si_6Li_6 [257], carbon nitride (C_2N) [258] and boron nitride [259]. In all these materials, the doped superalkalis serve as the source of excess electrons, which bring forth the large NLO responses of them. Sometimes, two superalkali units were simultaneously combined with a central atom or group to design NLO molecules, such as OLi_3-M-Li_3O (M = Li, Na and K) complexes [260] and compounds M_3-NO_3-M'_3 (M, M' = Li, Na, K) [261]. Misra and coworkers [262] proposed a class of endofullerene complexes with considerable NLO responses by encapsulating superalkalis (FLi_2, OLi_3 and NLi_4) into C_{60} and then interacting with superhalogens.

In addition, it is also well-known that some of superalkalis also contain the diffuse excess electrons and, therefore, exhibit high NLO responses themselves

[263-268]. Srivastava et al. [263] studied the NLO behaviour of Li_nF (n = 2–5) superalkali clusters, and found that these clusters may also possess alkalide and/or electride characteristics due to the existence of excess electrons. The exceptionally large hyperpolarizability of Li_2F and its electride characteristics are particularly highlighted. Ahsin et al. [264-268] revealed the remarkable NLO properties of bimetallic superalkalis [264] NM_3M' (M, M' = Li, Na, K) [265], M_2OCN and M_2NCO (M = Li, Na, K) [266], oxacarbon superalkali $C_3X_3Y_3$ (X = O, S and Y = Li, Na, K) [267] and germanium-based superalkalis [268]. Sun and coworkers also reported the significant NLO responses of the nonmetallic superalkalis [69] and magnetic superalkalis [76] of M@3^6adz type designed by them.

6.3.2 Hydrogen Storage Materials

As mentioned earlier, the positive charges of superalkali cations are highly polarized and separately distributed on each alkali metal ligand that in turn act as the effective binding sites for the molecular hydrogen through electrostatic interactions. Pan et al. [269] firstly confirmed this by assessing the hydrogen trapping capability of some superalkali cations (see Figure 6.19). Most of these studied superalkali cations are found to show the potential to become effective hydrogen storage materials with high gravimetric weight percent owing to the charges on the Li centres. More recently, Giri and coworkers [270] reported the H_2 storage capacity of Li-doped five member aromatic heterocyclic superalkali complexes.

In 2014, Wang et al. [41] proposed superalkali Li_2F-coated C_{60} materials with enhanced hydrogen binding affinity H_2 for hydrogen storage. They found that the orbital interactions play a dominant role in this system and eventually $68H_2$ molecules can be stably stored by a $C_{60}(Li_2F)_{12}$ cluster with a well binding energy of 0.12 eV/H_2 (see Figure 6.20). It is concluded that superalkalis would be beneficial to enhance interactions between hydrogen and hosts and, thus, the hydrogen storage capacities for solid sorbents can be greatly improved.

In 2019, Saedi et al. [42] have studied superhalogen (Al_{13})–superalkali (M_3O, M = Li, Na and K) assemblages as versatile materials for hydrogen storage. They found that the Al_{13}–M_3O systems with three alkali metal M atoms have higher hydrogen storage performance than the Al_{13}–M systems. Gao et al. [43] proposed that superalkali NM_4 clusters can be used to develop high-performance hydrogen storage materials by performing a trial to decorate NLi_4 onto the 1D graphene nanoribbon. They found that the charges on Li atoms were successfully transferred to the pristine monolayer, forming partial electronic field around each Li atom to polarize the adsorbed hydrogen molecules, which enhances the electrostatic interactions between the Li atoms and hydrogen. Each NLi_4 cluster can adsorb at most 16 hydrogen molecules, resulting in the high total capacity of hydrogen storage (11.2 wt%) of the designed novel material.

Similar results were also reported by Chen and coworkers [44] in the same year. They found that the decoration of NLi_4 on grapheme not only solves the aggregation of metal atoms, but also provides more adsorption sites for hydrogen. Thus, each NLi_4 unit can adsorb up to 10 H_2 molecules, and the NLi_4-decorated graphene

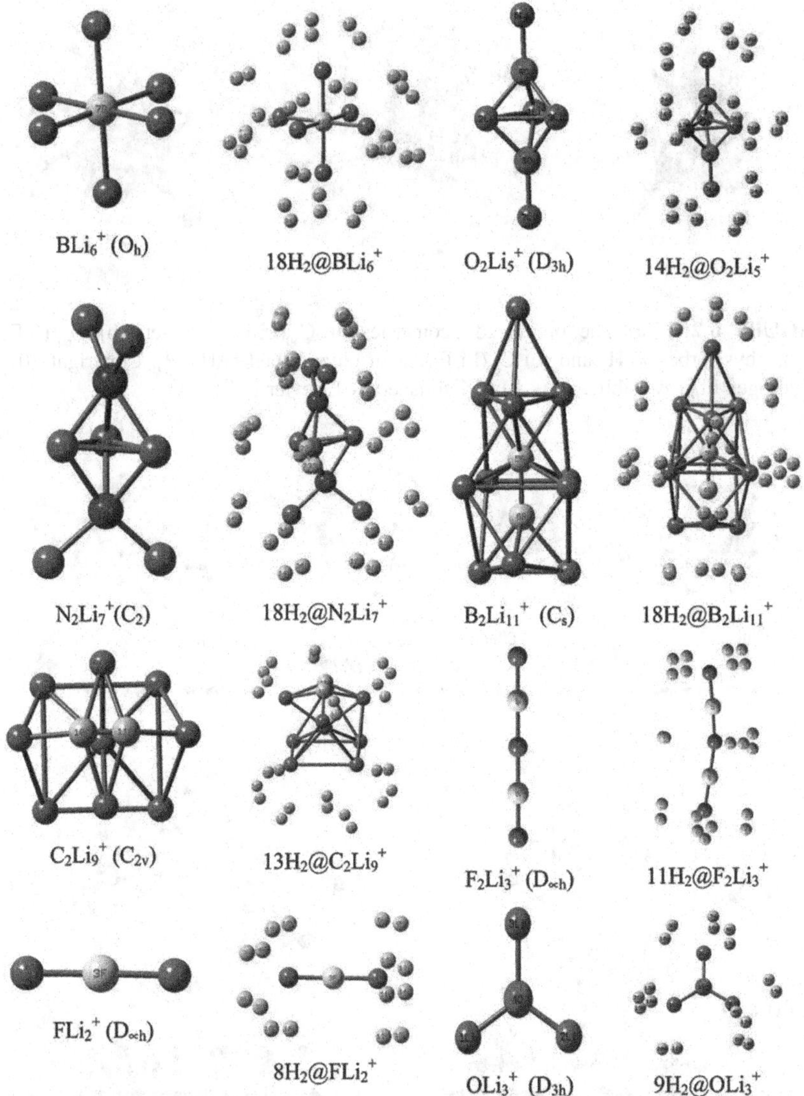

FIGURE 6.19 Optimized structures of studied superalkali cations and their hydrogen trapped analogues [269]. Copyright 2012 the Owner Societies.

can reach a hydrogen storage capacity of 10.75 wt% with an average adsorption energy -0.21 eV/H_2. Besides, the adsorption strengths fall in the ideal window for reversible hydrogen storage at ambient temperatures. Hence, the NLi_4-decorated graphene can be promising hydrogen storage material with high reversible storage capacities. They also used superalkali NLi_4 to decorate *h*-BN and investigate the hydrogen storage properties by density functional theory [45]. The H_2 molecules

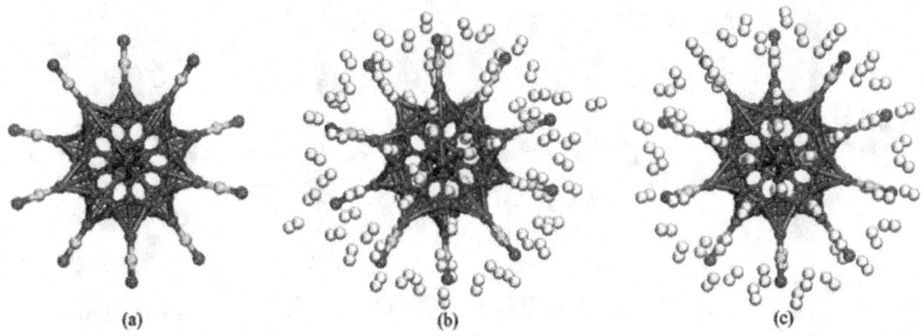

FIGURE 6.20 (a) The optimized geometries of $C_{60}(Li_2F)_{12}$ cluster, (b) $C_{60}(Li_2F)_{12}$ with physisorbed $80H_2$ and (c) $C_{60}(Li_2F)_{12}$ with physisorbed $68H_2$ [41]. Copyright 2014, Hydrogen Energy Publications, LLC. Published by Elsevier Ltd.

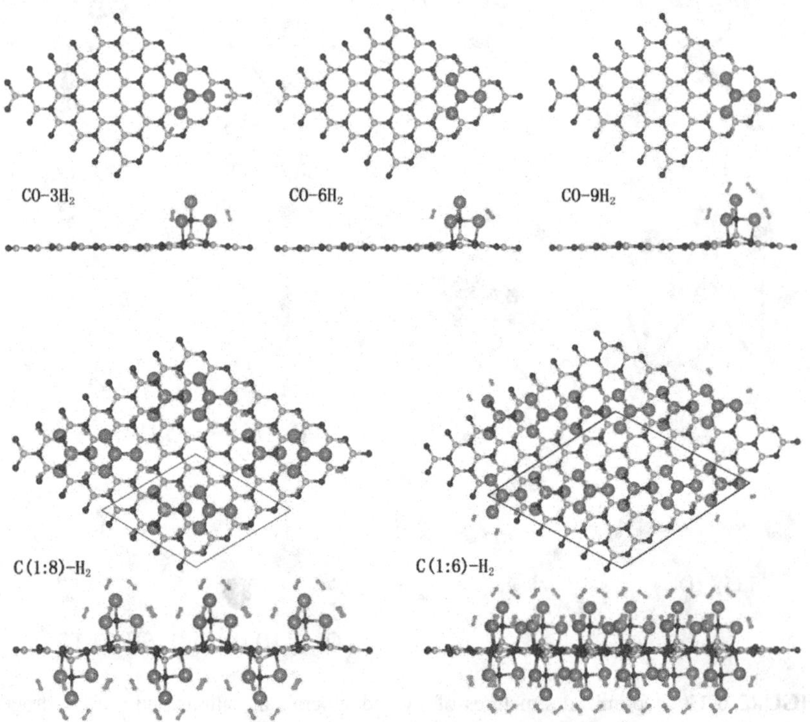

FIGURE 6.21 Optimised configurations for adsorbing nine H_2 molecules on each NLi_4 unit in the configurations C(1:8) and C(1:6) [45].

adsorbed on the bottom Li^+ cations are strongly polarized due to the small radius of Li^+, and the H_2 molecules attached to the top Li^+ are adsorbed by feeding back the excess electron up to the antibonding σ^* orbitals of the H_2 molecules. As shown in Figure 6.21, each NLi_4 can absorb nine H_2 molecules with adsorption energies about 0.20 eV/H_2. The adsorption strengths fall in the ideal window for reversible uptake-release at ambient temperatures. The hydrogen storage capacity at the decoration density of NLi_4/BN = 1:6 reaches 9.40 wt%. Their studies suggest that NLi_4-decorated 2D materials can be potential candidates for hydrogen storage at room temperature.

In a similar vein, the superalkali NLi_4 has been also doped onto the γ-graphyne (GY) [46] and β_{12}-borophene [47] to achieve promising hydrogen storage materials. Their results demonstrate that NLi_4 can be tightly bound onto the surface of GY, forming a $2NLi_4$/GY complex that can be adsorbed by 24 H_2 molecules with adsorption energy of -0.167 eV/H_2. The hydrogen storage gravimetric density reaches 6.78 wt%, which is higher than the target of 6.5 wt% from U.S. Department of Energy (DOE). Their *ab initio* molecular dynamic (AIMD) simulations indicate that the adsorbed H_2 molecules quickly escape from the $2NLi_4$/GY complex under room temperature of 300 K, confirming that NLi_4-decorated γ-graphyne can serve as an outstanding H_2 storage media. They further detected the hydrogen storage properties of NLi_4-decorated β_{12}-borophene [47]. The designed NLi_4-decorated β_{12}-borophene can adsorb up to 24H_2 molecules with an ideal H_2 adsorption energy of -0.176 eV/H_2 and, thus, its hydrogen uptake density achieves 7.66 wt%. Jason et al. [271] designed a novel H_2 storage architecture by decorating graphene-like haeckelite (r57) sheets with NLi_4. Exceptionally high H_2 storage capacities of 10.74% and 17.01% have been achieved for one-sided (r57-NLi_4) and two-sided (r57–$2NLi_4$) coverage of r57 sheets, respectively. Under maximum hydrogenation, the average H_2 adsorption energies have been found as -0.32 eV/H_2, which is ideal for reversible H_2 storage applications.

6.3.3 Gas-Trapping Materials

In view of the net positive charges on Li^+ of superalkali cations, such species have been also used to capture stable molecules, such as noble gas [272,273], volatile organic compounds [274], polar molecules [275] and N_2 [276]. Chattaraj and Merino [272] proved the noble-gas-trapping ability of the star-shaped $C_5Li_7^+$ and $O_2Li_5^+$ superalkali clusters by using *ab initio* and density functional theory (DFT). The stability of noble-gas-loaded clusters is analysed in terms of dissociation energies, reaction enthalpies and conceptual DFT-based reactivity descriptors. Xe-binding ability of these clusters is also studied by them [273]. Both $C_5Li_7^+$ and $O_2Li_5^+$ clusters can bind with maximum 12 Xe atoms. The electron transfer from Xe atoms to Li centres plays a crucial role in binding.

In 2018, the superalkali F_2Li_3 was used to capture volatile organic compounds by Meloni and coworkers [274]. The interactions between Li_3F_2 and four volatile organic compounds (VOCs), namely methanol, ethanol, formaldehyde and acetaldehyde, were assessed in their study. Stronger interactions are observed between Li_3F_2 and aldehydes than alcohols. The smaller aldehyde shows a larger binding energy (BE) with Li_3F_2 than the bigger aldehyde. However, alcohol clusters do not show this trend due to their weak interactions (low BEs). Chattaraj and coworkers [275] reported that the M_3O^+ (M = Li, Na, K) supported by pristine, B-doped and

BN-doped graphene nanoflakes (GR, BGR and BNGR, respectively) can sequestrate polar molecules, *viz.*, CO, NO and CH_3OH, in a thermodynamically more favourable way than GR, BGR and BNGR. In the adsorbed state, the CO, NO and CH_3OH molecules, in general, attain an "active" state as compared to their free counterparts.

In 2020, the possibility of the superalkali cation Li_3^+ for capturing N_2 and its behaviour in gaseous nitrogen have been theoretically studied by Yu et al. [276]. The evolution of structures and stability of the $Li_3^+(N_2)_n$ (n = 1–7) complexes shows that the N_2 molecules tend to bind to different vertices of the Li_3^+ core, and that Li_3^+ might have the capacity to capture up to twelve nitrogen molecules in the first coordination shell. They proved that Li_3^+ keeps its superatom identity in the lowest-lying $Li_3^+(N_2)_n$ (n = 1–4) complexes. The change in the Gibbs free energies of possible fragmentation channels also indicates the thermodynamic stability of Li_3^+ in the $(N_2)_n$ clusters when n ≤ 4. They also found that the superalkali cation Li_3^+ has better capacity than heavy alkali metal cations in capturing N_2 molecules, since it has a larger binding energy with N_2 than Na^+ and K^+ ions.

6.3.4 Catalytic Activation and Transformation of Stable Molecules

With high reducing ability, the neutral superalkali species have been also used to activate stable gas molecules, such as CO_2 [7,9,11–14,17,18,133,136,277], N_2 [8,19], SO_2 [15], CO [16] and nitrogen oxides (NO_x) [10]. After being activated by superalkalis, the greenhouse gas CO_2 can be further transformed into high-value chemicals, such as methanol fuel [21] and carboxylic acid [18], and the most abundant gas N_2 can be transformed into ammonia [20]. These studies demonstrate that superalkalis may be also potential cluster catalysts for the activation and transformation of small stable molecules.

In 2017, Zhao et al. [133] proved that the ability of superalkalis to transfer an electron easily can be used to activate a CO_2 molecule by transforming it from a linear to a bent structure. In the same year, the reduction of CO_2 by Li_3F_2 superalkali via electron transfer was also proposed by Park and Meloni [7]. The selectivity of Li_3F_2 toward CO_2 has been also evaluated by performing the same calculations for the most abundant atmospheric gas molecule N_2, which reveals a very small chemical affinity of Li_3F_2 for N_2. One year later, Srivastava [9] studied the interaction of CO_2 with FLi_2, OLi_3 and NLi_4 superalkalis, and found that this interaction leads to stable superalkali-CO_2 complexes in which CO_2 can be successfully reduced to CO_2^- anion with a bent structure due to electron transfer from superalkalis. They also proposed that CO_2 can be further reduced to CO_2^{2-} in case of the anionic complex such as $(FLi_2\text{-}CO_2)^-$. Thus, FLi_2 superalkali is also capable of double-electron reduction of CO_2. Thereafter, the activation of CO_2 by the superalkalis adsorbed on pyridinic graphene surfaces [11], planar superalkali C_6Li_6 [12], N_4Mg_6M (M = Li, Na, K) superalkalis [13], superalkali-doped fullerene [14] and $Al_{12}P$ [21] have been also reported. The recent studies on the single-electron reduction and activation of CO_2 by various kinds of superalkalis have been reviewed by Srivastava and coworkers [17].

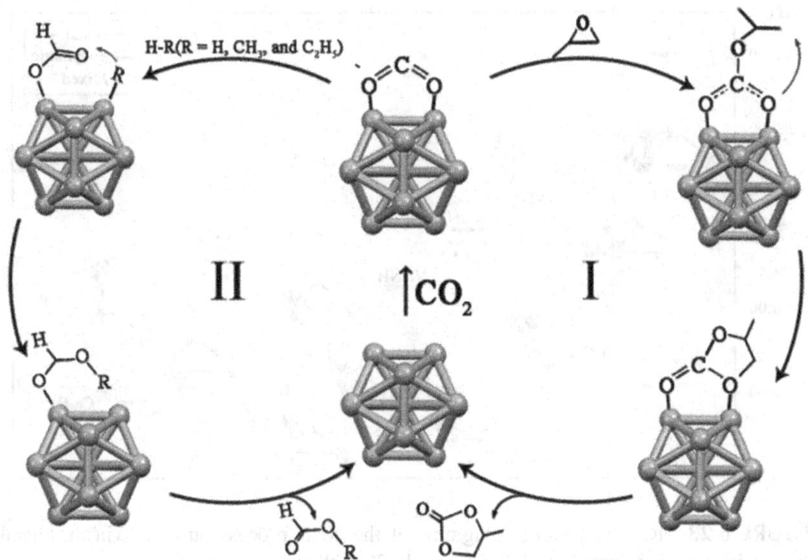

FIGURE 6.22 Proposed catalytic cycles for (I) cycloaddition of CO_2 with epoxides and (II) radical reactions between CO_2 and HR (R = H, CH_3 and C_2H_5) catalysed by $Al_{12}P$ [21]. Copyright 2020 Wiley-VCH GmbH.

In view of the excellent performance of superalkalis in activating CO_2, they may be efficient catalysts for the conversion of CO_2 into fuels and value-added chemicals to relieve the growing energy crisis and global warming. With the assistance of DFT calculations, Sun and coworkers [21] found that, different from other $Al_{12}X$ (X = Be, Al and C) superatoms, the alkali-metal-like superatom $Al_{12}P$ prefers to combine with CO_2 via a bidentate double oxygen coordination, yielding a stable $Al_{12}P(\eta^2\text{-}O_2C)$ complex containing an activated radical anion of CO_2 (*i.e.*, $CO_2^{\bullet-}$). As a result, this compound could not only participate in the subsequent cycloaddition reaction with propylene oxide, but also initiate the radical reaction with hydrogen gas to form high value chemicals, revealing that $Al_{12}P$ can play an important role in catalysing these conversion reactions (see Figure 6.22). They optimistically anticipated that this work could guide the experimentalists to discover such an additional superatom catalyst for CO_2 transformation, and open up a new research field of superatom catalysis.

The capability of the superalkali Li_3F_2 to activate dinitrogen was firstly presented by Meloni and coworkers [8] in 2018. The complete dissociation of N_2 was confirmed through visualized molecular orbitals and bond order calculation. The N-N bond is weakened by the addition of Li_3F_2 superalkali units. In the same year, Goel and coworkers [19] predicted the mononuclear superalkali BLi_6 to be a suitable catalyst for conversion of N_2 to ammonia with a limiting potential of 19.13 kcal/mol. The possibility of grafting the cluster on the support surface such as graphene and boron nitride sheet has also been explored. The study also found

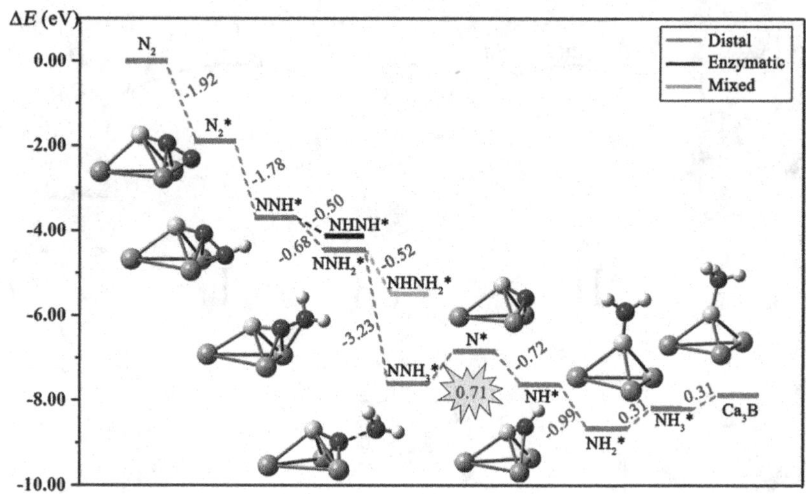

FIGURE 6.23 Reaction energy diagrams of the NRR process on Ca_3B via the distal, enzymatic and mixed pathways [20]. Copyright 2021 the Owner Societies.

that like the isolated cluster, the supported BLi_6 cluster is a promising catalyst for N_2 activation.

More recently, Sun and coworkers [20] have designed a new alkali-metal-like superatom Ca_3B with low VIE of 4.29 eV and high stability. This well-designed superatom exhibits unique geometric and electronic features, which can fully activate N_2 into N_2^{2-} via a "double-electron transfer" mechanism, and then convert the activated N_2 into NH_3 through a distal reaction pathway with a small energy barrier of 0.71 eV (see Figure 6.23). This work maybe intrigue more endeavour to design specific superatoms as excellent catalysts for the chemical adsorption and reduction of N_2 to NH_3.

In 2018, Srivastava [10] has found that three typical superalkalis, *i.e.*, FLi_2, OLi_3 and NLi_4 can reduce the major air pollutants NO_x into NO_x^- anion via single-electron transfer. He also found that the size of superalkalis plays a crucial role in the single-electron reduction of NO_x. Srivastava [16] also studied the capability of these three superalkalis to reduce carbon monoxide (CO) into CO^-. Das and coworkers [15] performed a systematic study on the reduction of SO_2 by using FLi_2, OLi_3 and NLi_4. The accumulation of negative charge on SO_2 molecule in the formed superalkali (SA)-SO_2 complexes suggested that SO_2 can be reduced with these superalkalis.

6.3.5 Cluster-Assembled Materials

6.3.5.1 Superatom Compounds, Supersalts and Nanocrystals

As a special kind of superatom, superalkalis can behave as alkali metal atoms and maintain their structural and electronic integrities, much in the same way as

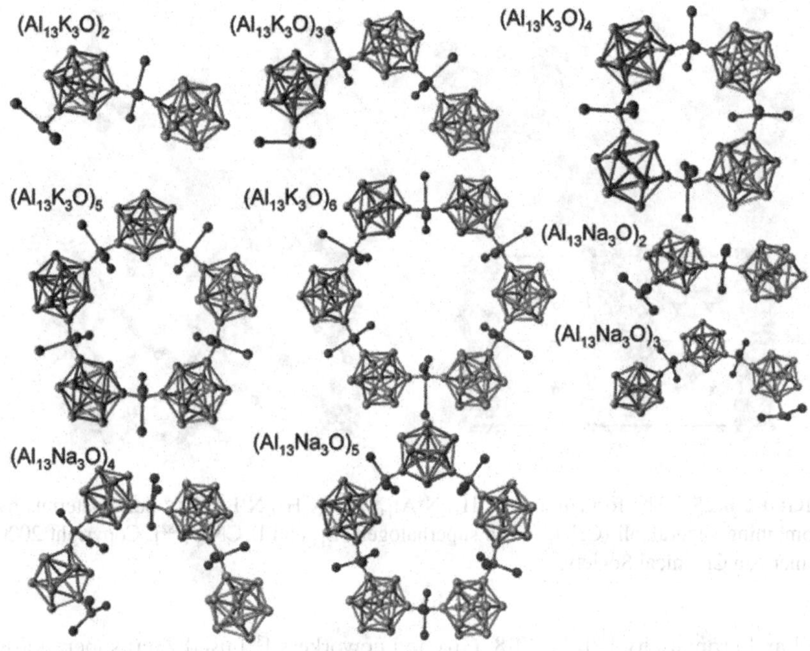

FIGURE 6.24 Assemblies of (A) $Al_{13}K_3O$ and (B) $Al_{13}Na_3O$ molecules [207]. Copyright 2007 American Chemical Society.

ordinary atoms, when they are assembled into extended nanostructures or used as building blocks of nano-materials [207]. In 2007, the superalkalis Na_3O and K_3O were used to combine with superhalogen Al_{13} to obtain stable superatom assemblies (see Figure 6.24), where the alkali metal atoms (Li-Cs) has not been promising because the size of alkali atom is small as compared to the larger Al_{13} unit. Thereafter, superalkalis have been utilized as new building blocks for a large number of nanomaterials with highly tunable properties in the past decades [26–54].

Initially, these ionic compounds formed by combining superalkalis and (super) halogen are named "superatom compounds" [207]. The charge transfer in such compounds results in their large NLO responses as mentioned above [26,27,208]. The interaction between superalkalis and superhalogens also brings forth some superatom compounds with intriguing characteristics, such as $(Li_3)^+(SH)$ (SH = LiF_2, BeF_3 and BF_4) with the aromaticity and the characters of electrides and alkalides [278], and $Li_3^+N_4^{2-}M_3^+$ (M = Li, Na and K) with the out-of-plane sigma-aromaticity and enhanced π-aromaticity [279].

In 2014, Jena and coworkers [33] begin to use the term "supersalt" to represent these compounds synthesized with superalkalis and superhalogens as building blocks. In the same year, Misra and coworkers reported a series of supersalts with Li_3X_3 (X = halogen atoms) type [34,35,280]. Such supersalts contain superalkali cation Li_2X^+ and suphalogen anion LiX_2^-, which shows both of alkalide characters

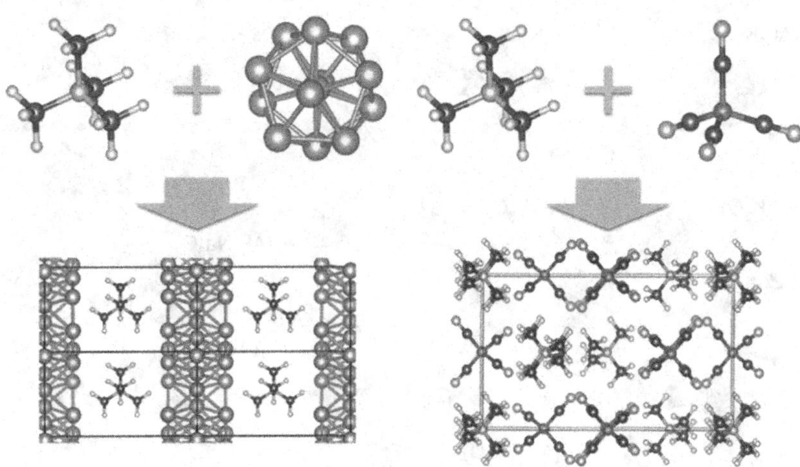

FIGURE 6.25 The formation of $(CH_3)_4N^+Al_{13}^-$ and $(CH_3)_4N^+B(CN)_4^-$ bulk materials by combining superalkali $(CH_3)_4N$ with superhalogen Al_{13} and $B(CN)_4$ [282]. Copyright 2020 American Chemical Society.

[34] and aromaticity [35]. In 2018, Giri and coworkers [36] used Zintl superalkalis as building blocks to achieve a series of novel supersalts. In the same year, Meloni and coworkers [281] explored the design of hypersalts starting from the hyperhalogen Li_3F_4 plus a Li atom and the hyperalkali Li_4F_3 plus an F atom. Milovanović [48] studied two approaches to generate the Li_nCl_n clusters, namely the polymerization of LiCl fragments and the combination of superalkalis and superhalogen clusters by examination of the lattice energy and the average Li-Cl bond length in rectangular Li_nCl_n ($n \leq 60$) clusters. They concluded that 50 LiCl are enough to mostly resemble the structure and stability of the bulk LiCl, which was regarded as a kind of nanocrystal formed by supersalts.

In 2020, as shown in Figure 6.25, Jena and coworkers [282] combined the nonmetallic superalkali $(CH_3)_4N$ with the superhalogen Al_{13} to construct the $[(CH_3)_4N^+][Al_{13}^-]$ crystal. However, they found that the icosahedral Al_{13} clusters began to coalesce while the $(CH_3)_4N$ superalkalis maintained their geometries. However, the results are different when the anion is chosen to be a nonmetal superhalogen, *e.g.*, the bison anion $B(CN)_4^-$. The resulting $[(CH_3)_4N^+][B(CN)_4^-]$ is found to be a charge-transfer supersalt, where both $(CH_3)_4N^+$ and $B(CN)_4^-$ maintain their individual structures, even at room temperature (300 K). This finding suggests a competition between the octet rule stabilizing the nonmetallic molecular ion $B(CN)_4^-$ vs the "jellium rule" stabilizing the metallic molecular ion Al_{13}^-, which could be the key for the design and synthesis of building blocks for cluster-assembled materials.

A family of stable ferroelectric/ferroelastic supersalts, PnH_4MX_4 (Pn = N, P; M = B, Al, Fe; X = Cl, Br) composed of superalkali PnH_4 and superhalogen MX_4 ions (see Figure 6.26) were predicted by Gao et al. in 2021 [37]. Unlike traditional

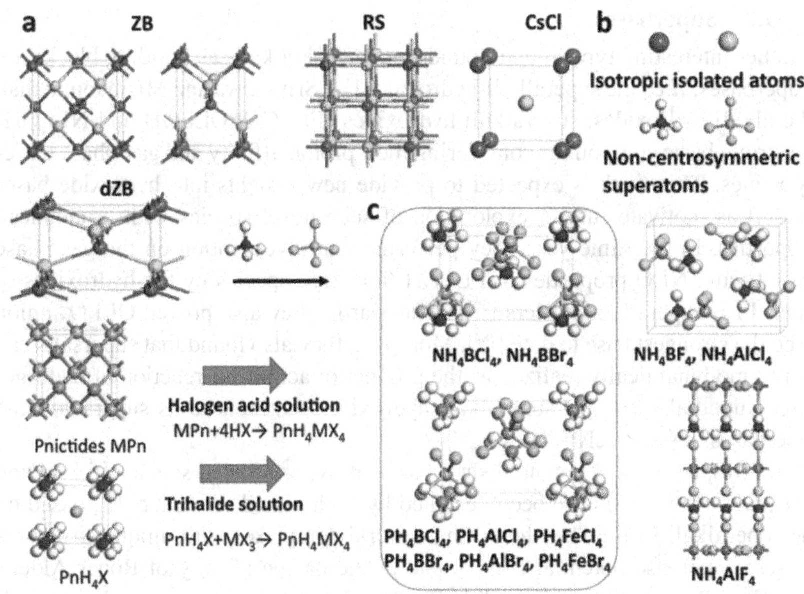

FIGURE 6.26 Design of supersalts PnH_4MX_4. (a) Structures of ZB, dZB, RS and CsCl phase, where dark green and yellow spheres, respectively, denote cations and anions. (b) Noncentro-symmetric superalkali and superhalogen compared with isotropic isolated atoms. (c) Several typical structures of supersalts PnH_4MX_4, which may be synthesized by reactions $MPn+4HX \rightarrow PnH_4MX_4$ or $PnH_4X + MX_3 \rightarrow PnH_4MX_4$. White, pink, blue, purple and light green spheres denote H, B, N, P and halogen atoms, respectively [37].

ferroelectric/ferroelastic materials, the cluster-ion based supersalts possess ultra-low switching barriers and can endure large ion displacements and reversible strain. In particular, PH_4FeBr_4 shows triferroic coupling of ferroelectricity, ferroelasticity and antiferromagnetism with controllable spin directions via either ferroelastic or 90-degree ferroelectric switching.

In 2022, Sikorska [49] studied the assembly of atomically-precise cluster solids with atomic precision by combining the electron-donating superalkali and electron-accepting superhalogen. The charge transfer between neutral molecular clusters in the designed solid results in the intercluster electrostatic attraction as a driving force for co-assembly. By the analysis of binding energy between superatomic counterparts, charge transfer and the relative size of the clusters, the resulting structures are analysed to be either molecular crystals or superatomic lattices. It is demonstrated that the constructed $[N_4Mg_6M]^+[AlX_4]^-$ (M = Li, Na and K; X = F and Cl) compounds form the same close-packed superatomic lattice structure through halogen bonding, with subtle differences in the orientation of the superatoms. These salts may also form molecular crystals where clusters are held to one another by electrostatic interactions. These results emphasize how the structure of superatomic solids can be tuned upon single atom substitution

6.3.5.2 Superbases

Another interesting type of compounds using superalkalis as building blocks are "superbases," *i.e.*, the superalkali-hydroxides [37]. Srivastava and Misra found that, like alkali hydroxides, superalkali-hydroxides (FLi_2OH, OLi_3OH and NLi_4OH) are strong basic compounds considering their proton affinity and gas-phase basicity values. This work is expected to provide new insights into hydroxide bases as well as motivate further exploration of such novel species with pronounced properties. In the same year, they performed an investigation on the gas phase basicity and NLO properties of FLi_nOH (n = 2–5) species by the hydroxides of small FLi_n superalkali clusters [283]. Afterwards, they also proved OLi_3O^- anion to be the strongest base to date [284]. Moreover, they also found that supersalts can be thermodynamically realized as the product of acid–base reaction of hydrogenated superhalogens and superalkali hydroxides that behave as superacids and superbases, respectively [285].

In 2018, a series of potential superbases of hyperlithiated species Li_3F_2O and $Li_3F_2(OH)_n$ (n = 1, 2) have been designed by Meloni and coworkers [39] based on the superalkali Li_3F_2. They found that several complexes with unique structures possess superbase potential comparable to the proton affinity of Roger Alder's canonical Proton Sponge. In 2021, Pandey [40] proposed novel polynuclear K- and Na-based superalkali hydroxides as superbases with stronger basic character (larger gas-phase proton affinity and gas-phase basicity values) than the LiOH and Li-related species. Design and synthesis of such theoretically examined superbases may pave alternative routes for the experimentally rewarding applications.

6.3.5.3 Superalkali-Based Perovskites

AMX_3-type perovskites (A and M are cations, but X is an anion) attracted great attention as promising absorbers for solar cells, due to their easy solution-based synthesis, low cost, variable compositions, controllable layers, tunable optical band gaps, high power conversion efficiency of 23.7% and other excellent photoelectric properties [50]. In 2017, Paduani et al. [286] reported that the band gap of perovskites can be tuned by introducing appropriate superalkali moieties at the cationic *A*-sites in the $CsPbI_3$-type structure. The computed band gap is 0.36 eV (direct) and 0.41 eV (indirect) for $[Li_3O]PbI_3$ and $[Li_3S]PbI_3$, respectively, indicating the drastic changes in the shape of both valence and conduction bands, as compared to $CsPbI_3$. The introduction of superalkali cations produces extra electronic states close to the Fermi level which arise from the formation of delocalized energy states, where a strong hybridization is identified between Pb and Li *s*-states near the top of the valence band. This can promote the hole mobility and increase the exciton diffusion length at longer wavelengths.

Unfortunately, the two band gaps of $[Li_3O]PbI_3$ and $[Li_3S]PbI_3$ are obviously lower than the band gap of 1.73 eV of $CsPbI_3$ inorganic perovskite, and even lower than the optimal band gap 1.1–1.4 eV for a single-junction solar cell, indicating that these two perovskites are unsuitable for solar cells [50]. To obtain the perovskite with the suitable band gaps and tunable electronic structures, cubic $(Li_3O)M(BH_4)_{3-x}Br_x$ (M = Ge, Sn and Pb; x = 0–3) perovskites are designed by introducing

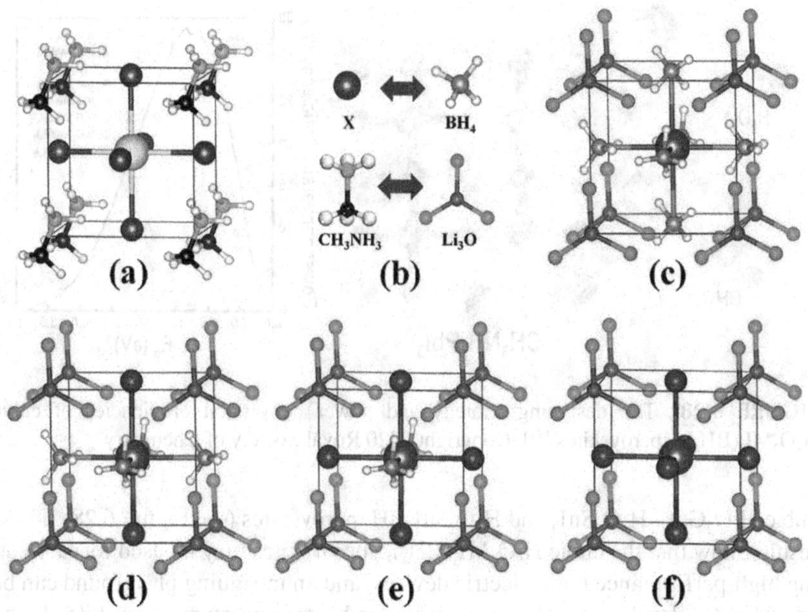

FIGURE 6.27 (a) The model of cubic $CH_3NH_3PbX_3$ (X = Cl, Br and I) perovskites. (b) halogen atoms (X), molecule (CH_3NH_3), superalkali (Li_3O) and superhalogen (BH_4). (c–f) The models of cubic $(Li_3O)M(BH_4)_3$, $(Li_3O)M(BH_4)_2Br$, $(Li_3O)M(BH_4)Br_2$ and $(Li_3O)MBr_3$ (M = Ge, Sn and Pb) perovskites, respectively. Atomic colors: H (white), Li (green), B (orange), C (black), N (cyan), O (red), Z (brown), Br (blue), M (gray) and Pb (yellow) [50]. Copyright 2019 Elsevier B.V.

both superalkali Li_3O and superhalogen BH_4 (see Figure 6.27) [50]. The calculated results show that the $(Li_3O)M(BH_4)_{3-x}Br_x$ perovskites show suitable tolerance factors and negative formation energies, tunable direct band gap from 0.97 to 2.42 eV, and small electron effective masses as well as the different charge densities for conduction band minimum and valence band maximum states. Their calculated results also suggest that cubic $(Li_3O)Ge(BH_4)Br_2$ and $(Li_3O)Pb(BH_4)_2Br$ perovskites can show the power conversion efficiency of 23.12% for the loss-inpotential of 0.5 eV. Interestingly, the power conversion efficiencies of the $(Li_3O)M(BH_4)_{3-x}Br_x$ perovskites are first increased then decreased with the increase of Br atom.

To expand the family of ABX_3-type perovskites, the nonmetallic superalkali cation $H_5O_2^+$ [65] and superhalogen BH_4^- are introduced into cubic $CsPbI_3$ perovskite by Peng et al. [51] to obtain novel hybrid organic-inorganic perovskites (HOIPs) (see Figure 6.28). The obtained cubic $H_5O_2MI_x(BH_4)_{3-x}$ (M = Ge, Sn; x = 0, 1, 2, 3) perovskites have the wide range of material properties, *i.e.*, stable dynamics performance under normal temperature and pressure, suitable tolerance factors (0.86 ~ 1.00), tunable direct band gaps (0.96 ~ 3.28 eV) and small effective hole and electron masses (0.17 ~ 0.70 m_e). Moreover, the power conversion efficiency is over 31% for single-junction perovskite solar cells based on the

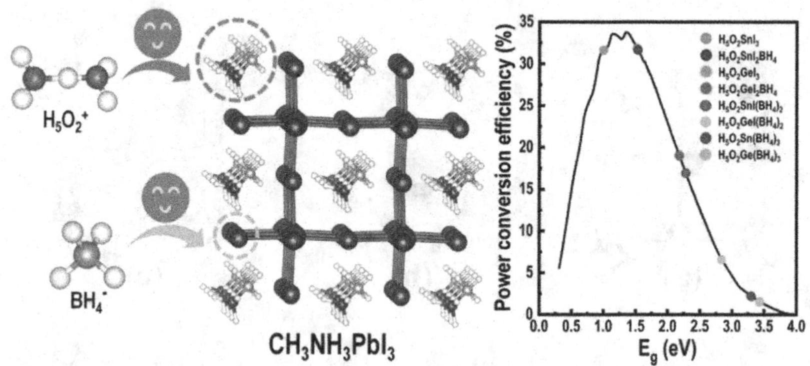

FIGURE 6.28 The designing strategy and power conversion efficiencies of cubic $H_5O_2MI_x(BH_4)_{3-x}$ perovskites [51]. Copyright 2020 Royal Society of Chemistry.

cubic $H_5O_2GeI_3$, $H_5O_2SnI_3$ and $H_5O_2SnI_2BH_4$ perovskites (see Figure 6.28). These results show that the cubic $H_5O_2MI_x(BH_4)_{3-x}$ perovskites may be used for fabricating high-performance photoelectric devices, and an intriguing playground can be provided for developing other new stable lead-free superatom perovskites based on various superalkalis and superhalogens.

Similarly, a new class of functional HOIPs has been theoretically designed based on the superalkali cations (*e.g.*, $H_5O_2^+$, $C_2H_5OH_2^+$ and CH_3SH^+) by Wu and coworkers [52]. Their computations suggest that many of the designed HOIPs possess not only direct bandgaps with values within the optimal range for solar light absorbing but also more desirable optical absorption spectra than that of prevailing $MAPbI_3$, where their ferroelectric polarizations also benefit photovoltaics. In 2021, Sikorska et al. [53] used the bimetallic superalkali cations ($LiMg^+$, $NaMg^+$, $LiCa^+$ and $NaCa^+$) to replace the Cs^+ cation in the $CsPbBr_3$ material. The bimetallic superalkali cation substitution introduces extra electronic states close to the Fermi level, and decreases the bandgap of the perovskite. In particular, the bandgaps of MgLi–$PbBr_3$ (1.35 eV) and MgNa–$PbBr_3$ (1.06 eV) are lower than the bandgap of $CsPbBr_3$ (2.48 eV) and within the optimal bandgap (*i.e.*, 1.1–1.4 eV) for single-junction solar cells, indicating that these inorganic perovskites are promising candidates for high-efficiency solar cells. In the same year, Zhou et al. [54] used the $H_5O_2^+$ cations as the building blocks to construct quasi-2D lead-free $[C_6H_5(CH_2)_2NH_3]_2H_5O_2Sn_2Br_7$ perovskites. Their calculated results show that the quasi-2D perovskite has negative formation energy, small effective hole and electron masses, stable dynamics performance, suitable exciton binding energy and direct band gap and H_5O_2 superalkali cations that don't agglomerate. Moreover, they proposed that this quasi-2D perovskite may be applied to red-light-emitting diodes.

6.3.5.4 Other Superalkali-Based Materials

Superalkalis can also be used to assemble the hydrides [287], endofullerenes [288], endoborospherene [289], sandwich-like compounds [290], alkalides [291], electrides

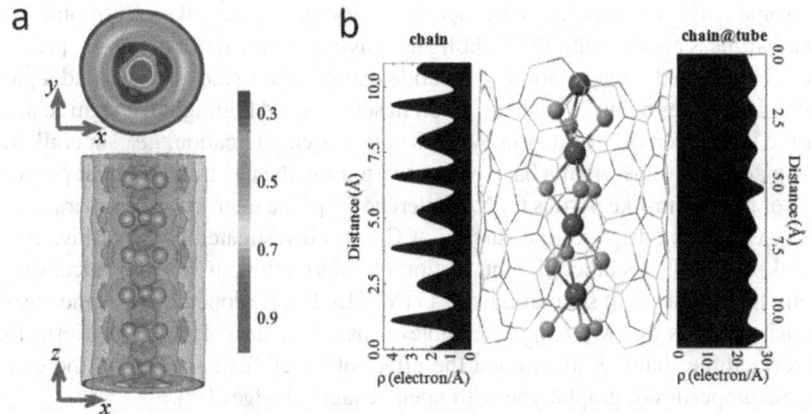

FIGURE 6.29 (a) Calculated electron localization function (ELF) of the 1D Li_3O. The localized electrons with ELF ≥ 0.7 are found as shells around the chain on the (Li_3) planes. (b) Calculated charge density integrated along the chain axis for both the 1D Li_3O and Li_3O@BNNT. Peaks are present at the positions of the localized electrons. Oxygen atoms are in red and lithium in cyan [[292]]. Copyright 2018 The Royal Society of Chemistry.

[[292]], superalkali-halogen clusters [[293]] and ion pairing in low-polarity solvents [[294]] with various interesting functions. For example, hydrogenated superalkalis (SAHs), including FLi_2H, OLi_3H and NLi_4H have potential to be used as strong bases, NLO molecules, reducing agents and building blocks of supersalts [[287]]. Wang et al. designed novel metal-[metal oxide]-nonmetal sandwich-like superalkali compounds, *i.e.*, H- and T-shaped $Li_3OMC_5H_5$ (M = Be, Mg and Ca), where the H-shaped structure exhibits electride characteristics, and the T-shaped structure with lithium anion exhibits alkaline characteristics [[290]].

Fang et al. [[292]] proposed that superalkalis can be used as building blocks of one-dimensional hierarchical inorganic electrides, characterized by the presence of localized electrons with high-charge density along a one-dimensional chain in 2018 (Figure 6.29). Embedding this one-dimensional chain inside a boron-nitride nanotube (BNNT) produces an electride with 1D hierarchical structure and improved stability and properties. For example, Li_3O@BNNT possesses ultra-low work function, making it an ideal material capable of activating CO_2. In addition, Li_3O electride is antiferromagnetic, with the anionic electrons carrying the magnetic moment. Thus, superalkalis as building blocks of electrides can bridge cluster science and solid-state chemistry. This study demonstrates that using superalkali, combined with nanoassembly, is a viable strategy for rational design of new electrides with novel properties.

In 2019, imidazole and benz-imidazole based different NHC ligands have been designed to synthesize a Cu(I)@NHC complex by Parida et al. [[294]] Their calculated results demonstrate that all the used ligands belong to superalkalis, and can form Cu(I)@NHC complexes with a trinuclear Cu_3 core having both σ- and

π-aromaticity. Aromaticity and other reactivity parameters like electrophilicity and hardness change with different ligand environments. Riedel et al. [295] revisited the classical interpretation of alkalide nature. Their results go beyond a picture of alkalides being a "gaslike" anion in solution and highlight the significance of the interaction of the alkalide with its complex countercation, *i.e.*, superalkali.

In addition, superalkalis have been used to tune the electronic-optical properties of porphyrin-like porous $C_{24}N_{24}$ fullerene [296]. The interaction of superalkalis FLi_2, OLi_3 and NLi_4 with the surface of C_{60} was investigated by Srivastava [297]. Sheikhi et al. [298] studied the interaction of chlormethine (CM) anticancer drug with the complexes of superalkalis M_3O (M = Li, Na, K) doped $B_{12}N_{12}$ nanocages, which suggests this M_3O-$B_{12}N_{12}$ complexes as a new drug delivery platform for chlormethine. Jiang et al. studied the effect of superalkali species on the electronic properties of graphdiyne with open hexagonal edges [299].

In 2021, George et al. [300] studied the Diels–Alder cycloaddition reaction between 1,3-cyclohexadiene and a series of C_{60} fullerenes with encapsulated (super)alkali/(super)halogen species ($Li^+@C_{60}$, $Li_2F^+@C_{60}$, $Cl^-@C_{60}$ and $LiF_2^-@C_{60}$) by means of DFT calculations. As compared to the parent C_{60} fullerene, the significant enhancement in reactivity was found for cation-encapsulating $Li^+/Li_2F^+@C_{60}$ complexes. Chattaraj and co-workers [301] have highlighted that superhalogens and superalkalis have the potential to enhance the efficacy of redox reactions. Jha et al. [302] have theoretically investigated the effect of confinement on the behaviour of superalkali and superhalogen species by taking different fullerene cages, including C_{70}, C_{80} and C_{90}. It has been observed that while the electron affinity of superhalogen decreases, ionization energies (IE) of superalkali increases. Steric effect is predominant between host and guest in smaller fullerenes whereas van der Waals interaction gradually increases in bigger fullerenes.

6.4 SUMMARY AND OUTLOOK

Superalkalis, as novel excellent reducing matters, are of great importance in chemistry. Therefore, such species have attracted more and more attention and achieved significant advances in the past 40 years. In this chapter, we have briefly summarized the previous investigations on the design and applications of superalkalis. Up to now, a great number of superalkalis of various types have been proposed. These achievements indicate that the potential of creating superalkalis is limitless and, thereby, will intrigue further interest to design more diverse superalkalis by using new strategies beyond the existing ones. Despite the remarkable progress in the theoretical design of superalkalis, the experimental works on such species are limited to simple ones by using mass spectrometric technology. Hence, more and more attention should be paid to further synthesize and characterize the available superalkalis theoretically designed in references or other newly designed ones in the future. To reveal this goal, the stability of superalkalis should be addressed in future design. Ligand-protected clusters with lower IEs than alkali-metal atoms seem to be the most promising type of superalkalis that can be synthesized by experimentalists. The realization of superalkali synthesis will be of great significance to open the door to new applications of superalkalis.

Nowadays, although tremendous achievements have been made for the potential application of superalkalis, their practical application in chemistry still remains a challenge. For example, though some superalkalis have been experimentally detected in the gas phase, scarce solid materials with superalkalis as building blocks have been synthesized up to now. In this regard, more efforts are expected to be devoted to synthesizing superalkali-assembled materials, such as supersalts and superatomic crystals. To realize this, superalkalis need to be produced in large quantities, and, once synthesized, they must not coalesce. Synthesizing superalkalis by coating metal core with suitable organic ligands or metal-encapsulating superalkalis with high stability may help to resolve this problem. Second, the instability of the diffuse excess electrons in the electrides and alkalides hinders their experimental synthesis and, thus, limits their practical application as NLO materials. Hence, it is important to balance the stability and NLO responses of superalkali-based NLO materials. Furthermore, depositing these superalkali clusters on various substrates may pave the way to practically utilize them as novel catalysts, hydrogen storage materials, drug delivery systems, noble gas-trapping agents, photovoltaic materials and so on. As a result, further extending the practical applications of superalkalis is still acquired in the future.

REFERENCES

1. Lide, D. R., Ed.; *CRC Handbook of Chemistry and Physics*; 84th ed.; CRC Press: Boca Raton, FL, 2003.
2. Gutsev, G. L., Boldyrev, A. I. *Chem. Phys. Lett. 92*, 262, 1982.
3. Zintl, E., Morawietz, W. *Z. Anorg. Allg. Chem. 236*, 372, 1938.
4. Jansen, M. *Angew. Chem. Int. Ed. 88*, 411, 1976.
5. Jansen, M. *Z. Anorg. Allg. Chem. 435*, 13, 1977.
6. Liu, H., Klein, W., Sani, A., Jansen, M. *Phys. Chem. Chem. Phys. 6*, 881 2004.
7. Park, H., Meloni, G. *Dalton Trans. 46*, 11942, 2017.
8. Park, H., Meloni, G. *ChemPhysChem 19*, 256, 2018.
9. Srivastava, A. K. *Int. J. Quantum Chem.* e25598, 2018.
10. Srivastava, A. K. *Chem. Phys. Lett. 695*, 205, 2018.
11. Ramondo, F., Leonzi, I., Meloni, G. *ChemPhysChem 20*, 3251, 2019.
12. Srivastava, A. K. *Int. J. Quantum Chem. 119*, 2019.
13. Sikorska, C., Gaston, N. *J. Chem. Phys. 153*, 144301, 2020.
14. Meloni, G., Giustini, A., Park, H. *Front. Chem. 9*, 712960, 2021.
15. Sarkar, S., Debnath, T., Das, A. K. *Comput. Theor. Chem. 1202*, 113317, 2021.
16. Srivastava, A. K. *J. Mol. Graph. Model. 102*, 107765, 2021.
17. Srivastava, H., Srivastava, A. K. *Front. Phys. 10*, 2022.
18. Ma, J.-F., Ma, F., Zhou, Z.-J., Tao Liu, Y. *RSC Adv. 6*, 84042, 2016.
19. Riyaz, M., Goel, N. *Comput. Theor. Chem. 1130*, 107, 2018.
20. Zhang, X.-L., Ye, Y.-L., Zhang, L., Li, X.-H., Yu, D., Chen, J.-H., Sun, W.-M. *Phys. Chem. Chem. Phys. 23*, 18908, 2021.
21. Zhang, X.-L., Zhang, L., Ye, Y.-L., Li, X.-H., Ni, B.-L., Li, Y., Sun, W.-M. *Chem. Eur. J. 27*, 1039, 2021.
22. Khanna, S., Jena, P. *Phys. Rev. Lett. 69*, 1664, 1992.

23. Khanna, S., Jena, P. *Phys. Rev. B 51*, 13705, 1995.
24. Luo, Z., Castleman, A. W. *Acc. Chem. Res. 47*, 2931, 2014.
25. Reber, A. C., Khanna, S. N. *Acc. Chem. Res. 50*, 255, 2017.
26. Li, Y., Wu, D., Li, Z.-R. *Inorg. Chem. 47*, 9773, 2008.
27. Yang, H., Li, Y., Wu, D., Li, Z. R. *Int. J. Quantum Chem. 112*, 770, 2012.
28. Wang, J.-J., Zhou, Z.-J., Bai, Y., Liu, Z.-B., Li, Y., Wu, D., Chen, W., Li, Z.-R., Sun, C.-C. *J. Mater. Chem. 22*, 9652, 2012.
29. Sun, W. M., Fan, L. T., Li, Y., Liu, J. Y., Wu, D., Li, Z. R. *Inorg. Chem. 53*, 6170, 2014.
30. Sun, W. M., Wu, D., Li, Y., Li, Z. R. *Dalton Trans. 43*, 486, 2014.
31. Sun, W.-M., Li, X.-H., Wu, D., Li, Y., He, H.-M., Li, Z.-R., Chen, J., Li, C. *Dalton Trans. 45*, 7500, 2016.
32. Sun, W. M., Li, Y., Li, X. H., Wu, D., He, H. M., Li, C. Y., Chen, J. H., Li, Z. R. *ChemPhysChem 17*, 2672, 2016.
33. Giri, S., Behera, S., Jena, P. *J. Phys. Chem. A 118*, 638, 2014.
34. Srivastava, A. K., Misra, N. *New J. Chem. 38*, 2890, 2014.
35. Srivastava, A. K., Misra, N. *Mol. Phys. 112*, 1, 2014.
36. Reddy, G. N., Kumar, A. V., Parida, R., Chakraborty, A., Giri, S. *J. Mol. Model. 24*, 306, 2018.
37. Gao, Y., Wu, M., Jena, P. *Nat. Commun. 12*, 1331, 2021.
38. Srivastava, A. K., Misra, N. *New J. Chem. 39*, 6787, 2015.
39. Winfough, M., Meloni, G. *Dalton Trans. 47*, 159, 2018.
40. Pandey, S. K. *ACS Omega 6*, 31077, 2021.
41. Wang, K., Liu, Z., Wang, X., Cui, X. *Int. J. Hydrogen Energ. 39*, 15639, 2014.
42. Saedi, L., Dodangi, M., Mohammadpanaardakan, A., Eghtedari, M. *J. Cluster Sci. 31*, 71, 2019.
43. Gao, P., Li, J.-w., Wang, G. *Int. J. Hydrogen Energ. 46*, 24510, 2021.
44. Qi, H., Wang, X., Chen, H. *Int. J. Hydrogen Energ. 46*, 23254, 2021.
45. Wang, X., Qi, H., Ma, L., Chen, H. *Appl. Phys. Lett. 118*, 093902, 2021.
46. Zhang, Y.-F., Liu, P.-P., Luo, Z.-H. *FlatChem 36*, 100429, 2022.
47. Zhang, Y., Liu, P. *Int. J. Hydrogen Energ. 47*, 14637, 2022.
48. Milovanović, M. *J. Comput. Chem. 42*, 1895, 2021.
49. Sikorska, C., Gaston, N. *Phys. Chem. Chem. Phys. 24*, 8763, 2022.
50. Zhou, T., Zhang, Y., Wang, M., Zang, Z., Tang, X. *J. Power Sources 429*, 120, 2019.
51. Peng, H., Tang, R., Deng, C., Li, M., Zhou, T. *J. Mater. Chem. A 8*, 21993, 2020.
52. Yang, Q., Wu, M., Zeng, X. C. *Research 2020*, 1, 2020.
53. Sikorska, C., Gaston, N. *J. Chem. Phys. 155*, 174307, 2021.
54. Zhou, T., Kuang, A. *Nanoscale 13*, 13152, 2021.
55. Sun, W. M., Wu, D. *Chem. Eur. J. 25*, 9568, 2019.
56. Rehm, E., Alexander, I. Boldyrev, Schleyer, P. V. R. *Inorg. Chem. 31*, 4834, 1992.
57. Tong, J., Li, Y., Wu, D., Li, Z. R., Huang, X. R. *J. Chem. Phys. 131*, 164307, 2009.
58. Tong, J., Li, Y., Wu, D., Li, Z.-R., Huang, X.-R. *J. Phys. Chem. A 115*, 2041, 2011.
59. Tong, J., Li, Y., Wu, D., Wu, Z.-J. *Inorg. Chem. 51*, 6081, 2012.
60. Tong, J., Wu, Z., Li, Y., Wu, D. *Dalton Trans. 42*, 577, 2013.
61. Tong, J., Wu, D., Li, Y., Wang, Y., Wu, Z. *Dalton Trans. 42*, 9982, 2013.
62. Liu, J. Y., Wu, D., Sun, W. M., Li, Y., Li, Z. R. *Dalton Trans. 43*, 18066, 2014.
63. Liu, J.-Y., Li, R.-Y., Li, Y., Ma, H.-D., Wu, D. *J. Cluster Sci. 29*, 853, 2018.
64. Sun, W.-M., Wu, D., Kang, J., Li, C.-Y., Chen, J.-H., Li, Y., Li, Z.-R. *J. Alloy. Comp. 740*, 400, 2018.
65. Hou, N., Li, Y., Wu, D., Li, Z.-R. *Chem. Phys. Lett. 575*, 32, 2013.
66. Srivastava, A. K. *New J. Chem. 43*, 4959, 2019.

67. Srivastava, A. K. *J. Mol. Graph. Model. 88*, 292, 2019.
68. Srivastava, A. K. *Chem. Phys. Lett. 721*, 7, 2019.
69. Ye, Y.-L., Pan, K.-Y., Ni, B.-L., Sun, W.-M. *Front. Chem. 10*, 853160, 2022.
70. Sun, W.-M., Li, Y., Wu, D., Li, Z.-R. *J. Phys. Chem. C 117*, 24618, 2013.
71. Parida, R., Reddy, G. N., Ganguly, A., Roymahapatra, G., Chakraborty, A., Giri, S. *Chem. Commun.*, 2018.
72. Guo, J. C., Feng, L. Y., Zhang, X. Y., Zhai, H. J. *J. Phys. Chem. A 122*, 1138, 2018.
73. Srivastava, A. K. *Mol. Phys. 118*, e1730991, 2020.
74. Paduani, C. *J. Phys. Chem. A 122*, 5037, 2018.
75. Parida, R., Reddy, G. N., Inostroza-Rivera, R., Chakraborty, A., Giri, S. *J. Mol. Model. 25*, 218, 2019.
76. Sun, W.-M., Cheng, X., Wang, W.-L., Li, X.-H. *Organometallics 41*, 2406, 2022.
77. Reddy, G. N., Giri, S. *Phys. Chem. Chem. Phys. 18*, 24356, 2016.
78. Srivastava, A. K. *Mol. Phys. 116*, 1642, 2018.
79. Srivastava, A. K. *Mol. Phys.* 1, 2019.
80. Srivastava, A. K. *Chem. Phys. Lett. 778* 138770, 2021.
81. Abegg, R. Z. *Anorg. Chem. 39*, 330, 1904.
82. Lewis, G. N. *J. Am. Chem. Soc. 38*, 762, 1916.
83. Langmuir, I. *J. Am. Chem. Soc. 41*, 868, 1919.
84. Yokoyama, K., Haketa, N., Hashimoto, M., Furukawa, K., Tanaka, H., Kudo, H. *Chem. Phys. Lett. 320*, 645, 2000.
85. Yokoyama, K., Haketa, N., Tanaka, H., Furukawa, K., Kudo, H. *Chem. Phys. Lett. 330*, 339, 2000.
86. Neškoviić, O. M., Veljkoviić, M. V., Veliičkoviić, S. R., Petkovska, L. T., Perić-Grujiić, A. A. *Rapid Commun. Mass Spectrom. 17*, 212, 2003.
87. Veličković, S., Djordjević, V., Cvetićanin, J., Djustebek, J., Veljković, M., Nešković, O. *Rapid Commun. Mass Spectrom. 20*, 3151, 2006.
88. Veličković, S. R., Koteski, V. J., Belošević Čavor, J. N., Djordjević, V. R., Cvetićanin, J. M., Djustebek, J. B., Veljković, M. V., Nešković, O. M. *Chem. Phys. Lett. 448*, 151, 2007.
89. Kudo, H., Wu, C. H., Ihle, H. R. *J. Nucl. Mater. 78*, 380, 1978.
90. Wu, C. H., Kudo, H., Ihle, H. R. *J. Chem. Phys. 70*, 1815, 1979.
91. Schleyer, P. V. R., Würthwein, E.-U., Pople, J. A. *J. Am. Chem. Soc. 104*, 5839, 1982.
92. Gutowskit, M., Simons, J. *J. Phys. Chem. 98*, 8326, 1994.
93. Lievens, P., Thoen, P., Bouckaert, S., Bouwen, W., Vanhoutte, F., Weidele, H., Silverans, R. E., Navarro-Vázquez, A., Schleyer, P. V. R. *J. Chem. Phys. 110*, 10316, 1999.
94. Neukermans, S., Janssens, E., Tanaka, H., Silverans, R. E., Lievens, P., Yokoyama, K., Kudo, H. *J. Chem. Phys. 119*, 7206, 2003.
95. Yokoyama, K., Tanaka, H., Kudo, H. *J. Phys. Chem. A 105*, 4312, 2001.
96. Wang, D., Graham, J. D., Buytendyk, A. M., Bowen, K. H. *J. Chem. Phys. 135*, 164308, 2011.
97. Zein, S., Ortiz, J. V. *J. Chem. Phys. 135*, 164307, 2011.
98. Viallon, J., Lebeault, M. A., Lépine, F., Chevaleyre, J., Jonin, C., Allouche, A. R., Aubert-Frécon, M. *Eur. Phys. J. D 33*, 405, 2005.
99. Schleyer, P. V. R., Wurthwein, E.-U., Kaufmann, E., Clark, T. *J. Am. Chem. Soc. 105*, 5930, 1983.
100. Otten, A., Meloni, G. *Chem. Phys. Lett. 692*, 214, 2018.
101. Peterson, K. I., Dao, P. D., Castleman, A. W. *J. Chem. Phys. 79*, 777, 1983.
102. Rechsteiner, C. E., Buck, R. P., Pedersen, L. *J. Chem. Phys. 65*, 1659, 1976.
103. Vituccio, D. T., Herrmann, R. F. W., Golonzka, O., Ernst, W. E. *J. Chem. Phys. 106*, 3865, 1997.

104. Veličković, S. R., Veljković, F. M., Perić-Grujić, A. A., Radak, B. B., Veljković, M. V. *Rapid Commun. Mass Spectrom. 25*, 2327, 2011.
105. Veljković, F. M., Djustebek, J. B., Veljković, M. V., Veličković, S. R., Perić-Grujić, A. A. *Rapid Commun. Mass Spectrom. 26*, 1761, 2012.
106. Veličković, S. R., Dustebek, J. B., Veljković, F. M., Veljković, M. V. *J. Mass Spectrom. 47*, 627, 2012.
107. Veljković, F. M., Djustebek, J. B., Veljković, M. V., Perić-Grujić, A. A., Veličković, S. R. *J. Mass Spectrom. 47*, 1495, 2012.
108. Đustebek, J., Milovanović, M., Jerosimić, S., Veljković, M., Veličković, S. *Chem. Phys. Lett. 556*, 380, 2013.
109. Milovanović, B., Milovanović, M., Veličković, S., Veljković, F., Perić-Grujić, A., Jerosimić, S. *Int. J. Quantum Chem. 119*, e26009, 2019.
110. Mitić, M., Milovanović, M., Veljković, F., Perić-Grujić, A., Veličković, S., Jerosimić, S. *J. Alloys Compd. 835*, 155301, 2020.
111. Hou, G.-L., Wang, X.-B. *Chem. Phys. Lett. 741*, 137094, 2020.
112. Dao, P. D., Peterson, K. I., Castleman, A. W. *J. Chem. Phys. 80*, 563, 1984.
113. Wurthwein, E.-U., Schleyer, P. V. R., Pople, J. A. *J. Am. Chem. Soc. 106*, 6973, 1984.
114. Goldbach, A., Hensel, F., Rademann, K. *Int. J. Mass Spectrom. Ion Proc. 148*, L5, 1995.
115. Hampe, O., Koretsky, G. M., Gegenheimer, M., Huber, C., Kappes, M. M., Gauss, J. R. *J. Chem. Phys. 107*, 7085, 1997.
116. Zein, S., Ortiz, J. V. *J. Chem. Phys. 136*, 224305, 2012.
117. Tong, J., Li, Y., Wu, D., Wu, Z.-J. *Chem. Phys. Lett. 575*, 27, 2013.
118. Srivastava, A. K. *Chem. Phys. Lett. 759*, 138049, 2020.
119. Zhang, Z., Chen, H. *Phys. Lett. A 383*, 125952, 2019.
120. Ariyarathna, I. R. *Phys. Chem. Chem. Phys. 23*, 16206, 2021.
121. Li, Y., Wu, D., Li, Z. R., Sun, C. C. *J. Comput. Chem. 28*, 1677, 2007.
122. Saunders, M. *J. Comput. Chem. 25*, 621, 2004.
123. Xi, Y. J., Li, Y., Wu, D., Li, Z. R. *Comput. Theor. Chem. 994*, 6, 2012.
124. Hou, D., Wu, D., Sun, W. M., Li, Y., Li, Z. R. *J. Mol. Graph. Model. 59*, 92, 2015.
125. Liu, J. Y., Xi, Y. J., Li, Y., Li, S. Y., Wu, D., Li, Z. R. *J. Phys. Chem. A 120*, 10281, 2016.
126. Ekardt, W. *Phys. Rev. B 29*, 1558, 1984.
127. Knight, W. D., Clemenger, K., Heer, W. A. D., A. Saunders, Chou, M. Y., Cohen, M. L. *Phys. Rev. Lett. 52*, 2141, 1984.
128. Bergeron, D. E., Roach, P. J., Jr., A. W. C., Morisato, T., Khanna, S. N. *Science 304*, 84, 2004.
129. Bergeron, D. E., Roach, P. J., Jr., A. W. C., Jones, N. O., Khanna, S. N. *Science 307*, 231, 2005.
130. Reveles, J. U., Khanna, S. N., Roach, P. J., Castleman, A. W. *Proc. Natl. Acad. Sci. U. S. A. 103*, 18405, 2006.
131. Sun, W.-M., Wu, D., Li, X.-H., Li, Y., Chen, J.-H., Li, C., Liu, J.-Y., Li, Z.-R. *J. Phys. Chem. C 120*, 2464, 2016.
132. Alexandrova, A. N., Boldyrev, A. I. *J. Phys. Chem. A 107*, 554, 2003.
133. Zhao, T., Wang, Q., Jena, P. *Nanoscale 9*, 4891, 2017.
134. Akutsu, M., Koyasu, K., Atobe, J., Hosoya, N., Miyajima, K., Mitsui, M., Nakajima, A. *J. Comput. Chem. A 110*, 12073, 2006.
135. Molina, B., Soto, J. R., Castro, J. J. *J. Comput. Chem. C 116*, 9290, 2012.
136. Sun, W.-M., Zhang, X.-L., Pan, K.-Y., Chen, J.-H., Wu, D., Li, C.-Y., Li, Y., Li, Z.-R. *Chem. A Eur. J. 25*, 4358, 2019.

137. Bera, P. P., Sattelmeyer, K. W., Saunders, M. H. F. S., Schleyer, P. V. R. *J. Phys. Chem. A 110*, 4287, 2006.
138. Sergeeva, A. P., Averkiev, B. B., Zhai, H. J., Boldyrev, A. I., Wang, L. S. *J. Chem. Phys. 134*, 224304, 2011.
139. Alexandrova, A. N., Boldyrev, A. I. *J. Chem. Theory Comput. 1*, 566, 2005.
140. Lu, T. *Molclus Program, Version 1.6*, www.keinsci.com/research/molclus.html (accessed November 10th, 2022).
141. Wang, H., Zhang, Y., Zhang, L., Wang, H. *Front. Chem. 8*, 589795, 2020.
142. Wang, X., Wang, H., Luo, Q., Yang, J. *J. Chem. Phys. 157*, 074304, 2022.
143. Hou, N., Wu, D., Li, Y., Li, Z. R. *J. Am. Chem. Soc. 136*, 2921, 2014.
144. Hückel, E. *Z. Phys. 70*, 204, 1931.
145. Hirsch, A., Chen, Z., Jiao, H. *Angew. Chem. Int. Ed. 39*, 3915, 2000.
146. Chen, D. L., Tian, W. Q., Lu, W. C., Sun, C. C. *J. Chem. Phys. 124*, 154313, 2006.
147. Lau, J. T., Hirsch, K., Klar, P., Langenberg, A., Lofink, F., Richter, R., Rittmann, J., Vogel, M., Zamudio-Bayer, V., Möller, T., Issendorff, B. V. *Phys. Rev. A 79*, 053201, 2009.
148. Shibuta, M., Ohta, T., Nakaya, M., Tsunoyama, H., Eguchi, T., Nakajima, A. *J. Am. Chem. Soc. 137*, 14015, 2015.
149. Wade, K. *J. Chem. Soc. D 792–793*, DOI: 10.1039/C29710000792, 1971.
150. Wade, K. *Advances in Inorganic Chemistry and Radiochemistry*; H. J. Emeléus and A. G. Sharpe, Eds.; Academic Press: Pittsburgh, 1976.
151. Mingos, D. M. P. *Acc. Chem. Res. 17*, 311, 1984.
152. Mingos, D. M. P. *Chem. Soc. Rev. 15*, 31, 1986.
153. Kudo, H., Hashimoto, M., Yokoyama, K., Wu, C. H., Dorigo, A. E., Bickelhaupt, F. M., Schleyer, P. V. R. *J. Phys. Chem. 99*, 6477, 1995.
154. Hashimoto, M., Yokoyama, K., Kudo, H., Wu, C. H., Schleyer, P. V. R. *J. Phys. Chem. 100*, 15770, 1996.
155. Tanaka, H., Yokoyama, K., Kudo, H. *J. Chem. Phys. 113*, 1821, 2000.
156. Tanaka, H., Yokoyama, K., Kudo, H. *J. Chem. Phys. 114*, 152, 2001.
157. Anusiewicz, I. *Aust. J. Chem. 63*, 1573, 2010.
158. Anusiewicz, I. *J. Theor. Comput. Chem. 10*, 191, 2011.
159. Świerszcz, I., Anusiewicz, I. *Mol. Phys. 109*, 1739, 2011.
160. Li, X., Kuznetsov, A. E., Zhang, H.-F., Boldyrev, A. I., Wang, L.-S. *Science 291*, 859, 2001.
161. Boldyrev, A. I., Wang, L.-S. *Chem. Rev. 105*, 3716, 2005.
162. King, R. B. *Chem. Rev. 101*, 1119, 2001.
163. Chen, Z., King, R. B. *Chem. Rev. 105*, 3613, 2005.
164. Schleyer, P. V. R., Maerker, C., Dransfeld, A., Jiao, H., Hommes, N. J. R. V. E. *J. Am. Chem. Soc. 118*, 6317, 1996.
165. Banjade, H. R., Deepika, Giri, S., Sinha, S., Fang, H., Jena, P. *J. Phys. Chem. A 125*, 5886, 2021.
166. Sevov, S. C., Goicoechea, J. M. *Organometallics. 25*, 5678, 2006.
167. Karttunen, A. J., Fassler, T. F., Linnolahti, M., Pakkanen, T. A. *ChemPhysChem 11*, 1944, 2010.
168. Xue, D., Wu, D., Chen, Z., Li, Y., Sun, W., Liu, J., Li, Z. *Inorg. Chem. 60*, 3196, 2021.
169. Giri, S., Reddy, G. N., Jena, P. *J. Phys. Chem. Lett.* 800, 2016.
170. Monkhorst, H. J. *Chem. Commun.* 1111, 1968.
171. Guo, J.-C., Wu, H.-X., Ren, G.-M., Miao, C.-Q., Li, Y.-X. *Comput. Theor. Chem. 1083*, 1, 2016.

172. Guo, J.-C., Feng, L.-Y., Dong, C., Zhai, H.-J. *Phys. Chem. Chem. Phys. 21*, 22048–22056, 2019.
173. Wang, Z.-X., Schleyer, P. V. R. *Science 292*, 2465, 2001.
174. Guo, J. C., Tian, W. J., Wang, Y. J., Zhao, X. F., Wu, Y. B., Zhai, H. J., Li, S. D. *J. Chem. Phys. 144*, 244303, 2016.
175. Guo, J.-C., Cheng, Y.-X., Wu, X.-F. *Comput. Theor. Chem. 1180*, 112824, 2020.
176. Sun, W.-M., Li, X.-H., Li, Y., Liu, J.-Y., Wu, D., Li, C.-Y., Ni, B.-L., Li, Z.-R. *J. Chem. Phys. 145*, 194303, 2016.
177. Paduani, C., Rappe, A. M. *J. Phys. Chem. A 120*, 6493, 2016.
178. Srivastava, A. K. *Chem. Phys. 524*, 118, 2019.
179. Srivastava, A. K., Misra, N., Tiwari, S. N. *SN Appl. Sci. 2*, 307, 2020.
180. Srivastava, A. K., Srivastava, H., Tiwari, A., Misra, N. *Chem. Phys. Lett. 790*, 139352, 2022.
181. Srivastava, H., Srivastava, A. K. *Struct. Chem. 34*, 617, 2022.
182. Nowiak, G., Skurski, P., Anusiewicz, I. *J. Mol. Model. 22*, 87, 2016.
183. Faron, D., Skurski, P., Anusiewicz, I. *J. Mol. Model. 25*, 24, 2019.
184. Srivastava, A. K., Tiwari, S. N., Misra, N. *Int. J. Quantum Chem. 118*, e25507, 2017.
185. Moran, D., Stahl, F., Eluvathingal D. Jemmis, III, H. F. S., Schleyer, P. V. R. *J. Phys. Chem. A 106*, 5144, 2002.
186. Wang, H., Jia, J.-F., Wu, H.-S. *Chin. J. Chem. 24*, 1509, 2006.
187. Tkachenko, N. V., Sun, Z. M., Boldyrev, A. I. *ChemPhysChem 20*, 2060, 2019.
188. Dye, J. L. *Acc. Chem. Res. 42*, 1564, 2009.
189. Zhong, R.-L., Xu, H.-L., Li, Z.-R., Su, Z.-M. *J. Phys. Chem. Lett. 6*, 612, 2015.
190. Sun, W.-M., Li, X.-H., Li, Y., Ni, B.-L., Chen, J.-H., Li, C.-Y., Wu, D., Li, Z.-R. *ChemPhysChem 17*, 3907, 2016.
191. Sun, W.-M., Li, X.-H., Wu, J., Lan, J.-M., Li, C.-Y., Wu, D., Li, Y., Li, Z.-R. *Inorg. Chem. 56*, 4594, 2017.
192. Sun, W.-M., Ni, B.-L., Wu, D., Lan, J.-M., Li, C.-Y., Li, Y., Li, Z.-R. *Organometallics 36*, 3352, 2017.
193. Sun, W.-M., Cheng, X., Ye, Y.-L., Li, X.-H., Ni, B.-L. *Organometallics 41*, 412, 2022.
194. Tkachenko, N. V., Rublev, P., Boldyrev, A. I., Lehn, J.-M. *Front. Chem. 10*, 880884, 2022.
195. Yan, J., Teo, B. K., Zheng, N. *Acc. Chem. Res. 51*, 3084, 2018.
196. Du, Y., Sheng, H., Astruc, D., Zhu, M. *Chem. Rev. 120*, 526, 2020.
197. Kang, X., Li, Y., Zhu, M., Jin, R. *Chem. Soc. Rev. 49*, 6443, 2020.
198. Chauhan, V., Sahoo, S., Khanna, S. N. *J. Am. Chem. Soc. 138*, 1916, 2016.
199. Chauhan, V., Reber, A. C., Khanna, S. N. *J. Am. Chem. Soc. 139*, 1871, 2017.
200. Chauhan, V., Reber, A. C., Khanna, S. N. *Nat. Commun. 9*, 2357, 2018.
201. Reber, A. C., Bista, D., Chauhan, V., Khanna, S. N. *J. Phys. Chem. C, 123*, 8983, 2019.
202. Wang, J., Zhao, Y., Li, J., Huang, H. C., Chen, J., Cheng, S. B. *Phys. Chem. Chem. Phys. 2*, 1062, 2019.
203. Li, J., Zhao, Y., Bu, Y.-F., Chen, J., Wei, Q., Cheng, S.-B. *Chem. Phys. Lett. 754*, 137709, 2020.
204. Li, J., Cui, M., Yang, H., Chen, J., Cheng, S. *Chin. Chem. Lett. 3*, 1117, 2022.
205. Pandey, S. K., Arunan, E., Das, R., Roy, A., Mishra, A. K. *Front. Chem. 10*, 1019166, 2022.
206. Sun, W.-M., Li, C.-Y., Kang, J., Wu, D., Li, Y., Ni, B.-L., Li, X.-H., Li, Z.-R. *J. Phys. Chem. C. 122*, 7867, 2018.
207. Reber, A. C., Khanna, S. N., Castleman, A. W. *J. Am. Chem. Soc. 129*, 10189, 2007.
208. Omidvar, A. *Inorg. Chem. 57*, 9335, 2018.
209. Dawes, S. B., Ward, D. L., Huang, R. H., Dye, J. L. *J. Am. Chem. Soc. 108*, 3534, 1986.

210. Dye, J. L., Ceraso, J. M., Tak, M. L., Barnett, B. L., Tehan, F. J. *J. Am. Chem. Soc. 96*, 608, 1974.
211. Tehan, F. J., Barnett, B. L., Dye, J. L. *J. Am. Chem. Soc. 96*, 7203, 1974.
212. Bai, Y., Zhou, Z.-J., Wang, J.-J., Li, Y., Wu, D., Chen, W., Li, Z.-R., Sun, C.-C. *J. Phys. Chem. A 117*, 2835, 2013.
213. Zhao, X., Yu, G., Huang, X., Chen, W., Niu, M. *J. Mol. Model. 19*, 5601, 2013.
214. Tu, C., Yu, G., Yang, G., Zhao, X., Chen, W., Li, S., Huang, X. *Phys. Chem. Chem. Phys. 16*, 1597, 2014.
215. Li, Z., Yu, G., Zhang, X., Huang, X., Chen, W. *Phys. E 94*, 204, 2017.
216. Lin, Z., Lu, T., Ding, X.-L. *J. Comput. Chem. 38*, 1574, 2017.
217. Ullah, F., Kosar, N., Arshad, M. N., Gilani, M. A., Ayub, K., Mahmood, T. *Opt. Laser Technol. 122*, 105855, 2020.
218. Shakerzadeh, E., Tahmasebi, E., Solimannejad, M., Chigo Anota, E. *Appl. Organomet. Chem. 33*, e4654, 2019.
219. Kosar, N., Mahmood, T., Ayub, K., Tabassum, S., Arshad, M., Gilani, M. A. *Opt. Laser Technol. 120*, 105753, 2019.
220. Ullah, F., Kosar, N., Ayub, K., Gilani, M. A., Mahmood, T. *New J. Chem. 43*, 5727, 2019.
221. Ullah, F., Kosar, N., Ayub, K., Mahmood, T. *Appl. Surf. Sci. 483*, 1118, 2019.
222. Noormohammadbeigi, M., Shamlouei, H. R. *J. Inorg. Organomet. P. 28*, 110, 2017.
223. Omidi, M., Sabzehzari, M., Shamlouei, H. R. *Chin. J. Phys. 65*, 567, 2020.
224. Li, Y., Ruan, M., Chen, H. *Mol. Phys. 119*, e1909161, 2021.
225. Raza Ayub, A., Aqil Shehzad, R., Alarfaji, S. S., Iqbal, J. *J. Nonlinear Opt. Phys. Mater. 29*, 2050004, 2021.
226. Bano, R., Arshad, M., Mahmood, T., Ayub, K., Sharif, A., Tabassum, S., Gilani, M. A. *Mat. Sci. Semicon. Proc. 143*, 106518, 2022.
227. Bano, R., Ayub, K., Mahmood, T., Arshad, M., Sharif, A., Tabassum, S., Gilani, M. *Dalton Trans. 51*, 8437, 2022.
228. Hou, N., Wu, Y.-Y., Wu, H.-S., He, H.-M. *Synthetic Met. 232*, 39, 2017.
229. Ullah, F., Ayub, K., Mahmood, T. *New J. Chem. 44*, 9822, 2020.
230. Kosar, N., Zari, L., Ayub, K., Gilani, M. A., Mahmood, T. *Surf. Interf. 31*, 102044, 2022.
231. Kosar, N., Zari, L., Ayub, K., Gilani, M. A., Arshad, M., Rauf, A., Ans, M., Mahmood, T. *Optik 171*, 170139, 2022.
232. Yi, X. G., Wang, Y. F., Zhang, H. R., Cai, J. H., Liu, X. X., Li, J., Wang, Z. J., Bai, F. Q., Li, Z. R. *Phys. Chem. Chem. Phys. 24*, 5690, 2022.
233. Ahsin, A., Ayub, K. *Mat. Sci. Semicon. Proc. 138*, 106254, 2022.
234. Mai, J., Gong, S., Li, N., Luo, Q., Li, Z. *Phys. Chem. Chem. Phys. 17*, 28754, 2015.
235. Li, B., Peng, D., Gu, F. L., Zhu, C. *ChemistrySelect 3*, 12782, 2018.
236. Li, X. *J. Mol. Liq. 277*, 641, 2019.
237. Hou, N., Du, F. Y., Feng, R., Wu, H. S., Li, Z. R. *Int. J. Quantum Chem. 121*, e26477, 2020.
238. Kosar, N., Shehzadi, K., Ayub, K., Mahmood, T. *Optik 218*, 165033, 2020.
239. Kosar, N., Shehzadi, K., Ayub, K., Mahmood, T. *J. Mol. Graph. Model. 97*, 107573, 2020.
240. Song, Y.-D., Wang, Q.-T. *Optik 220*, 164947, 2020.
241. Hussnain, M., Shehzad, R. A., Muhammad, S., Iqbal, J., Al-Sehemi, A. G., Alarfaji, S. S., Ayub, K., Yaseen, M. *J. Mol. Model. 28*, 46, 2022.
242. Chen, S., Xu, H. L., Sun, S. L., Zhao, L., Su, Z. M. *J. Mol. Model. 21*, 209, 2015.
243. Zhang, F.-Y., Xu, H.-L., Su, Z. *J. Phys. Chem. C 121*, 20419, 2017.
244. Song, Y.-D., Wang, L., Wang, Q.-T. *Optik 165*, 319, 2018.
245. Hou, N., Wei-Ming, S., Du, F.-Y., Wu, H.-S. *Optik 183*, 455, 2019.

246. Khan, A. U., Khera, R. A., Anjum, N., Shehzad, R. A., Iqbal, S., Ayub, K., Iqbal, J. *RSC Adv. 11*, 7779, 2021.
247. Song, Y.-D., Wang, L., Wang, Q.-T. *Optik 154*, 411, 2018.
248. Hanif, A., Kiran, R., Khera, R. A., Ayoub, A., Ayub, K., Iqbal, J. *J. Mol. Graph. Model. 107*, 107973, 2021.
249. Kiran, R., Khera, R. A., Khan, A. U., Ayoub, A., Iqbal, N., Ayub, K., Iqbal, J. *J. Mol. Struct. 1236*, 130348, 2021.
250. Omidvar, A. *Comput. Theor. Chem. 1198*, 113178, 2021.
251. Roy, R. S., Ghosh, S., Hatua, K., Nandi, P. K. *J. Mol. Model. 27*, 74, 2021.
252. Sajid, H., Mahmood, T. *Phys. E 134*, 114905, 2021.
253. Umar, A., Yaqoob, J., Khan, M. U., Hussain, R., Alhadhrami, A., Almalki, A. S. A., Janjua, M. R. S. A. *J. Phys. Chem. Solid. 169*, 110859, 2022.
254. Sajid, H., Ullah, F., Khan, S., Ayub, K., Arshad, M., Mahmood, T. *RSC Adv. 11*, 4118, 2021.
255. Song, Y.-D., Wang, Q.-T. *Optik 242*, 166830, 2021.
256. Bano, R., Arshad, M., Mahmood, T., Ayub, K., Sharif, A., Perveen, S., Tabassum, S., Yang, J., Gilani, M. A. *J. Phys. Chem. Solid. 160*, 110361, 2022.
257. Khaliq, F., Afzaal, A., Tabassum, S., Mahmood, T., Ayub, K., Khan, A. L., Yasin, M., Gilani, M. A. *Colloid. Surf. A 653*, 129985, 2022.
258. Ishfaq, T., Ahmad Khera, R., Zahid, S., Yaqoob, U., Aqil Shehzad, R., Ayub, K., Iqbal, J. *Comput. Theor. Chem. 1211*, 113654, 2022.
259. Shafiq, S., Shehzad, R. A., Yaseen, M., Ayub, K., Ayub, A. R., Iqbal, J., Mahmoud, K. H., El-Bahy, Z. M. *J. Mol. Struct. 1251*, 131934, 2022.
260. Srivastava, A. K., Misra, N. *Int. J. Quantum Chem. 117*, 208, 2017.
261. Yu, J., Xiao, W., Xin, J., Jin, R. *Comput. Theor. Chem. 1185*, 112853, 2020.
262. Srivastava, A. K., Kumar, A., Misra, N. *Chem. Phys. Lett. 682*, 20, 2017.
263. Srivastava, A. K., Misra, N. *J. Mol. Model. 21*, 305, 2015.
264. Ahsin, A., Ayub, K. *J. Nanostruct. Chem., 12*, 529, 2021.
265. Ahsin, A., Ayub, K. *J. Mol. Graph. Model. 109*, 108031, 2021.
266. Ahsin, A., Ayub, K. *Optik 227*, 166037, 2021.
267. Ahsin, A., Ayub, K. *J. Mol. Graph. Model. 106*, 107922, 2021.
268. Ahsin, A., Shah, A. B., Ayub, K. *RSC Adv. 12*, 365, 2022.
269. Merino, G., Chattaraj, P. M., Pan, S. *Phys. Chem. Chem. Phys. 14*, 10345, 2012.
270. Kanti Dash, M., Sinha, S., Sekhar Das, H., Chandra De, G., Giri, S., Roymahapatra, G. *Sustain. Energy Techn. 52*, 102235, 2022.
271. Jason J, I., Pal, Y., P, A., Bae, H., Lee, H., Ahuja, R., Hussain, T., Panigrahi, P. *Int. J. Hydrogen Ener. 47*, 33391, 2022.
272. Pan, S., Contreras, M., Romero, J., Reyes, A., Chattaraj, P. K., Merino, G. *Chem. Eur. J. 19*, 2322, 2013.
273. Pan, S., Jalife, S., Romero, J., Reyes, A., Merino, G., Chattaraj, P. K. *Comput. Theor. Chem. 1021*, 62, 2013.
274. Park, H., Meloni, G. *ChemPhysChem 19*, 2266, 2018.
275. Chakraborty, D., Chattaraj, P. K. *Phys. Chem. Chem. Phys. 18*, 18811, 2016.
276. Yu, D., Wu, D., Liu, J.-Y., Li, Y., Sun, W.-M. *Phys. Chem. Chem. Phys. 22*, 26536, 2020.
277. Kumar, R., Kumar, A., Srivastava, A. K., Misra, N. *Mol. Phys. 119*, e1841311, 2020.
278. Wang, F. F., Li, Z. R., Wu, D., Sun, X. Y., Chen, W., Li, Y., Sun, C. C. *ChemPhysChem 7*, 1136, 2006.
279. Ma, F., Li, R. Y., Li, Z. R., Chen, M. M., Xu, H. L., Li, Z. J., Wu, D., Li, Z. S. *J. Mol. Struct. Theochem. 913*, 80, 2009.

280. Srivastava, A. K., Misra, N. *RSC Adv. 4*, 41260, 2014.
281. Price, C., Winfough, M., Parka, H., Meloni, G. *Dalton Trans. 47*, 13204, 2018.
282. Huang, C., Fang, H., Whetten, R., Jena, P. *J. Phys. Chem. C 124*, 6435, 2020.
283. Srivastava, A. K., Misra, N. *RSC Adv. 5*, 74206, 2015.
284. Srivastava, A. K., Misra, N. *Chem. Phys. Lett. 648*, 152, 2016.
285. Srivastava, A. K., Misra, N. *Chem. Phys. Lett. 644*, 1, 2016.
286. Paduani, C., Rappe, A. M. *PCCP 19*, 20619, 2017.
287. Srivastava, A. K., Misra, N. *J. Mol. Model. 22*, 122, 2016.
288. Srivastava, A. K., Pandey, S. K., Misra, N. *Chem. Phys. Lett. 71*, 655, 2016.
289. Stasyuk, A. J., Sola, M. *Phys. Chem. Chem. Phys. 19*, 21276, 2017.
290. Wang, Y. F., Chen, W., Yu, G. T., Li, Z. R., Sun, C. C. *Int. J. Quantum Chem. 110*, 1953, 2010.
291. Srivastava, A. K., Misra, N. *Chem. Phys. Lett. 639*, 307, 2015.
292. Fang, H., Zhou, J., Jena, P. *Nanoscale 10*, 22963, 2018.
293. Srivastava, A. K., Misra, N. *J. Mol. Model. 21*, 147, 2015.
294. Parida, R., Das, S., Karas, L. J., Wu, J. I. C., Roymahapatra, G., Giri, S. *Inorg. Chem. Front. 6*, 3336, 2019.
295. Riedel, R., Seel, A. G., Malko, D., Miller, D. P., Sperling, B. T., Choi, H., Headen, T. F., Zurek, E., Porch, A., Kucernak, A., Pyper, N. C., Edwards, P. P., Barrett, A. G. M. *J. Am. Chem. Soc. 143*, 3934, 2021.
296. Shakerzadeh, E., Tahmasebi, E., Solimannejad, M., Chigo Anota, E. *Appl. Organomet. Chem. 33*, e4654, 2019.
297. Srivastava, A. K. *Mol. Phys. 120*, 2021.
298. Sheikhi, M., Kaviani, S., Azarakhshi, F., Shahab, S. *Comput. Theor. Chem. 1212*, 113722, 2022.
299. Jiang, S., Shen, M., Jameh-Bozorghi, S. *Mol. Phys. 118*, e1734679, 2020.
300. George, G., Stasyuk, A. J., Sola, M. *Dalton Trans. 50*, 16214, 2021.
301. Nambiar, S. R., Jana, G., Chattaraj, P. K. *Chem. Phys. Lett. 762*, 138131, 2021.
302. Jha, R., Giri, S., Chattaraj, P. K. *Comput. Theor. Chem.* 113491, 2021.

7 Mass Spectrometry and "Superalkali" Clusters

Suzana Velickovic and Filip Veljkovic

7.1 HYPERVALENT AND/OR "SUPERALKALI" CLUSTERS

Originally, clusters of the type M_nX (M-alkali metal, X-nonmetal, $n > 2$) were called hypervalent molecules because they have more than 8 electrons in the valence shell and resembled already known compounds such as ICl_3, BrF_3 and IF_7. The term hypervalent as well as the first definition were proposed by Musher in 1969. According to Musher, hypervalent molecules are formed from central atoms of the groups V–VIII with other than the lowest valence, *i.e.*, 3, 2, 1 and 0 [1]. As can be seen, this definition does not include compounds that have a metal atom in their composition, so it can be considered that Wu, Kudo and Ihle have opened a new chapter in the study of hypervalent compounds with the discovery of the Li_3O [2]. Schleyer's group confirmed the existence of Li_3O in their theoretical work and predicted a large number of new hypervalent compounds with metals [3–6].

It turned out that the term hypervalent is too broad, so the term hyperalkaline molecules was introduced when an alkali metal is present in the composition of the compound, and depending on which alkali metal is present in the molecule in question, further distinctions can be made: hyperlithium (CLi_5, CLi_6 [7], Li_4N [8], Li_3O [2], Li_4O [9], Li_2F [10], Li_3S, Li_4S [11], Li_4P [12], Li_2CN [13] and $Li_n(OH)$ [14]), hypersodium (Na_3O, Na_4O [15,16], Na_2CN [17]) and hyperpotassium (K_3O, K_4O [15], K_2CN [18]). Other metals can also form this type of compounds, for example there are hypermagnesium (Mg_2O, Mg_3O and Mg_4O) [19,20] and hyperaluminum (Al_3O, Al_4O) [21] molecules. In the literature, the general term hypermetals (hypermetallation) can be found for the mentioned clusters. A group of hypersilicon molecules such as Si_2O, Si_3O [22] has also been proposed.

The first experimental results indicated that some hypervalent molecules are more stable than octet or stoichiometric molecules (in the gas phase), which further stimulated interest in this type of compound. Theoretical calculations revealed that the above molecules do not have a well-defined spatial arrangement, *i.e.*, they have several energetically close isomers, which is why they were initially referred to as "floppy" molecules or hypervalent clusters or, more generally, nonstoichiometric clusters. The hypervalent electronic structure means that the "excess" of electrons is delocalized between the lithium atoms, making the lithium "cage" or "network" positively charged (Li_n^+) and leaving the nonmetal negatively charged (F^-). The electrostatic attraction between the two ions mentioned

DOI: 10.1201/9781003384205-7

above makes these compounds stable. Additional theoretical studies of Li_nX and Li_nF_{n-1} type lithium halides have revealed a large number of isomers, only some of which exhibit the hypervalent structure described. However, the chemical bonding presented in this simple way proved to be a very useful explanation for the structure of the "superalkali" clusters [6]. It should be noted that the term and concept of hypervalency have been criticised, even by Schleyer and Gillespie. Gillespie wrote in 2002, "Since there is no fundamental difference between the bonds in hypervalent and nonhypervalent (Lewis octet) molecules, there is no reason to continue using the term hypervalent" [23]. Therefore, the term hypervalent is rarely used today for this type of cluster.

In parallel with Kudo and Schleyer, Honea and co-workers proposed a new approach to heterogeneous alkali metal clusters of the type M_nX_{n-1}, $n \geq 2$. Honea [24,25] described M_nX_{n-1} clusters as an ionic core $M_nX_{n-1}^+$ to which an electron is bound. They have made the classification of the given clusters on the basis of their values of ionization energies, intensity of peaks in the mass spectrum, and according to the theoretical assumptions calculated on the subject of the structure of the given clusters. On this basis, three groups of clusters are distinguished: cubic-ionic clusters, clusters with an F-centre and nanocubic clusters. Cubic clusters have a preserved cubic crystal lattice, where the excess electrons are weakly bound to the surface of the crystal. The abundance of these clusters is low, and their ionization energy is very low (about 1.88 eV). F-centre clusters contain a cubic lattice in which the excess electrons are localized in the anion cavity. They belong to the stable clusters, *i.e.*, they occur in large numbers. The ionization energy is between 2.8eV and 3.6eV. In nanocubic crystals, the excess electron is localized on a metal ion and neutralizes it, which allows easy dissociation of the atom in question and low abundance of this type of clusters. The ionization energy is higher than in the other two cluster types (> 5eV).

A slightly modified classification of the M_nX_{n-1} cluster was made by authors Landman, Scharf and Jortner (the abbreviation LSJ is used in the literature) [26–28]. According to the LSJ theory, these are ion clusters containing nM^+ and mX- (n > m) ions, with the excess electrons localized in different ways. There are three ways to localize excess electrons in nonstoichiometric clusters: the excess electrons can be localized in the cavity of the anion, as in the case of defects of the F-centre, which is found in most cases in the crystals of alkali halides, but is not expected in small clusters (n ~ 10); the cluster may look as if it has no preferred location where the excess electrons are located; the excess electrons may be bound to a metal ion (*e.g.*, Na^+), which is neutralized in this way, and are the result of the easy dissociation of such a metal atom.

The authors used the term nonstoichiometric for M_nX_{n-1} cluster. They considered that nonstoichiometric clusters are found at the transition between two extreme states of condensed matter, the metals and the insulators. Figure 7.1 shows the schematic of the energy levels of stoichiometric $(MX)_n$ clusters and nonstoichiometric clusters with excess electrons (M_nX_{n-1}). For example, in the stoichiometric $(NaF)_n$ cluster, the minimum energy for the transition of an electron from

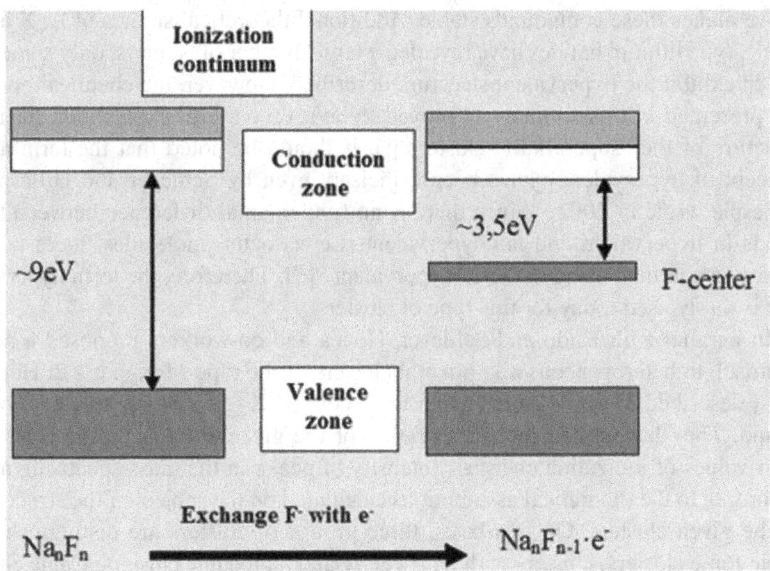

FIGURE 7.1 Schematic of energy levels of stoichiometric $(MX)_n$ clusters and nonstoichiometric clusters (M_nX_{n-1}), M-alkaline metal and X-halogen.

the 2p orbital of the F ion, located at the top of the valence zone, to the empty 3s orbital of Na^+ from the conduction zone is about 9eV, when a fluorine anion is replaced by an unpaired electron, the 2p orbital shifts toward the conduction band, reducing the energy gap to about 3.5 V [25]. This is the reason why non-stoichiometric clusters are conductors and stoichiometric clusters are insulators.

Hebant and Picard assumed that small lithium chloride clusters form as a layer between liquid lithium and the eutectic LiCl–KCl mixture during industrial electrorefining [29–31]. The reactivity of metals and halogen elements is a key issue in the above process, and the reactivity of metals is explained by the presence of M_2X type molecules, where M–Li, Na, K, Rb and X–F, Cl, Br, J. The authors used the term subhalides for the above clusters [8,32,33].

Simultaneously with the experimental discovery of the first hypervalent clusters, Boldyrev and Gutsev proved theoretically, using M_nX, M-Na, Li, X–F, Cl, O, S, N, P ($n \geq 2$) as examples, that these compounds have ionization energies lower than the ionization energy of an alkali atom. They named this group of compounds "superalkali." The same group of researchers discovered that MX_n type clusters have greater electron affinity than the halogen elements, and this group of compounds was named "superhalogen." The aforementioned authors not only discovered the key properties of these clusters, but also continue to actively work on the discovery of new molecules that exhibit "superalkali/superhalogen" properties.

The study of "superalkali" and "superhalogen" clusters has attracted the attention of researchers for four decades. It has been experimentally confirmed that "superalkali" clusters indeed have ionization energies lower than the ionization energies of alkali metals [34–42]. There are also numerous theoretical publications on the applications of "superalkali" clusters, some of which are listed below. One of the most important discoveries in the field of this type of clusters, made by Reber, Castelman and Khana, is the fact that "superalkali" and "superhalogen" clusters can mimic the chemical properties of atoms of the periodic table. This means that these clusters can combine with each other or with other atoms without losing their structural and electronic identity. It is believed that this type of cluster represents the third dimension of the periodic table. Together, "superalkali" or "superhalogen" can form new types of clusters called superatoms. These superatoms can serve as potential building blocks for new materials with unique properties [43,44]. Paduani et al. have calculated that by combining "superalkali" Li_3O, Li_4O and Na_8 clusters with Gd or V, compounds with excellent magnetic properties can be formed [45,46]. Li et al. have theoretically shown that a series of "superalkali (super)halogen" compounds, BLi_6-X (X = F, LiF_2, BeF_3, BF_4), can be used as building blocks for the synthesis of novel materials with unusual nonlinear optical properties [47].

"Superalkali" clusters have very low ionization energies and can therefore be excellent reducing reagents for nitrogen oxides (NO_x, x = 1 and 2), nitrogen molecules (N_2) and carbon dioxide (CO_2) [48–50]. This process allows the conversion of these molecules, which are harmful to the environment, into useful products [48]. For example, Srivastava found that the "superalkali" Li_2F can easily transfer electrons from NO_n to NO_{n-} [49]. Roy et al. predicted that the combination of "superalkali" Li_2, Li_4, Li_6 and Li_8 clusters and N_2 would lead to the activation of diatomic nitrogen [51]. Park and Meloni calculated that in the $(Li_3F_2)_6$-N_2 complex, the addition of electrons from "superalkali" Li_3F_2 clusters into the empty N_2 MOs leads to complete bond cleavage in diatomic N_2. They have also shown that the "superalkali" Li_3F_2 clusters can easily reduce CO_2 to CO_2^- (although CO_2 has no positive electron affinity) [48]. Wang et al. have found that hydrogen storage capacities for solid sorbents (such as C_{60}) can be significantly improved by using "superalkali" Li_2F cluster. It is believed that the positively charged metal "network" of "superalkali" clusters can be connected to hydrogen through electrostatic interactions, and "superalkali" clusters have the potential to become an effective hydrogen storage material [52].

All the above theoretical discoveries are a great challenge and guide for future experimental research.

7.2 PRODUCTION OF CLUSTERS

In general, mass spectrometry has played a crucial role in the field of clusters. At the beginning, the sources for clusters were special chambers, such as a seeded supersonic nozzle source, a gas aggregation source, a laser vaporization source, a pulsed arc cluster ion source, *etc.* [53]. Clusters of different size and composition were formed in the aforementioned chambers under certain conditions. The first task in the study of the clusters obtained is their separation according

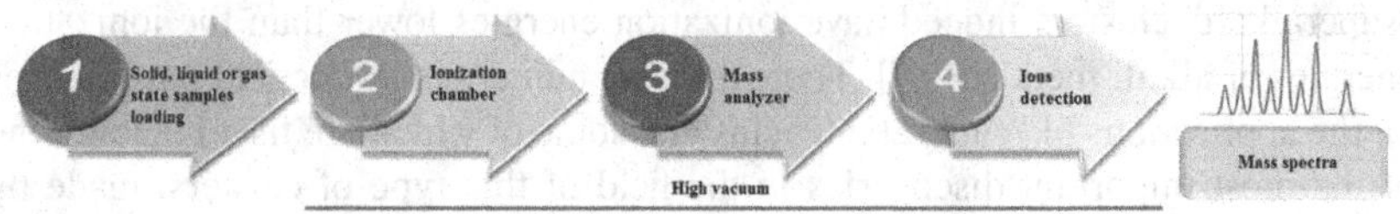

FIGURE 7.2 Schematic representation of the main components of the mass spectrometry instrument.

to size. For this purpose, mass analysers included in mass spectrometers were used. Mass analysers separate only ions. For this reason, the obtained clusters were ionized by electron impact, and then the separation was performed on the basis of the ratio of mass and charge (*m/z*). The preparation of clusters, their ionization and separation by size (or composition) are interrelated processes, so the standard experimental setup for the study of clusters should include the above parts, which are practically the main components of mass spectrometers (Figure 7.2).

Commercial mass spectrometers, often modified, are instruments for the generation and characterization of "superalkali" clusters. To date, "superalkali" clusters occur only in the gas phase. In commercial mass spectrometers, generation and ionization often occur together in the ionization chambers. Numerous ionization methods are used in cluster research: electron impact (EI), laser desorption ionization (LDI), electrospray ionization (ESI), fast atom/ion bombardment (FAB/FIB), surface ionization (SI) or thermal ionization (TI), field desorption/ionization (FDI) and spark source mass spectrometry (SS). Electron impact is the first mass spectrometric method that has been used in the field of clusters. However, electron impact is not used as a source of cluster formation, but only for ionization of neutral clusters formed by other means. Other mass spectrometric methods mentioned above can provide information about the conditions of cluster formation and the stability of the distinguished species.

This part lists studies of some groups of authors who found in their work that the clusters of the type $((MX)_nM)^+/M_nX^+$ form easily in mass spectrometers. Moreover, their results have shown that these clusters are very abundant in the mass spectra compared to MX^+ and M^+ ions, indicating their surprising stability. However, it should be mentioned that these authors focused on the study of the vaporization of alkali salt mixtures in their mass spectrometers and on the intensity ratio of the obtained clusters, while the chemical bonds in the obtained clusters or the "superalkali" nature of these clusters were not studied. For example, the M_2X type clusters were detected using a spark source mass spectrometer (SSMS). In this experiment, samples were prepared as a suspension/slurry by mixing equimolar proportions of the salt pairs (LiX/NaX; MF/MCl, where X = CI, F and M = Li, Na) with high purity carbon in ethanol and water and immediately frozen in a dry ice/acetone bath. These samples were pressed into rod-shaped electrodes (0.062 inch diameter) mounted in split indium cups and sparked. The results showed that the intensity ratios were as follows: $Na_2F^+/Na_2Cl^+ = 2.39$; $Li_2F^+/Li_2Cl^+ = 25.0$; $Li_2Cl^+/Na_2Cl^+ = 2.79$; $Li_2F^+/Na_2F^+ = 1.50$ [[54]]. Equimolar binary mixtures of NaCl/NaF, LiCl/LiF, LiCl/NaCl and LiF/NaF in aqueous solution applied to nickel and cobalt emitters were analysed by field desorption mass spectrometry (FDMS). FDMS is a soft ionization technique in which the sample from solution is deposited on the surface of a field anode (nickel, cobalt or tungsten wire). Ionization/desorption of the molecules is achieved by applying a high voltage of about 10 kV between the wire and a remote counter electrode while the emitter wire is heated by a resistor. In this case, the following pair wise orders were observed: $Li_2Cl^+ > Na_2Cl^+$, $Li_2Cl^+ > Li_2F^+$, $Na_2Cl^+ > Na_2F^+$ [[55,56]]. It was also found that the M_2X type clusters can be obtained from binary mixtures of alkali halides MX (M = Li, Na; X = F, Cl) using secondary ion mass spectrometry with a quadrupole mass analyser. Secondary ion mass spectrometry (SIMS) is a technique in which the sample surface is sprayed with a focused primary ion beam (*e.g.*, $^{40}Ar^+$, Xe^+, SF_5^+, oxygen ($^{16}O^-$, $^{16}O^{2-}$) and the emitted secondary ions are collected and analysed. Four equimolar binary mixtures were prepared from solids: NaCl/NaF, LiCl/LiF, LiCl/NaCl, LiF/NaF; these bulk mixtures were dried overnight in a vacuum oven at 353 K; the fine powders of the samples were deposited on indium foil for analysis. The order of ionic intensity of the clusters was $Li_2F^+ > Li_2Cl^+ > Na_2F^+ > Na_2Cl^+$. However, this order may be affected by instrumental parameters of the mass spectrometer and by variations in composition due to sample preparation [[57]]. It was found that $Cs^+(CsI)_n$ cluster ions of extremely high mass (*m/z* > 18,000) could be detected with a high performance secondary ion mass spectrometer equipped with a double focusing mass analyser. These clusters were produced by xenon ion bombardment (at 4.0–4.7 keV) onto a CsI crystal. The cubically shaped clusters were found to have higher stability than others, *i.e.*, the magic numbers of the $Cs^+(CsI)_n$ clusters are n = 4, 6, 9, 13, 16, 22, 25, 37, 44, 46 and 62 [[58]]. The fast atom bombardment mass spectra of solid alkali chlorides (Na, K, Cs) and fluorides (Na, K, Rb, Cs) contain the peaks of the clusters $[M(MX)_n]^+$ and the small peak of M^+ and $[M_2]^+$ (M = metal). Fast atom bombardment (FAB) is a soft ionization technique in which a beam of high-energy atoms strikes the surface of a sample and generates ions. These results show that for alkali fluorides, the ion intensity of the $[M(MX)_n]^+$ clusters (X-F and Cl) decreases with increasing "n" (n was 2–7) [[59,60]]. In this work, the authors propose that the mechanism of cluster formation involves three steps: (i) after each projectile impact, an expanding cloud is generated whose main component is the neutral XY species; (ii) although the lifetime of the cloud is very short (tens of ps), a sufficient number of collisions can occur in this time interval to generate the $(XY)_n$ clusters, where n can be up to ten; (iii) competition for the formation of large clusters favours the most stable cluster ions [[61]].

Electro-spray ionization and MALDI are two of the newest ionization methods in mass spectrometry. In 2002, Fenn, the inventor of ESI-MS, Tanaka (for the development of MALDI mass spectrometry) and Wuthrich (for work in NMR spectroscopy) shared the fourth Nobel Prize in Mass Spectrometry. In electro-spray ionization mass spectrometry (ESI), a sample solution is passed through a capillary tube. A strong electric field at the tip of the tube nebulizes the sample solution into electrically charged droplets. Ions are then released from the droplets and transported from the ionization source (atmospheric pressure region) to a mass analyser (vacuum region) of the mass spectrometer. ESI is a soft ionization in the sense that no fragmentation occurs during ionization, even the very weak noncovalent interactions are preserved in the gas phase of the analyte. In ESI MS the ions are very often multiply charged, so the *m/z* values of the resulting ions become lower and fall within the mass ranges of all current mass analysers [62,63]. The alkali halide clusters of the type ($M(MX)_n$ and $X(MX)_n$) have been studied extensively by ESI MS for several reasons: identification of the ESI mechanism; investigation of the hydration energies and entropies for MXM^+ ($NaFNa^+$, $NaClNa^+$, $NaBrNa^+$, $NaINa^+$, $NaNO_2Na^+$, $NaNO_3Na^+$, KFK^+, $KBrK^+$, KIK^+, $RbIRb^+$, $CsICs^+$, $NH_4BrNH_4^+$ and $NH_4INH_4^+$); investigation of the nucleation of alkali metal chlorides in methanol at the cluster level; study of the influence of solvation energy on the formation of alkali metal clusters, study of the gas-phase stability of these clusters (observation of magic numbers); in addition, the protonated water clusters and the salt clusters of NaF, KF, NaI, KI, RbI, CsI, $CsNO_3$ and Cs_2CO_3 can be excellent calibration standards for positive and negative ions [64–67].

The ESI MS is the method of choice to study the conditions for the formation of multiply charged clusters and their stability. For example, in the case of LiCl with a concentration of 0.05 M (50% water and 50% methanol solution), 246 different cluster ions are obtained, of which 167 are in the positive mode and 79 are in the negative mode. Singly charged $Li(LiCl)_n^+$ clusters in positive mode were detected for $n_p = 1–51$, doubly charged $Li_2(LiCl)_n^{2+}$ clusters for np = 31–101, and triply charged $Li_3(LiCl)_n^{3+}$ clusters for $n_p = 109–152$. In the negative mode, singly charged clusters $(LiCl)_nCl$- were detected for $n_n = 1–36$, doubly charged clusters $(LiCl)_nCl_2^{2-}$ for $n_n = 29–71$, while triply charged clusters $(LiCl)_nCl_3^{3-}$ were not detected. Studies have shown that multiply charged clusters are also obtained by salts such as NaCl, KCl, RbCl and CsCl. For example, the ESI mass spectra of NaCl clusters contained $(MCl)_nM^{x+}$ for $x = 1$, $n_p = 1–67$, $n_n = 1–55$, for $x = 2$, $n_p = 26–128$, $n_n = 25–104$ and for $x = 3$, $n_n = 79–102$ [64,66,68].

The quadrupole mass filter, used in commercial and academic laboratories for more than half a century, was patented by Wolfgang Paul (Nobel Prize winner in Physics, 1989). ESI MS is equipped with three quadrupole analysers (triple quadrupole), where the middle quadrupole has a role and traps for ions, *i.e.*, a space where ion fragmentation takes place. Quadrupole analysers consist of four electrodes (*i.e.*, rods with a radius of 5–15 mm and a length of 50 to 500 mm) arranged parallel to each other and equidistant from a central axis extending in the *z*-direction. One pair of rods has the potential $U+V\cos(\omega t)$, the other pair has the opposite potential $-U+V\cos(\omega t)$, where U is the voltage DC, V is the maximum amplitude of the high-frequency voltage RF, ω is the angular frequency, and t is the time (Figure 7.3).

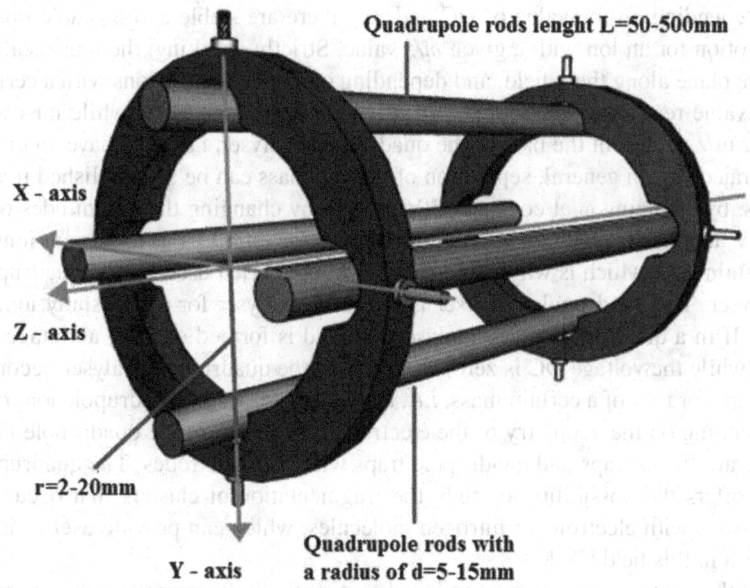

FIGURE 7.3 Schematic of a quadrupole mass analyser [69].

The quadrupole analyser selects ions according to the *m/z* ratio by combining a static DC voltage (U) and a time-dependent, *i.e.*, RF voltage (Vcos(ωt)). The equations of motion of ions are:

$$\frac{d^2x}{d^2t} = -\left(\frac{e}{m}\right)\frac{[U + V\cos(\omega t)]}{r_o^2}x \tag{7.1}$$

$$\frac{d^2y}{d^2t} = -\left(\frac{e}{m}\right)\frac{[U + V\cos(\omega t)]}{r_o^2}y \tag{7.2}$$

$$\frac{d^2z}{d^2t} = 0 \tag{7.3}$$

$$\frac{a_u}{q_u} = \frac{2U}{V} \tag{7.4}$$

where the coefficients a_u and q_u are given by the expressions:

$$a_u = \frac{8eU}{mr_0^2\omega} \tag{7.5}$$

$$q_u = \frac{4eV}{mr_0^2\omega} \tag{7.6}$$

Depending on the value of DC and AC, there are stable and unstable ranges of motion for an ion with a given *m/z* value. Strictly speaking, the ions oscillate in the plane along the x-field, and depending on the U/V ratio, ions with a certain m/z value reach the detector, *i.e.*, they have a stable trajectory, while ions with other m/z values hit the bars of the quadrupole analyser, *i.e.*, they have an unstable trajectory. In general, separation of ions by mass can be accomplished in two ways: by changing ω at constant U/V ratio or by changing the magnitudes of U and V at constant ω. It can be said that the voltage ratio U/V is a filter for ions of a certain size, which is why the term mass filter is often used for the quadrupole analyser. The quadrupole analyser is the ideal analyser for electrospray ionization. If in a quadrupole mass analyser the field is formed only by a voltage RF AC (while the voltage DC is zero, U = 0), then the quadrupole analyser becomes a "trap" for ions of a certain mass, *i.e.*, a quadrupole "trap" (quadrupole ion trap). Depending on the geometry of the electrodes used to form the quadrupole field, there are linear traps and quadrupole traps with ring electrodes. The quadrupole trap offers the possibility to study the fragmentation of clusters that occurs in collisions with electrons or nitrogen molecules, which can provide useful information in this field [70,71].

In the 1960s, a laser was introduced as a method of generating ions for mass spectrometry. However, the amount of ions produced was not satisfactory until the laser microprobe mass analyser (LAMMA) was developed, in which the laser beam is focused on a point on the sample with a diameter of less than 1 μm. Initially, a ruby UV laser at 347 nm was used, and later a Nd: YAG laser at 266 nm (far UV). Clusters M_2X were formed directly from the solid state of the alkali halide using the LAMMA method [72,73]. Laser mass spectrometry reached its peak with the development of the MALDI method, which is now recognized as an excellent analytical technique for large biologically important macromolecules, organic polymers, and polar organic molecules. The MALDI MS is not usually used to obtain clusters. However, laser methods (laser desorption, laser vaporization and laser ablation) have proven to be very effective methods in the field of clusters. It is important to point out that the laser photoionization method provides very accurate values for the ionization energy of clusters, whereas the neutrals of clusters are usually obtained by laser vaporization/desorption/ablation of the corresponding target.

This book contains a chapter on photoionization, so the details of the method mentioned are not presented here. However, some examples of obtaining alkyl halide clusters by laser vaporization are given. A large number of clusters of different sizes are formed by laser vaporization mass spectrometry (LVMS) [74–76]. For example, the experimental setup for the preparation of Na_nF_{n-p} clusters by laser vaporization consists of two different chambers. In the first chamber, these clusters are produced by laser vaporization (Nd:YAG laser at 532 nm, 10-ns pulse) of a sodium rod into which helium is injected with a small amount of SF_6 (the SF_6 content ranged from 0.4 to 0.8%, depending on the stoichiometry to be produced). The clusters are ionized in the second chamber with one or two dye lasers pumped by an eximer laser (XeCl, 15 ns pulse). In this experiment, Na_nF_{n-1} (n up to 42), Na_nF_{n-2} (n up to 25), Na_nF_{n-3} (n up to 40) and Na_nF_{n-5} (n up to 30) clusters were detected whose ionization potentials were below 4eV. The experimental IPs agreed well with those of the theoretical calculations of Honea and Landaman [77–81]. The LVMS of the surface of a powdered sample or a single crystal of NaCl, NaI, CsCl, CsI salt in a freely expanding supersonic jet of helium gas can produce clusters $M(MX)^+$, M = Na, Cs, X = Cl, I. Two lasers were used in this experiment, an excimer laser (ArF at 193 nm, with 80–200 mJ in each 17-ns pulse) and a neodymium-doped yttrium aluminum garnet (Nd:YAG) laser (532 nm, with 50–100 mJ, and at 266 nm 5–10 mJ). The following cluster sizes were detected: Na_nCl cluster (n = 1–200), Na_nI cluster (n = 1–125), Cs_nCl cluster (n = 1–125) and Cs_nI cluster (n = 1–100) [82].

From a fundamental point of view, it is interesting to observe how, with the increase of the number of metal atoms in heterogeneous M_nX clusters (X = electronegative nonmetal), the transition from purely ionic to metallic clusters occurs. For this purpose, the change in ionization energy of Li_nO and Li_nC (n = 3–70) was studied. Laser vaporization combined with He gas condensation was used to form Li_nO and Li_nC clusters. In small ionic lithium clusters (n < 10), the value of ionization energy oscillates with the increase of the number of lithium atoms, *i.e.*, it does not show a well-defined trend, indicating that the nonmetal in these clusters has a greater influence on the nature of chemical bonding. In large heterogeneous lithium clusters, the "excess" metal leads to an increase in the metal fraction of the cluster where delocalized electrons move in a small volume, and the influence of the nonmetals is negligible, leading to a decrease in the ionization energy approaching the value of the lithium-metal work rate (2.38 eV).

An interesting phenomenon is the discontinuity of the value of the ionization energy with the increase of the number of lithium atoms. The ionization energy for Li_nO clusters exhibits a discontinuity at n = 20 and 40. Similarly, the ionization energy for Li_nO decreases at n = 10, 22 and 42 and for Li_nC clusters at n = 24 and 44 [83–85].

The polylithium clusters Li_3F_2 and Li_4F_3 were obtained by laser ablation mass spectrometry. In the experiment, a mixture of LiF and Li_3N salts is pressed into the shape of a disk at a ratio of 1:1, resulting in a sample with a diameter of 10 mm and a thickness of 20 mm. A Nd:YAG laser with a wavelength of 532 nm and a pulse energy of about 5 mJ was used for sample ablation. Clustering of the vaporized species was performed in a special channel in a pulsed helium stream under a pressure of 3.5 at, and the obtained clusters were expanded in a vacuum chamber. The neutral clusters were ionized using a pulsed Nd:YAG laser with a wavelength of 266 nm and a pulse energy of about 38 mJ. The signal in the mass spectrum was obtained by accumulating 100 shots of the ablation laser at a repetition rate of 10 Hz. Experimental parameters such as the frequency of helium injection, the ablation laser pulse, and the ionization laser pulse were chosen to achieve the maximum ion intensity of the Li_2F^+ cluster. Interestingly, no Li_nN-type clusters were detected during laser ablation of LiF/Li_3N. Based on the results of vaporization experiments of Li_3N salt from Knudsen cell and theoretical calculations, it was assumed that Li_3N molecules have high affinity to react with

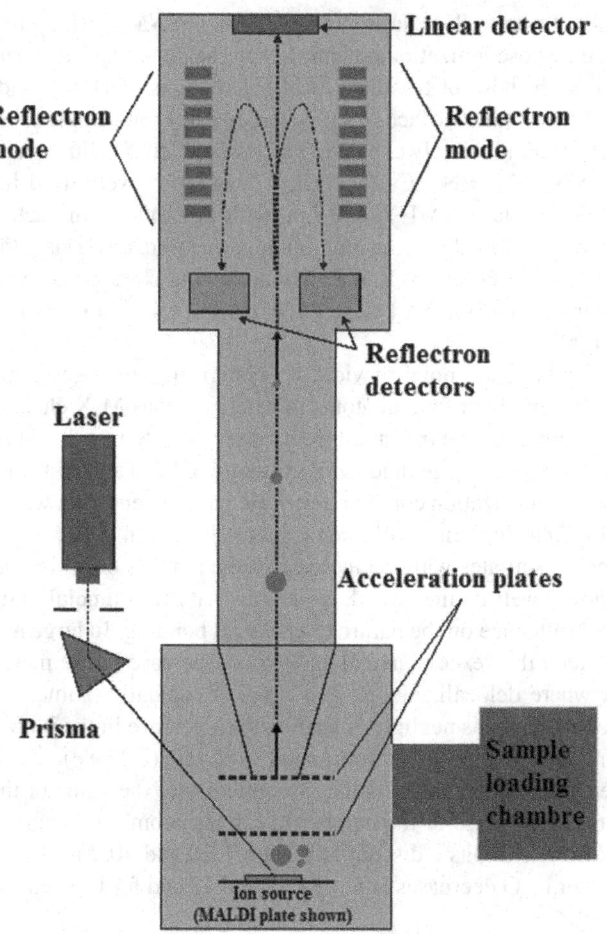

FIGURE 7.4 Schematic of a time-of-flight mass analyser.

each other and the resulting compound decomposes into nitrogen and lithium very quickly in an exothermic reaction, which prevents the formation of lithium clusters with nitrogen [10].

Time-of-flight (TOF) mass analysers (Figure 7.4) are used in the LDI mass spectrometer.

The basic idea of separation is quite simple and consists of a few steps: Ions produced in the ion source have different masses, and ions of different masses have different velocities of motion, *i.e.*, they travel a given distance L in different times. Lighter ions, *i.e.*, ions with lower mass, reach the detector faster than heavier ions, *i.e.*, ions with higher mass. This assumption is described by basic formulas.

$$v = \frac{L}{t} \tag{7.7}$$

$$t = \frac{L}{v} \tag{7.8}$$

$$E_k = \frac{mv^2}{2} \tag{7.9}$$

$$v = \sqrt{\frac{2E_k}{m}} \tag{7.10}$$

where v = velocity, m = mass of ions, t = time, E_k = kinetic energy.

In this case, the resulting mass spectrum would have very broad peaks and the accuracy of the *m/z* determination would be low, *i.e.*, unusable. The problem is that ions of the same mass acquire different amounts of energy during ionization, *i.e.*, ions of the same mass have different kinetic energies. Practically, this means that the flight time of the ion to the detector depends not only on the mass, but also on the kinetic energy, *i.e.*, on a certain initial velocity, which is different for the same mass. The solution to this problem is to accelerate the ions in a field with a high voltage of about 20 kV, which gives all ions the same kinetic energy. The kinetic energy of the ion can be expressed as follows:

$$E_k = \frac{mv^2}{2} = zU = E_e \tag{7.11}$$

so, the velocity of the ion is:

$$v = \left(\frac{2zU}{m}\right)^{1/2} \tag{7.12}$$

the time taken by the ions to travel the distance L is

$$t^2 = \frac{m}{z}\frac{L^2}{2U} \tag{7.13}$$

The most important problem of the first TOF analysers was their low mass resolution. The resolution of a mass spectrometer is:

$$R = \frac{m}{\Delta m} = \frac{t}{\Delta t} = \frac{L}{2\Delta x} \tag{7.14}$$

where m and t are the mass and time of flight of the ion, and Δm and Δt are the peak widths measured at the 50% level of the mass and time scales, respectively. L is the flight distance and Δx is the thickness of an ion packet approaching the detector. It can be seen that the resolution increases with increasing length of the analyser tube. However, it should be noted that a tube that is too long will decrease the performance of the analyser due to ion losses when colliding with other molecules or due to angular dispersion. On the other hand, the equation states that decreasing the voltage will increase the resolution, but low voltage values will decrease the sensitivity of the instrument. The optimum values for the length of the analyser tube are between 1 m and 2 m, and the value of the accelerating voltage, *i.e.*, the voltage across the plate, is usually in the range of 15 kV to 20 kV. This mode of operation of the time-of-flight analyser is called continuous extraction, in which the problem of resolution is not adequately solved. Delayed extraction mode or pulsed ion extraction (Delayed Extraction Mode) is one way to solve this problem. In delayed ion extraction, the ions are first allowed to expand into a field-free region in the source, and after a certain delay (hundreds of nanoseconds to a few microseconds), a voltage pulse is applied to extract the ions outside the source. Nowadays, all LDI MS operate with a delayed extraction mode, and as an additional tool to achieve better resolution and greater accuracy, there is also a reflectron mode. In reflectron mode, a set of electrodes (usually ring or grid shaped) is inserted into the analyser tube. The electrodes create an electrostatic, decelerating reflector field that causes faster ions (of mass *m*) to penetrate deeper into the field, travel a longer distance, and allow slower ions (of the same mass *m*) to catch up. The delayed extraction mode and the reflectron mode have made TOF analysers a powerful tool for analytical chemistry of organic and inorganic molecules [[86]].

7.3 SURFACE IONIZATION MASS SPECTROMETRY

In surface ionization, also known as thermal ionization, positive and negative ions are generated by heating the sample on a metal surface. The first experimental measurements of ion emission interacting with a heated metal filament were made in 1903 by Richardson. Langmuir, Kingdon, and Ives studied the phenomenon of desorption of cesium atoms in the form of positive ions from the surface of a heated tungsten wire; they called this phenomenon "surface ionization." In the same year, Saha began research on the relationship between the number of adsorbed and ionized atoms of a heated metal surface, which would later form the backbone of the application of surface ionization to the determination of ionization energy, electron affinity, work function, dissociation energy, activation energy, and so on. Taylor is the first to study the behaviour of a molecular beam on a heated metal surface. About 10 years after the discovery of positive surface ionization, Morgulis pointed out the possibility of formation of negative ions in the process of surface ionization. Sutton and Mayer confirmed experimentally that atoms with a high affinity for the electron can "capture" the electron leaving the metal filament by heating and form a negative ion. Zandgber and Ionov performed detailed and systematic experiments on the vaporization and ionization of alkali atoms from alkali halide salts by the method of surface ionization [[87–95]]

The surface ionization method can be used to analyse samples in all three states of matter. The efficiency of surface ionization is two orders of magnitude higher than that of electron ionization. However, its major drawback is the fact that the ionization energy of the atoms or molecules to be analysed must be below 7 eV to produce positive ions (which limits this method to elements of the first and second groups of the periodic table and a small number of molecules). Negative ions can be detected if the electron affinity (electronegativity) of the atom/molecule under study is above 2 eV. Research has shown that surface ionization is a very suitable method for the precise measurement of the isotopic composition of atoms. Originally, the isotopic composition of uranium was measured by using surface ionization sources [96].

The basic laws governing the process of surface ionization are derived for the case when the sample is gaseous. A beam of neutral atoms in the gaseous state is directed onto the clean, single-crystal, heated surface of the metal filament. Part of the neutral atoms is desorbed, the other part is ionized. The number of neutral atoms falling on an ionizing surface in a unit time is denoted by N_M, the number of desorbed atoms (atoms evaporating from the surface) is denoted by n_M, and the number of ions generated is denoted by $n_M{}^+$. The ionization process is characterized by two physical quantities: Degree of ionization and ionization coefficient described by the following formulas:

$$\text{Degree of ionization } \alpha = n_{M^+} / n_M \tag{7.15}$$

$$\text{Ionization coefficient } \beta = n_{M^+} / N_M \tag{7.16}$$

The actual process of ionization on a heated metal surface was illustrated by Saha with a reaction of the type:

$$M \rightarrow M^+ + e \tag{7.17}$$

Using the expression for the equilibrium constant and the equation for the emission of electrons produced by heating a metal filament, Saha and Langmire defined the degree of ionization as follows:

$$\alpha = \frac{g^+}{g} e^{\frac{\phi - IE}{kT}} \tag{7.18}$$

g^+, g- statistical weights of ions and neutrons, IE = ionization energy of atoms, ϕ = ejection work of metal surface. This is the common equation most often found in publications. By analogy, the expression for the degree of ionization in the production of negative ions is:

$$\alpha^- = \frac{g^-}{g} e^{\frac{EA - \varphi}{kT}} \tag{7.19}$$

EA is the electron affinity of the atom.

In general, the number of ions produced in mass spectrometers is proportional to the current I^+ measured at the detector. The ion current is proportional to the number of ions desorbed from a surface unit in a unit time and is given by the expression

$$I^+ = eSn_M^+ \tag{7.20}$$

S = size of the surface from which the ions were desorbed.

Combining the expression for the ion current and the degree of ionization, we get:

$$I^+ = eS\frac{g^+}{g}N_M e^{\frac{\phi - IE}{kT}} = AN_M e^{\frac{\phi - IE}{kT}} \tag{7.21}$$

The negative ion current is:

$$I^- = eS\frac{g^-}{g}N_M e^{\frac{EA - \phi}{kT}} = AN_M e^{\frac{EA - \phi}{kT}} \qquad (7.22)^{[5]}$$

The number of positive and negative ions produced during surface ionization depends on the properties of the atoms and the metal surface on which the ionization occurs. When we speak of the properties of the atoms or molecules, we mean primarily the value of their ionization energy or electron affinity, but also the degree of volatility and dissociation energy of the compound under study. The requirements for the metal surface are as follows: Metals with low work function values and a high melting point are used. In addition, the surface of the metal must be homogeneous, *i.e.*, the work function must be the same over the entire surface, which is very difficult to achieve. The problem of homogeneity is solved by determining the work function of the metal filament for each set of measurements based on atoms with known ionization energy [[95]]. The problem of constraints on the choice of the sample to be ionized can be solved by choosing the appropriate coatings to be applied to the metal filament to change the value of the work function. The above equations are very simple to apply in the case of atoms, but for diatomic molecules the situation is more complex, MX. When a diatomic molecule falls on a heated metal filament, the following processes can occur: part of the molecule desorbs unchanged, part of the molecule dissociates into atoms, the resulting atoms can desorb, and part of them can be ionized. It should be emphasized here that the above reactions apply to the case when the MX sample is in gaseous state and this gas is directed to the heated metal surface where ionization takes place. The example of desorption process of the triatomic CaI_2 molecule by surface ionization using a source consisting of two filaments is interesting. The sample was placed on one filament (a vaporization filament), and the other filament (that served as an ionization filament) was placed parallel to this vaporization filament. Rhenium filaments with dimensions of 8.35 × 0.7 mm and thickness of 0.04 mm were used, and the spacing was 10.4 mm. Before each experiment, the ionization filament must be heated to about 2,000 K to remove all impurities from the filament. The process of generating ions by surface ionization through vaporization of CaI_2 can be described by the following reactions:

$$MI_2 \rightarrow MI + I \rightarrow M + 2I \rightarrow M^+ + 2I + e^- \tag{7.23}$$

If the CaI_2 sample is imagined to evaporate and fall in unchanged from onto the ionization filament, the energy required forming the M^+ ion is the sum of the dissociation energy and the ionization energy.

An expression for the degree of ionization can be represented as follows

$$\frac{n_+}{n_0} = Be^{\frac{-E}{kT_{IF}}} \tag{7.24}$$

Here, E represents a more complex expression for the energy than in the case of equation 26, and can be described as follows:

$$E = (ED_1 + ED_2 + IE) - \phi \tag{7.25}$$

ED_1 is the dissociation energy MI_2 to MI; ED_2 = the dissociation energy MI; IE = the ionization energy of the atom M and ϕ = the work function of the rhenium metal filament (the ionization filament). In the case when MI falls on the ionization filament (it is assumed that the dissociation of MI_2 to MI took place on the sample vaporization filament), the expression for the energy is as follows:

$$E = (ED_2 + IE) - \phi \tag{7.26}$$

If the temperature of the vaporization filament is such that dissociation of MI_2 to MI and further dissociation of MI to atoms can occur, then the expression for E has only two terms:

$$E = IE - \phi \tag{7.27}$$

The authors found that the first term (ED1) in equation 25 dominates in the low temperature range of the evaporative filament, the second term (ED2) describes the process at some intermediate temperatures, and the third term (IE) dominates at high temperatures of the evaporative filament. When studying complex processes such as this, it is necessary to determine the dominant processes for different temperature intervals as well as for different geometries of sources of surface ionization, and on this basis to define the energy term (E) used in the Saha-Langmire equation. This shows that, in addition to determining the ionization energy, the Saha-Langmire equation can also be used to quantitatively

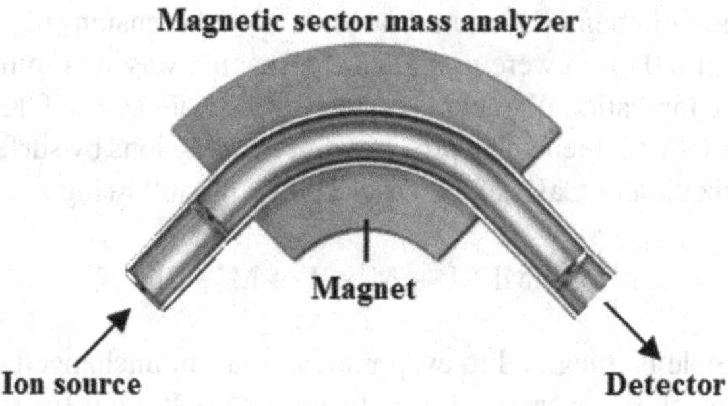

FIGURE 7.5 Schematic of a magnetic sector mass analyser.

describe the dissociation or activation energy, carefully distinguishing between the processes occurring in a given temperature interval [97,98].

The magnetic sector is a mass analyser (Figure 7.5) used in the surface ionization mass spectrometer. The sector field mass spectrometers are the oldest analysers. They were invented by Sir Joseph John Thomson (Nobel laureate in physics, 1906), who was the first to separate ions according to their respective *m/z* values by applying a magnetic and an electric field, and Francis William Aston, Thomson's student (Nobel laureate in chemistry, 1922), built the first double-sector mass spectrograph [99].

The separation principle is as follows: Ions enter the magnetic field at right angles, the magnetic induction vector B is perpendicular to the ion velocity vector ν, and the Lorenc force on an ion (with elementary charge e and mass m) in a magnetic sector is:

$$F_B = B \cdot e \cdot v \tag{7.28}$$

F_B = Lorentz force, ν = the velocity vector of the ion, B = the magnetic induction vector, e = elementary charge e and mass m (F_B perpendicular to both the magnetic field and the velocity vector of the ion itself, in the direction determined by the right-hand rule of cross products and the sign of the charge). The centripetal force (Fc) is:

$$F_C = \frac{mv^2}{r} \tag{7.29}$$

The action of the above two forces on the ions can be described by the expression:

$$\frac{mv}{e} = r \cdot B \tag{7.30}$$

The derived expression shows that the magnetic field performs separation based on the ion momentum (the product of m and the ion velocity ν). To achieve separation based on the ratio of mass to charge (*m/z*, where z is the total charge of the ion and n is the number of charges, *i.e.*, z = ne), it is necessary that the ions produced have the same velocity, *i.e.*, the same kinetic energy, when leaving the ion source. In practice, this is achieved by passing the ion through a potential difference (U) of the order of 3 kV (the value is constant for a given measurement), then the kinetic energy of the ion is described by the equation:

$$\frac{mv^2}{2} = zeU \tag{7.31}$$

Combining the last two equations, we obtain an expression that gives the dependence of *m/z* of the ion and the strength of the magnetic field:

$$\frac{m}{z} = \frac{r^2 B^2}{2U} \tag{7.32}$$

r = radius of curvature of the sector magnetic field shown in the figure (precisely fixed value).

The separation of ions depends directly on the value of the magnetic field strength, while the radius of the sector magnetic field and the potential difference are constant. In our experiment, a sector of the magnetic field of 90° was used, and the curve between the two magnets, which practically represents the path of the ion, has a radius of 32 cm. Magnetic analysers offer high resolution, sensitivity and reproducibility of results and can be easily combined with ion sources that supply a continuous ion current to the analyser, *e.g.*, electron ionization, surface ionization and Knudsen cell.

7.3.1 Surface Ionization MS and "Superalkali" Clusters

"Superalkali" clusters are perfect compounds for study by the surface ionization method, thanks to their low ionization energy. All results presented in this chapter were obtained by surface ionization in the positive mode. Xiao investigated the conditions for the generation of M_2X^+ cluster ions by alkali halides (NaCl, KCl, RbCl, CsF, CsCl, CsBr, CsI) in thermal or surface ionization mass spectrometry (TIMS or SIMS). The M_2X^+ clusters belong to the group of "superalkali," but the authors did not use this term in their work. Nevertheless, we believe that the experiments of Xiao should be included in this chapter because the aim of the work was to investigate for the first time the factors affecting the intensity of M_2X^+ clusters. In their experiments, the authors used three tantalum filaments (triple filament SIMS or TIMS), the dimensions of the metal filament were 7.5 × 0.76 × 0.025 mm. For the experiment, a graphite solution was first applied to the central filament, dried and then a solution of the sample (alkali metal salts) was applied. In the study of cesium clusters, different series of measurements were performed with cesium chloride, cesium bromide and cesium iodide salts. A solution of cesium carbonate salt was applied to the side filament. The experiment was designed to increase the temperature of the side filament independently of the central filament until the complete dissociation of the cesium carbonate, so that the intensity of cesium ions in the system is enormous. By heating the central filament with a current greater than 1 A (~433 K), a cluster of cesium chloride, Cs_2Cl^+, was detected after ten minutes. The exact mechanism of ion formation is difficult to explain. The results showed that the intensity of M_2Cl^+ clusters increased significantly in the presence of graphite on the central filament. Since Cs_2Cl^+ clusters were formed at the lowest current intensity, they were chosen as the model system. The authors assumed that part of the cesium chloride salt, CsCl, dissociates into neutral Cs° and Cl°, which are then ionized to Cs^+ and Cl^-. It is known that the clustering reaction is driven by high metal ion intensity. According to this model, the presence of Cs^+ ions leads to an ion-molecule reaction: $Cs^+ + CsCl \rightarrow Cs_2Cl^+$, resulting in the formation of clusters. The intensity of the M_2Cl^+ clusters first increases with time, reaches its maximum value in the period of 100 to 150 minutes, and then decreases until the end of 10 hours. The maximum intensity of Cs_2Cl^+ was reached at a current intensity of 1.6 A (~623 K). The maximum of the Rb_2Cl^+ cluster was detected at 1.85 A (~723 K) and that of K_2Cl^+ at 1.95 A (~753 K). When the current is further increased, the intensity of the cluster decreases sharply. The clusters of K_2Cl^+, Rb_2Cl^+, and Cs_2Cl^+ were detected over a period of 600 min at the indicated current intensities using the surface ionization method. Although Na_2Cl^+ clusters can be obtained very easily by the surface ionization method in these tests, the intensity of disodium chloride was very low [100].

The results also show that for a particular metal ion, *e.g.*, cesium, the cluster intensity increases with increasing anion radius from fluorine to iodine, *e.g.*, the relative cluster intensities are given in parentheses: Cs_2F^+ (0.17), Cs_2Cl^+ (1.00), Cs_2Br^+ (1.94), Cs_2I^+ (2.68). This experiment is also shown that the intensity of the M_2Cl cluster also increases with the increase of the radius of the anion from sodium to cesium, provided that the anion, in this case chlorine, is constant (Na_2Cl^+ (0.0009), K_2Cl^+ (0.084), Rb_2Cl^+ (0.24), Cs_2Cl^+ (1.00). This means that the formation of clusters is preferred for large alkali cations. In this experiment, bimetallic clusters were detected for the first time using the surface ionization method. The following clusters were detected: $KLiCl^+$, $RbLiCl^+$, $KNaCl^+$, $CsLiCl^+$, $RbNaCl^+$, $CsNaCl^+$, $RbKCl^+$, $CsRbCl^+$. The clusters were obtained by vaporization of the corresponding combination of alkali chlorides. The intensity is highest for the bimetallic salts of large cations such as cesium and rubidium, while the intensity is very low for the combination of small cations potassium, sodium and lithium, but again higher than the intensity of the disodium chloride cluster [100].

Special attention is paid in this work to the possibility of measuring the isotopic composition of chlorine. The change in the intensity of: M_2Cl^+, M_2Cl^+/M^+ and $^{37}Cl^-/^{35}Cl$- was followed as a function of time (at constant intensity of the current of the central filament) for the clusters K_2Cl^+, Rb_2Cl^+, Cs_2Cl^+. The results showed that the measurement of the chlorine isotope ratio $3^7Cl/^{35}Cl$ by measuring Cs_2Cl^+ is much more accurate than the measurement of the ratio Cs_2Cl^+/Cs^+, which is particularly important for monitoring geological processes. The isotopic ratio

TABLE 7.1
The temperature values for the central and side filaments in which the M_2Cl^+, M = Li, Na, K, Rb, Cs clusters were formed.

Clusters	$T_{central\ filament}$ (K)	$T_{side\ filament}$ (K)
Li_2Cl	1,450–1,620	1,215
Na_2Cl	1,220–1,550	1,050
K_2Cl	1,330–1,700	1,140
Rb_2Cl	1,420–1,650	1,161
Cs_2Cl	1,340–1,640	511

of chlorine is important for studying geochemical processes in aerosols, rocks, underwater waters and oceans [101–108].

Inspired by this experiment, "superalkali" clusters of M_2Cl^+, M = Li, Na, K, Rb, Cs, were obtained in our research using surface ionization MS. The sample, a solution of NaCl, KCl, RbCl and CsCl salt in ethanol, was placed on the side rhenium filaments (the temperature of the side filaments was constant during the experiment). The central rhenium filament was used without graphitizing agent. The temperature values for the central and side filaments in which the above clusters were detected are shown in Table 7.1.

In our experiments, the conditions for the preparation of lithium halide clusters (Li_nX, X-F, Cl, Br, I) were also investigated using the SIMS. The Li_2F cluster was chosen as a prototype to test the conditions for the preparation of "superalkali" Li_nX clusters. The influence of the chemical composition of the sample was investigated using two samples: LiF and a mixture of LiF/LiI salt (LiI is additional sources of Li^+ ions). Lithium iodide was chosen because the dissociation energy of lithium iodide is the lowest in the lithium halide group, so lithium can be easily obtained by heating the LiI salt. On the other hand, iodine is a monoisotopic element, which facilitates the interpretation of the obtained mass spectra. Also, iodine has a much higher mass (*m/z* 127) than fluorine (*m/z* 19), so lithium-iodine clusters are detected at much higher masses than lithium-fluorine clusters and are clearly separated in the mass spectrum. The influence of the geometry of the surface ionization source on the formation of clusters was also studied. The simplest source for surface ionization consists of a single metal filament. This source is most commonly used when working with gaseous samples. However, the use of a single filament to obtain clusters from an inorganic salt solution would mean that processes such as vaporization, salt dissociation, formation of new types of compounds, ionization of rewarded neutrals take place on the surface of a single filament, which leads to great confusion, especially when the goal is to work out the characterization of the obtained clusters. For this reason, a source with only

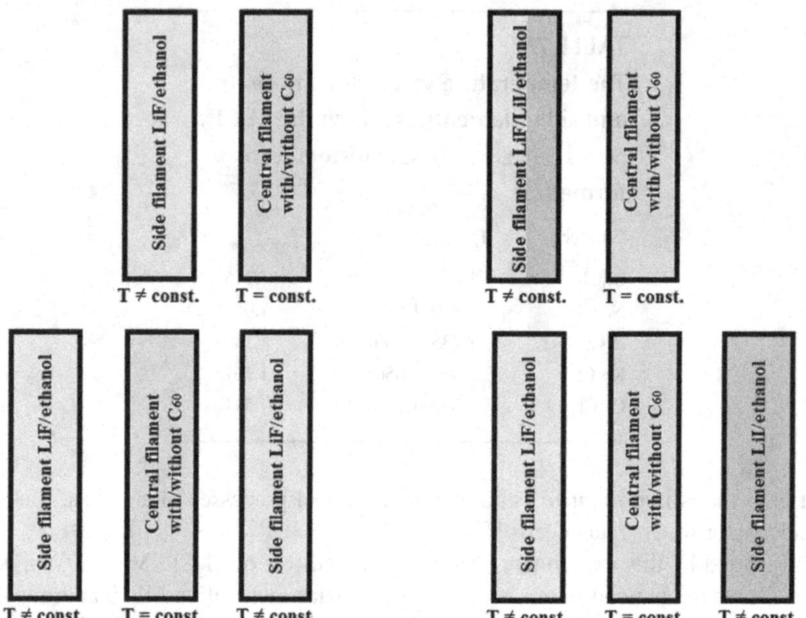

FIGURE 7.6 Schematic representation of the schedule for placing the samples on the filaments in the case of double and triple SIMS.

one metal filament was not used. The experiments were performed with the double and triple filament SIMS. Platinum, tantalum, rhenium and tungsten filaments are the most commonly used. In our experiments, rhenium filaments were used, which are 9 mm long, 1 mm wide and 0.15 mm thick. In the case of the double filament SIMS (the side filament and the central filament are at a 90° angle), the LiF solution was added to the side filament in the first series of measurements, while the LiF and LiI solutions were added individually in the second series of measurements (Figure 7.6). In the case of the triple filament source SIMS (with two parallel side filaments, while the central filament is horizontal and at an angle of 90° to the intervening side filaments), the LiF solution was added to both side filaments in the first series of measurements, and in the second series of measurements, the LiF solution was dropped on one side filament and the LiI solution was dropped on the other side filament of the source (Figure 7.6) [[34,109–111]].

For all measurements, the sample was a solution of the appropriate lithium salt in ethanol (ethanol was chosen because it facilitates drying of the sample). The sample was placed on the side filaments at room temperature and dried with a UV lamp outside the mass spectrometer. In these experiments, the temperature of the central filament was kept constant at 1,694 K, while the temperature of the vaporization filament was gradually increased. The fullerene C_{60} was used for graphitization of the central filaments. The experiment was designed to observe results

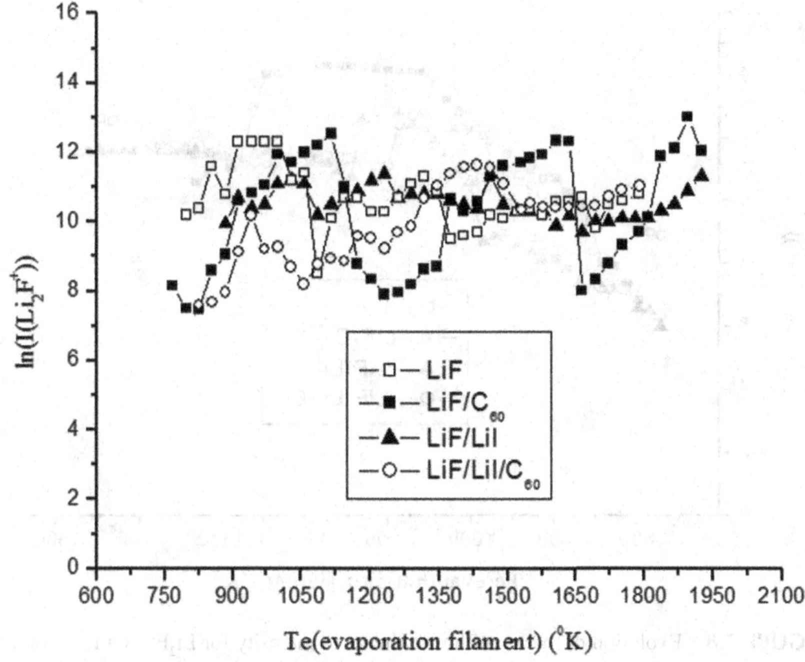

FIGURE 7.7 Evolution of the logarithm of the ionic intensity for Li_2F^+ as a function of the temperature of the side (vaporization) filament of the double filament set of SIMS: open square = sample was the solution of LiF salt; filled square = the LiF/C_{60}; filled triangle = sample was the solution of a mixture LiF and LiI; open circle = sample was LiF/LiI/C_{60}.

with pure central filaments and with graphitized central filaments. Experimental results showed that the distance between the two side strands is a crucial factor for the formation of superalkali clusters. It was found that when the distance between the side strands exceeds 4 mm, the detection of these clusters is much more difficult. The Figures 7.7 and 7.8 shows the evolution of the logarithm of the ion intensity for Li_2F^+ as a function of the temperature of the side filament of the double and triple filament SIMS, respectively.

The research results showed the following trends: when the sample was LiF, the cluster was detected in the side filament temperature interval of 792–1,786 K; the use of graphitized allows the detection of Li_2F^+ at lower temperatures, 761 K, and in a longer side filament temperature interval of 761–1,924 K; the LiF/LiI combination provides similar stability of cluster ions at high temperatures as C_{60}, but the onset of cluster formation requires a slightly higher temperature than in the previous two cases, 890–1,924 K; in the case of LiF/LiI the temperature interval for cluster detection is 830–1,783 K (the double filament set of SIMS, Figure 7.7).

The temperature intervals of the side filament in which Li_2F^+ was detected were: for LiF sample 882–1,449 K, for LiF/C_{60} 797–1,405 K, for LiF/LiI 771–1,491 K and

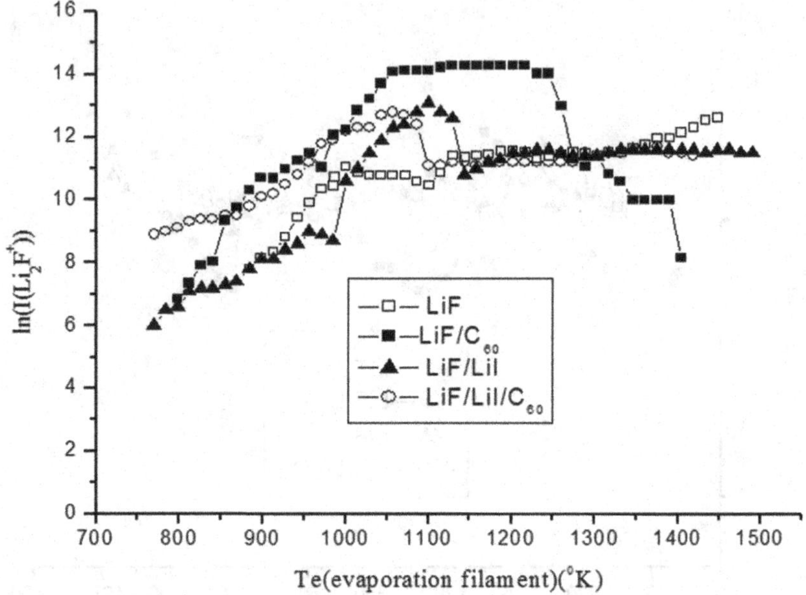

FIGURE 7.8 Evolution of the logarithm of the ionic intensity for Li_2F^+ as a function of the temperature of the side (vaporization) filament of the triple filament set of SIMS: open square = sample was the solution of LiF salt; filled square = sample was the LiF/C_{60}; filled triangle = sample was solution LiF and LiI; open circle = sample was LiF/LiI/C_{60}.

LiF/LiI/C_{60} 771–1,419 K (the triple filament set of SIMS, Figure 7.8) [34,112–114]. The intensity of the rewarded ions is slightly higher for the triple filament than for the double filament source. For the triple filament, the first region, in the temperature interval from about 700 K to about 1,000 K, is characterized by a linear increase in the intensity of Li_2F^+ ions with increasing temperature and occurs in all series of measurements. The second range, the range of signal saturation, temperature above 1,000 K, occurs when the system under study is LiF, LiF/LiI, LiF/LiI/C_{60}. The difference is that when the LiF/C_{60} system is studied, the signal of Li_2F^+ ions decreases after saturation. The presence of LiI allows more efficient formation of Li_2F^+ at lower temperatures than vaporization of LiF alone. Experimentally, the fullerene on the central filament was shown to increase the signal intensity of the Li_2F cluster in both sample types, the LiF salt and the LiF/LiI mixture. The fullerene on the central filament, *i.e.*, graphene, most likely leads to suppression of the dissociation process of the formed clusters. The main difference between these two series of measurements is that the intensity of Li_2F^+ ions in the double filament source does not show a defined trend with increasing temperature of the side filament, while two regions of the curve can be distinguished in the triple filament source. Therefore, a triple filament surface ionization source was chosen to obtain lithium halide clusters [109].

The clusters with 9 and 10 valence electrons, such as dilithium and trilithium Li_nX (n = 2, 3, X = F, Cl, Br, I), were obtained by this method. For Li_nCl and Li_nBr clusters, the sample was a mixture of LiCl/LiI and LiBr/LiI, respectively; Li_nI clusters were detected in both cases; in all experiments, the C_{60} was located on the central filament. The triple-filament surface ionization method was used to experimentally demonstrate the existence of Li_4F (with 11 valence electrons) for the first time. The results showed that C_{60} allowed a higher yield of Li_2X clusters, while the Li_3X and Li_4F clusters could not be detected without fullerene. Multinuclear Li_nF_{n-1} clusters with n = 3, 4, 5 and 6 were obtained by surface ionization through vaporization of LiF, LiI and BiF_3 salts, respectively [[34,109–111]].

7.3.2 Surface Ionization Mass Spectrometry and Ionization Energy

To determine the ionization energy by the SIMS, the Saha-Langmuir equation is used in the following form:

$$\ln I^{+} = C + (\phi - IE)/kT \tag{7.33}$$

The use of the Saha-Langmuir equation to calculate the ionization energy of a cluster requires that the value of the work function of the metal filament under the conditions in which the cluster is generated be known (since the use of tabular values would lead to a large error in the calculation of the ionization energy). It should be noted that the Saha-Langmuir equation is used both to determine the work function of the metal filament and to determine the ionization energy of the cluster. The problem of determining the work function of the filament can be solved by a reference ion. It is necessary that the reference ion be rewarded in the same temperature interval in which the observed clusters are formed, *e.g.*, $^7Li^+$ is commonly used as a reference ion for lithium halide clusters. In practice, the work function is determined based on the dependence of the intensity of the reference ion on temperature ($\ln I(M^+) = f(1/T)$, M^+ is the reference ion).

Based on the knowledge of the slope of the line $\ln I(M^+) = f(1/T)$ and the ionization energy of the reference ion, the expression

$$tg\alpha = (\phi - IE\left(M^{+}\right)/k \tag{7.34}$$

(k is the Boltzmann constant and is 8.6 × 10–5eV/K) is used to obtain the value of the work function.

The second problem that arises in determining the ionization energy is the formation of neutral beam clusters under the conditions of surface ionization. In practice, it is necessary to separate the process of generating a beam of neutral clusters from the process of ionization on a heated metal filament. The solution to this problem was found in the control of the temperature of the side filament. The temperature of the side filament, at which no ionization of the sample occurs, is determined experimentally (the central filament is not heated). The presence of neutral clusters in the system is checked by electron impact ionization. The neutral ions generated in this way fall on the central filament, which is gradually heated, and the change in intensity of the cluster ions as a function of temperature is observed. Finally, the ionization energy of the cluster is determined by the dependence of the intensity of the cluster ions on the temperature, $\ln I(M_2X^+) = f(1/T)$, the slope of this curve is calculated, which together with the calculated work function and using the Saha-Langmuir equation allows the value of the ionization energy of the cluster to be determined.

By combining the Saha-Langmuir equation for the reference ions and the Saha-Langmuir equation for the cluster ion (whose ionization energy we want to determine) we get the expression

$$\ln \frac{I\ (M^+)}{I\ (M_2X^+)} = \left(\frac{IE(M_2X^+) - IE\ (M^+)}{kT} \right) \tag{7.35}$$

$IE(M_2X)$ = ionization energy of the cluster, $IE(M)$ = ionization energy of the reference ion = known value, $I(M^+)$ = intensity of the reference ion, $I(M_2X^+)$ = intensity of the cluster ion, T = temperature of the central filament, k = Boltzmann constant. Based on the measurement of the intensity of the reference ion and the intensity of the cluster ion, the ionization energy of the cluster can be determined without knowing the value of the work function of the metal filament.

The results of the ionization energy for the clusters Li_nX, X = Cl, Br, I (n = 2, 3) obtained by the triple filament surface ionization by vaporization of the salt comb ne C_{60} are shown in Table 7.2. The same table also shows the ionization energies of polynuclear lithium clusters of the Li_nF_{n-1} type (the sample was $LiF/LiI/BiF_3$, the central filament graphitized with ination LiX/LiI, X = F, Cl, Br, I from the side filaments in the case where the central filament is coated with the fullerene C_{60}) [110–111].

The ionization energies of Li_3F_2 and Li_4F_3 clusters measured by the surface ionization method are in excellent agreement with the results obtained by the photoionization method (4.32 0.2 eV for Li_3F_2 and 4.300.2 eV for Li_4F_3) [112]. These are important data showing that the values obtained with the Saha-Langmuir

TABLE 7.2
The ionization energy for the clusters Li_nX and LinFn-1 clusters (X = F, Cl, Br, I, n =2, 3) obtained using the surface ionization MS.

Clusters	IE (eV)	Clusters	IE (eV)	Clusters	IE (eV)
Li_2F	3.8 ± 0.1	Li_3F	4.0 ± 0.1	Li_3F_2	4.2 ± 0.1
Li_2Cl	3.8 ± 0.1	Li_3Cl	4.0 ± 0.1	Li_4F_3	4.3 ± 0.1
Li_2Br	3.9 ± 0.1	Li_3Br	4.1 ± 0.1	Li_5F_4	/
Li_2I	4.0 ± 0.1	Li_3I	4.1 ± 0.1	Li_6F_5	4.1 ± 0.1

equation are very reliable when the conditions for the formation of neutral clusters are well defined.

7.4 KNUDSEN CELL MASS SPECTROMETRY

In the early 20th century, Danish physicist Martin Knudsen constructed a cell to measure vapor pressure [113]. The application of the Knudsen cell has gained importance since 1948 when the Russian scientist Ionov incorporated the Knudsen cell into the ion source of a mass spectrometer to study the vaporization process of alkali halides [114]. In the literature, Knudsen cell mass spectrometry is also referred to as high temperature mass spectrometry (HTMS) or Knudsen effusion cell mass spectrometry (KCMS) or mass spectrometry with an effusion cell or molecular beam mass spectrometry. The method consists in the following, a solid sample is introduced into the Knudsen cell, that is uniformly heated until equilibrium is attained between the condensed and vapor phases. Since the opening of the cell is many times smaller than its cross-section, when the pressure in the cell is such that the mean free path of the neutral is greater than the diameter of the opening, effusion of the neutral beam occurs. The neutral beam is directed into the ionization chamber, where ionization occurs by collision with electrons [115]. Under standard conditions, the Knudsen cell is located outside the ionization chamber, and there is a movable shatter between the ionization chamber and the cell, which makes it possible to determine the origin of the ions (whether they were generated in the Knudsen cell or in the ionization chamber). The generated ions are passed into the mass analyser, where separation takes place based on the relationship between the mass of the ions and their charge, *i.e.*, until identification.

The first step in quantitative analysis is to establish a functional dependence between the intensity of the detected ions, the partial pressure and the concentration. Since the effusion process does not disturb the thermodynamic equilibrium between the condensed and vapor phases, the ion intensity is proportional to its pressure (pi) as follows:

$$p_i = (k'TI_i / \sigma_i) \tag{7.36}$$

I_i = ion intensity obtained in the mass spectrometer, σ_i = ionization cross section of a given ion type, p_i = pressure of a given ion type (Pa), T = cell temperature, k' = a constant containing various correction factors (geometry of the system, apertures, *etc.*) and determined on the basis of a reference sample. In the Knudsen cell, the partial pressures of the components are below 10 Pa, so we can apply the equation of the ideal gas state pV = NRT (p = pressure, V = volume, N = number of molecules, R = gas constant, T = temperature), which in combination with the equation allows us to relate the ion intensity of the mass spectrum and their concentrations.

Considering the chemical reaction AB + C = AC + B, the equilibrium constant of the reaction can be written in the following form

$$K_P = P_{AC}P_B / P_CP_{AB} = (\sigma_{AB}\sigma_C / (\sigma_{AC}\sigma_B)(I_{AC}I_B / I_{AB}I_C) \tag{7.37}$$

By measuring the ionic intensity of each species and based on the table values for (σ_{AB} σC/σ_{AC} -σ_B), the equilibrium constant of the reaction (K_P) is calculated using the equation. By observing the change in the equilibrium constant as a function of the temperature of the cell, thermodynamic quantities such as free energy, enthalpy and entropy can be calculated based on the equations of the second and third laws of thermodynamics. The second law of thermodynamics is described by the van't Hoff equation:

$$\ln K_P = -\Delta H_T^0 / RT + C \quad (7.38)$$

where ΔH^0_T is the enthalpy of reaction. From the slope of the experimentally determined dependence $\ln K_p = f(1/T)$, the enthalpy of reaction can be determined very easily without knowing the entropy of the reactants or their absolute pressures. However, this method requires a large number of measurements to achieve satisfactory accuracy. The third law of thermodynamics establishes a link between the equilibrium constant (K_P) and the free energy of reaction (ΔG^0_T):

$$\Delta G_T^0 = -RT \ln K_P \quad (7.39)$$

and

$$\Delta G_T^0 = \Delta H_T^0 - T\Delta S_T^0 \quad (7.40)$$

where H^0_T and ΔS^0_T are the enthalpy and entropy of the reaction, respectively. The enthalpy of reaction calculated by knowing the free energy is a more accurate method [116].

7.4.1 The Knudsen Cell as a Source of "Superalkali" Clusters

In the field of homogeneous and heterogeneous alkali metal clusters, the Knudsen cell MS has played an important role from the discovery of said clusters to the present day. The first homogeneous lithium clusters, Li_n, n = 3–6, were obtained by this method [117,118]. The Knudsen effusion mass spectrometry was used to determine their heats of formation and atomization energies. In the early 1960s, Berkovic studied the process of formation of gaseous chemical species in lithium chloride vapours using a Knudsen cell MS and discovered Li_2Cl. The authors believed they had discovered an unusual molecule, but the goal of their work was to study the formation of dimers and trimers of lithium chloride [119].

The beginning of the study of "superalkali" clusters is represented by the work of Kudo and Wu. Their goal was a detailed study of the vaporization of Li_2O crystals, and mass spectrometry using the Knudsen cell was the ideal method because it allowed precise detection of the resulting compounds. Crystalline Li_2O is a potential candidate for tritium breeder material in thermonuclear fusion reactors, and it was of particular interest to study its behaviour at high temperatures. For the experiment, Kudo specially constructed a Knudsen cell mass spectrometer with a quadrupole mass analyser. The Knudsen cell was located outside the ionization chamber with a shatter between the ionization chamber and the Knudsen cell. The molecules produced in the Knudsen cell are ionized by electrons of energy 5 eV. Experiments were performed in a Knudsen cell made of molybdenum, tantalum, graphite and platinum. The results showed that Li_2O reacts with the molybdenum cell walls and drastically reduces the intensity of the observed ions. It was found that the highest yield of detected ions was obtained with the platinum cell. Upon vaporization of Li_2O crystals in the temperature range of 1,300–1,700 K, the following ion species were detected: LiO(g), Li_2O(g), Li_2O_2(g) and Li_3O(g). Among the mentioned ionic species, the Li_3O cluster attracted the most attention [2]. This is the first experiment to show that Li_3O, a compound with nine valence electrons, can exist as a neutral molecule. As an explanation for the formation of this molecule, they looked for the similarity of Li_3O with H_3O molecules. At this point, it should be emphasized that apart from the formal similarity, these two compounds are fundamentally different. Indeed, hypervalent H_3O and H_4O molecules are formed by weak van der Waals forces between H_2O molecules and H or H_2. However, the nature of chemical bonding in the Li_3O cluster is quite different [2]. Unlike Li_2O, Li_2C_2 crystals do not react with molybdenum cell walls. For this reason, a molybdenum cell with a volume of 1.5 cm^3 and an opening cross section of 0.3 mm was used to test the lithium and carbon cluster. The neutrals of the produced gas species were ionized with electrons of energy 13 eV, the following ionic species Li^+, Li_2^+, CLi_3^+, CLi_4^+, CLi_6^+ were detected [7]. The described apparatus was used to obtain a series of "superalkali" heterogeneous clusters, such as Li_3S, Li_4S, clusters were obtained by heating the Li_2S salt in a molybdenum cell in the temperature range of 950–1,240 K. In addition to the above clusters, Li^+, Li_2^+, LiS^+, Li_2S^+, LiS_2^+, $Li_2S_2^+$, S^+, S_2^+ positive ions were also detected [11]. Lithium-phosphorus clusters were obtained by vaporization of lithium triphosphate introduced into a Knudsen cell of molybdenum under argon atmosphere. The mass spectrum was obtained by ionization with electrons of energy 15 eV, identifying the following ions: Li^+, Li_2^+, P^+, P_2^+, P_3^+, LiP^+, Li_3P^+, LiP_2^+, $Li_2P_2^+$, Li_4P^+ (in the temperature range from 900 K to 1,357 K) [12]. The dependence of the calculated partial pressures on the reciprocal of the Knudsen cell temperature and the measured heat of formation for all detected species was monitored. The experimental enthalpy values agree with those obtained by other methods, which is a good indication that the KCMS method can be used to determine the enthalpies of unknown reactions.

The dissociation energies according to the reaction $Li_nA \rightarrow Li_{n-1}A + Li$ were also determined for the whole series of superalkali clusters (Table 7.3) [2–4,11–13,18].

This experimental setup allows the determination of the ionization energy of the obtained clusters using the so-called ionization efficiency curves (IEC). With the help of the IEC, *i.e.*, by measuring the cluster intensity as a function of the electron energy and using the extrapolated voltage difference method with reference values from IE (ions of alkali metals), the ionization energies of the M_nX clusters were determined [12]. In this way, it was experimentally demonstrated for the first time that the hypervalent clusters have "superalkali" character.

TABLE 7.3

Clusters obtained with a Knudsen cell of platinum and molybdenum, ionic species formed by vaporization of a particular alkali salt, dissociation energy of the separated clusters ($Li_nA \rightarrow Li_{n-1}A + Li$), temperature range of the cell.

Clusters	Sample	Energies disciation (kJ/mol)	Themperature Knudsen celln (K)	Ion detected by electron impact
Li_2CN	LiCN	137 ± 14	400–800	M-Li, Na, K
Na_2CN	NaCN	104 ± 14		M^+, M_2^+, M_3^+, MCN^+, M_2CN^+
K_2CN	KCN	82 ± 8		
Li_3S	Li_2S	138 ± 4.18	950–1,240	Li^+, Li_2^+, LiS^+, Li_2S^+, LiS_2^+, $Li_2S_2^+$, S^+, S_2^+, Li_3S^+, Li_4S^+
Li_4S		212 ± 13		
Li_4P	Li_3P	186 ± 24	960–1,168	Li^+, Li_2^+, P^+, P_2^+, P_3^+, LiP^+, Li_3P^+, LiP_2^+, $Li_2P_2^+$, Li_4P^+
CLi_6	Li_2C_2	—	900–1,357	Li^+, Li_2^+, CLi_3^+, CLi_4^+, CLi_6^+
Li_3O	Li_2O	212 ± 42	1300–1,600	Li^+, Li_2^+, Li_2O^+, $Li_2O_2^+$, Li_3O^+
Li_4O		197 ± 30		
Li_5O		121 ± 25		

In our laboratory we used a magnetic mass spectrometer modified by Veljkovic specifically for the study of "superalkali" clusters. The special feature of this mass spectrometer is the ionization chamber, where it is possible to combine electron impact, surface ionization and the Knudsen cell. The surface ionization, Knudsen cell, and electron impact ionization are arranged at a 90° angle. At the top of the ionization chamber there is an inlet for gaseous samples and an opening for a cryo cell [[120]]. The results of the surface ionization performed in this mass spectrometer have already been presented in the previous chapter.

Using the mass spectrometer mentioned above, a series of experiments were performed to investigate the possibility of cluster formation using the Knudsen cell. The Figure 7.9 schematically shows three ways of using the Knudsen cell in a mass spectrometer: when the cell is placed outside the ionization chamber (standard method, previously described experiments), the Knudsen cell in the ionization chamber when the heater is located on the outer wall of the cell, the Knudsen cell in the ionization chamber when the heater is located in the cell itself.

In the case where the Knudsen cell was placed in the ionization chamber, the opening of the Knudsen cell was at a distance of 2 mm from the electron beam. This should allow all clusters generated in the Knudsen cell to be ionized immediately. The problem with this mode could be the identification of clusters due to the overlap of m/z with ions normally present in the ionization chamber. A nickel Knudsen cell was used for these experiments. The dimensions of the cell were: Height 7 mm, outer diameter 6 mm and opening diameter 0.1 mm. The outer

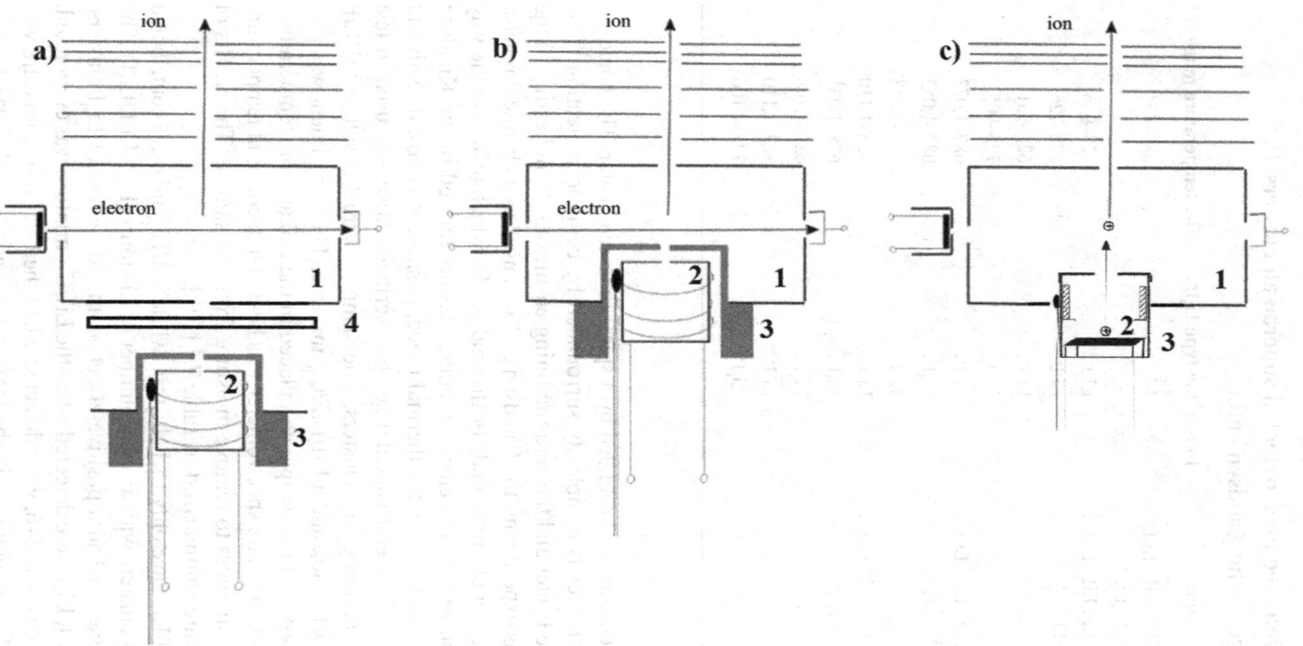

FIGURE 7.9 Schematically three ways of using the Knudsen cell in a mass spectrometer: (a) when the cell is placed outside the ionization chamber, shatter the heater is located on the outer wall of the cell (standard method), (b) the Knudsen cell in the ionization chamber when the heater is located on the outer wall of the cell, (c) the Knudsen cell in the ionization chamber when the heater is located in the cell itself; mass spectra obtained by: the electron impact for (a) and (b), the thermal mode for (c). 1 = Ionization chamber; 2 = heater; 3 = the Knudsen cell; 4 = shatter.

TABLE 7.4
List of samples, detected ions, detected superalkali clusters, the temperature range of the Knudsen cell.

Sample	Ions	Ions of the type Li_nI^+	The temperature range (K)
LiI	I^+, I_2^+, Li^+, LiI^+, $Li_2I_2^+$, $Li_3I_2^+$, $Li_4I_3^+$	Li_2I^+	681–854
LiI/C_{70}	I^+, I_2^+, Li^+, LiI^+, $Li_2I_2^+$, $Li_3I_2^+$, $Li_4I_3^+$	Li_2I^+,	628–946
		Li_3I^+,	628–726
		Li_4I^+,	628–931
		Li_6I^+	725–905
LiI/LiF	I^+, I_2^+, Li^+, LiI^+, $Li_2I_2^+$, $Li_3I_2^+$, $Li_4I_3^+$	Li_2I^+,	687–1,073
		Li_4I^+,	687–1,053
		Li_6I^+	687–1,053
LiI/LiF/C_{70}	I^+, I_2^+, Li^+, LiI^+, $Li_2I_2^+$, $Li_3I_2^+$, $Li_4I_3^+$	Li_2I^+,	628–1,110
		Li_3I^+,	628–1,031
		Li_4I^+,	628–1,108
		Li_5I^+,	882–1,110
		Li_6I^+	725–1,106

wall of the Knudsen cell is surrounded by a spiral tungsten heater. The Knudsen cell with external heater is completely surrounded by ceramic protection. This arrangement of the heater and the ceramic lining ensures uniform heating of the cell volume. In this experiment, the Knudsen cell was used as a chemical reactor that can be operated in electron mode or thermal mode, which differs in the way the clusters are ionized. In the electron mode, clusters formed in the Knudsen cell are ionized by electrons. In the thermal mode, positive ions flow directly out of the Knudsen cell. The experimental setup described above was used to test the possibility of obtaining Li_nI clusters. The samples were LiI salt, a LiI/LiF salt mixture (in a 7/1 ratio), and a LiI/LiF/C_{70} mixture. The experiment was performed in such a way that the sample was placed in the cell at room temperature. The Knudsen cell in the mass spectrometer was heated to a constant temperature of 150 °C for several hours to remove moisture from the sample. The results of these experiments are summarized in Table 7.4 [[35,121]].

Vaporization of LiI gave only Li_2I clusters, while LiI/LiF mixture contributed to the formation of clusters with an even number of Li atoms (Li_2I, Li_4I, Li_6I). The addition of fullerene C_{70} had a positive effect on the formation of Li_nI clusters. Thus, Li_2I, Li_3I, Li_4I, Li_6I were detected when the LiI/C_{70} mixture was evaporated, while the Li_nI clusters, n = 2–6, were detected when the LiI/LiF/C_{70} mixture was evaporated. The clusters obtained in the temperature interval 673–1,053 K were ionized by electrons, *i.e.*, electron mode. At temperatures around 1,073 K, there is a transition from the electron mode to the thermal mode of the Knudsen cell.

TABLE 7.5
Experimental values of Li_nI ionization energy (n = 2–6) and theoretical adiabatic and vertical values.

Clusters	Ionization energies experimental (eV)	Ionization energies adiabatic	Ionization energies vertical
Li_2I	4.69 ± 0.25	4.53[a] 4.41[b]	5.26[a] 5.19[b]
Li_3I	5.14 ± 0.25	5.07[a] 5.09[b]	5.56[a] 5.61[b]
Li_4I	4.86 ± 0.25	4.40[a] 4.37[b]	4.83[a] 4.78[b]
Li_5I	4.62 ± 0.25	4.49[c] 4.45[d]	4.96[c] 5.09[d]
Li_6I	4.96 ± 0.25	—	—

[a] UB3LYP/Li=cc-pVQZ,I=aug-cc-pVQZ-PP; [b] RQCISD(T)/Li=cc-pVTZ,I=cc-pVTZ-PP;
[c] B3LYP3/Li=aug-cc-pVTZ,I=aug-cc-pVTZ-PP; [d] RCCSD/Li=cc-pVTZ,I=cc-pVTZ-PP.

At temperatures above 1,073 K, the clusters in the Knudsen cell were obtained in the thermal mode, *i.e.*, positively charged clusters flowed from the Knudsen cell directly into the ionization chamber. In the electron mode, the ionization energy was determined using the so-called ionization efficiency curves, *i.e.*, the ion currents of the tested clusters and the reference ion were measured as a function of electron energy. The ionization efficiency curves for the observed clusters and the reference ion show a clear linear increase, which allows the application of the linear extrapolation method in the threshold region of the IEC. Using this method, the ionization energy of the observed cluster was determined from the intersection of the tangent of the linear growth curve and the baseline of the plot. The ionization energy of the reference ion (LiI^+) was used as a calibrator for the electron energy scale. The error in determining the ionization energy results from the energy width of the electrons used for ionization. The energy width of the electrons in the ionization chamber is determined using the helium efficiency curve (helium was chosen because it has a wide range between the first and second ionization energies) [122,123]. Nonstoichiometric Li_3I_2 and Li_4I_3 clusters were detected in all measurements, but in a very short time interval, so it was not possible to measure their ionization energy under these conditions. Table 7.5 shows the experimental values of Li_nI cluster ionization (n = 2–6) and the theoretical values of adiabatic and vertical ionization energies [35,121].

Using the Knudsen cell inserted into the ionization chamber of the magnetic sector mass spectrometer by heating the LiI/LiF mixture at a ratio of 1/7, heterogeneous clusters of Li_nF (n = 2–6) can be detected. However, it should be noted that Li_nF (n = 2–6) clusters are obtained only in the thermal mode. A typical mass spectrum is shown in Figure 7.10.

As can be seen in Figure 7.10, Li_2F^+ has the highest peak in the mass spectrum, while the Li_3F^+ peak has the lowest intensity. The peaks of Li_6F^+ and Li_4F^+ ions are slightly more intense than the peak of Li_5F^+. From the results shown, it can

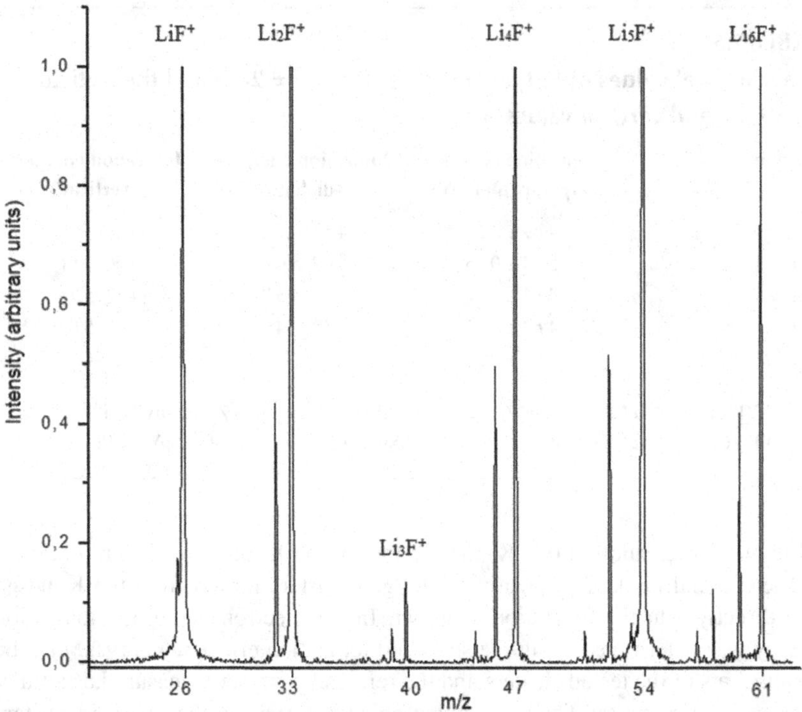

FIGURE 7.10 Mass spectra of Li_nF clusters (n = 2–6) obtained by vaporization of the LiI/LiF mixture (1/7) from the Knudsen cell inserted into the ionization chamber of the magnetic sector mass spectrometer.

be concluded that the method described is suitable for obtaining only positive ions from lithium fluoride clusters. However, their ionization energy could not be determined by the IEC method. However, because a good linear dependence of ion intensity as a function of temperature was found, the Saha-Langmuir equation was applied to determine the ionization energy of Li_nF clusters. As mentioned earlier, the Saha-Langmuir equation can be applied to determine ionization energies when a neutral molecule/cluster/atom falls on a heated metal and is ionized. The Saha-Langmuir equation can also be applied to determine the dissociation energy or activation energy of a particular process, provided, of course, that the formation mechanism of the observed ion is known. With this experimental setup in the thermal mode, it was not possible to determine exactly where the ionization of the clusters forming in the Knudsen cell occurs. Ionization of the forming clusters may occur on the cell walls or on the cell envelope, the temperature of which was not measured. Also, the possibility of formation of cluster ions by an ion-molecule reaction or a dissociation reaction should not be excluded [[124]].

In the context of this experiment, it is difficult to assume the mechanism of cluster formation, and regardless of the fact that the obtained values of ionization energy agree with theoretical calculations, the Saha-Langmuir equation should be used with great caution in such situations. It should be noted, however, that the Knudsen cell incorporated into the ionization chamber is a successful method for obtaining Li_nF^+ (thermal mode) and Li_nI (electronic mode) clusters, which can be used for future research on "superalkali" clusters. The Knudsen cell placed in the ionization chamber becomes an even more efficient source of positive cluster ions when a heater (rhenium filament) is placed directly in the cell (the external heater is completely removed). In this case, the temperature of the cell is not uniform throughout the inner cavity. Results show that heating the inorganic salts MX, M–Li or K, X–Cl, Br leads to the formation of a series of mononuclear Li_nX^+ and K_nX^+ (n = 2–6) and dinuclear $K_nBr_{n-1}^+$ (n = 3–5), $Li_nCl_2^+$ (n = 4–7) and $Li_nCl_{n-1}^+$ (n = 3–5) clusters, which were not obtained with the standard Knudsen cell mass spectrometry setting [[36,37,39,40,42]].

7.5 PERSPECTIVE IN THE FUTURE

Knowledge of nonstoichiometric "superalkali" clusters is an important part of research to synthesize new materials with predefined properties obtained under laboratory conditions. Thanks to a variety of different ionization methods, powerful mass analysers and detectors, mass spectrometry offers numerous possibilities not only for the production of clusters (with precisely defined size and composition), but also for the verification of theoretical assumptions for the application of "superalkali" clusters. Surface ionization and the Knudsen cell, although methods discovered early in the last century, may play an important role in future research. The cases where SIMS is the source of the cluster and the means to determine the ionization energy of the cluster were considered here. In some future research, surface ionization mass spectrometry (since it ionizes only compounds with low ionization energy below 7eV) could be a simple method to determine the ionization energy of "superalkali" clusters formed with another source, such as specially designed cells.

The Knudsen cell is an efficient chemical reactor for "superalkali" clusters and can be a useful tool for depositing well-defined cluster layers on various surfaces. Changing the dimensions, the way of heating, combining several Knudsen cells, combining the Knudsen cell with surface ionization – all these and many other modifications provide significant opportunities for efficient production of new types of "superalkali" clusters.

ACKNOWLEDGEMENT

This chapter was financially supported by Ministry of Science, Technological Development and Innovation of the Republic of Serbia, Contract Nos. 451–03–47/2023–01/200017.

REFERENCES

1. J. I. Musher, The chemistry of hypervalent molecules, *Angew. Chem. Int. Ed*, 1969, 8, 54–68. https://doi:10.1002/anie.196900541

2. H. Kudo, C. H. Wu, H. R. Ihle, Mass-spectrometric study of the vaporization of Li_2O(s) and thermochemistry of gaseous LiO, Li_2O, Li_3O, and Li_2O_2, *J. Nucl. Mater.*, 1978, 78, 380, 380–389. https://doi.org/10.1016/0022-3115(78)90460-9
3. P. V. R. Schleyer, E. U. Würthwein, J. A. Pople, Effectively hypervalent first-row molecules. 1. Octet rule violations by OLi_3 and OLi_4, *J. Am. Chem. Soc.*, 1982, 104, 21, 5839–5841. https://doi.org/10.1021/ja00385a072
4. P. V. R. Schleyer, E. U. Würthwein, E. Kaufmann, T. Clark, J. A. Pople, effectively hypervalent molecules. 2. Lithium carbide (CLi_5), lithium carbide (CLi_6), and the related effectively hypervalent first row molecules, $CLi_{5-n}H_n$ and $CLi_{6-n}H_n$, *J. Am. Chem. Soc.*, 1983, 105, 18, 5930–5932. https://doi.org/10.1021/ja00356a045
5. W. J. Here, L. Radom, P. V. R. Schleyer, J. A. Pople, *Ab Initio Molecular Orbital Theory*, John Wiley & Sons, New York, 1986, 425.
6. P. V. R. Schleyer, *New Horizons of Quantum Chemistry*, ed. by P. O. Lowdin, B. Pullman, Reidel Publisher, Dordrecht, 1983, 95–109.
7. H. Kudo, Observation of hypervalent CLi_6 by Knudsen-effusion mass spectrometry, *Nature*, 1992, 355, 432–434. https://doi.org/10.1038/355432a0
8. G. L. Gutsev, A. I. Boldyrev, DVM Xα calculations on the electronic structure of "superalkali" cations, *Chem. Phys. Lett.*, 1982, 92, 3, 262–266. https://doi.org/10.1016/0009-2614(82)80272-8
9. C. H. Wu, The stability of the molecules Li_4O and Li_5O, *Chem. Phys. Lett.*, 1987, 139, 3–4, 357–359. https://doi.org/10.1016/0009-2614(87)80571-7
10. K. Yokoyama, N. Haketa, M. Hashimoto, K. Furukawa, H. Tanaka, H. Kudo, Production of hyperlithiated Li_2F by a laser ablation of LiF–Li_3N mixture, *Chem. Phys. Lett.*, 2000, 320, 5–6, 645–650. https://doi.org/10.1016/s0009-2614(00)00260-8
11. H. Kudo, K. Yokoyama, C. H. Wu, The stability and structure of the hyperlithiated molecules Li_3S and Li_4S: An experimental and *ab Initio* study, *J. Chem. Phys.*, 1994, 101, 5, 4190. https://doi.org/10.1063/1.467469
12. H. Kudo, K. F. Zmbov, Observation of gaseous Li_4P: A hypervalent molecule, *Chem. Phys. Lett.*, 1991, 187, 1–2, 77–80. https://doi.org/10.1016/0009-2614(91)90487-T
13. H. Kudo, M. Hashimoto, K. Yokoyama, C. H. Wu, A. E. Dorigo, F. M. Bickelhaupt, P. V. R. Schleyer, structure and stability of the Li_2CN molecule: An experimental and *ab initio* study, *J. Phys. Chem.*, 1995, 99, 17, 6477–6482. https://doi.org/10.1021/j100017a030
14. H. Tanaka, K. Yokoyama, H. Kudo, Structure and energetics of $Li_n(OH)_{n-1}$ (n=2–5) clusters deduced from photoionization efficiency curves, *J. Chem. Phys.*, 2001, 114, 1, 152. https://doi.org/10.1063/1.1329645
15. P. D. Dao, K. I. Peterson, A. W. Castleman Jr., The photoionization of oxidized metal clusters, *J. Chem. Phys.*, 1984, 80, 1, 563–564. https://doi.org/10.1063/1.446431
16. E. U. Würthwein, P. V. R. Schleyer, J. A. Pople, Effectively hypervalent molecules. 3. Hypermetalation involving sodium: ONa_3, ONa_4, $HONa_2$, and $HONa_3$, *J. Am. Chem. Soc.*, 1984, 106, 23, 6973–6978. https://doi.org/10.1021/ja00335a018
17. M. Hashimoto, K. Yokoyama, H. Kudo, C. H. Wu, P. V. R. Schleyer, structure and stability of the hypervalent Na_2CN molecule: An experimental and *ab initio* study, *J. Phys. Chem.*, 1996, 100, 39, 15770–15773. https://doi.org/10.1021/jp960829t
18. H. Kudo, M. Hashimoto, K. Yokoyama, C. H. Wu, P. V. R. Schleyer, Thermodynamic stability and optimized structures of hypervalent molecules M_2CN (M = Li, Na, K), *Thermochim. Acta*, 1997, 299, 1–2, 113–121. https://doi.org/10.1016/S0040-6031(97)00146-9
19. A. I. Boldyrev, I. L. Shamovsky, P. V. R. Schleyer, Ab initio prediction of the structure and stabilities of the hypermagnesium oxide molecules: Mg_2O, Mg_3O, and Mg_4O, *J. Am. Chem. Soc.*, 1992, 114, 16, 6469–6475. https://doi.org/10.1021/ja00042a027

20. A. I. Boldyrev, J. Simons, P. V. R. Schleyer, Ab initio study of the hypermagnesium Mg_2O^+ and Mg_3O^+ cations, *Chem. Phys. Lett.*, 1995, 233, 3, 266–273. https://doi.org/10.1016/0009-2614(94)01443-Y
21. A. I. Boldyrev, P. V. R. Schleyer, Ab initio prediction of the structures and stabilities of the hyperaluminum molecules: Al_3O and square-planar Al_4, *J. Am. Chem. Soc.*, 1991, 113, 24, 9045–9054. https://doi.org/10.1021/ja00024a003
22. A. I. Boldyrev, J. Simons, Ab initio study of the silicon oxide (Si_2O and Si_3O) molecules, *J. Phys. Chem.*, 1993, 97, 22, 5875–5881. https://doi.org/10.1021/j100124a016
23. R. Gillespie, The octet rule and hypervalence: Two misunderstood concepts, *Coord. Chem. Rev.*, 2002, 233–234, 53–62. https://doi.org/10.1016/S0010-8545(02)00102-9
24. E. Honea, M. Homer, P. Labastie, R. Whetten, Localization of an excess electron in sodium halide clusters, *Phys. Rev. Lett.*, 1983, 63, 4, 394–397. https://doi.org/10.1103/PhysRevLett.63.394
25. E. Honea, M. Homer, R. Whetten, Electron binding and stability of excess-electron alkali halide clusters: Localization and surface states, *Phys. Rev. B*, 1993, 47, 12, 7480–7493. https://doi.org/10.1103/PhysRevB.47.7480
26. U. Landman, D. Scharf, J. Jortner, Electron localization in alkali-halide clusters, *Phys. Rev. Lett.*, 1985, 54, 16, 1860–1863. https://doi.org/10.1103/PhysRevLett.54.1860
27. U. Landman, D. Scharf, J. Jortner, Electron excitation dynamics, localization and solvation in small clusters, *J. Phys. Chem.*, 1987, 91, 19, 4890–4899. https://doi.org/10.1021/j100303a005
28. J. Jortner, D. Scharf, U. Landman, Elemental and molecular clusters, *Springer Ser. Mat. Science*, 1988, 6, 148.
29. P. Hebant, G. Picard, Conformational and thermodynamic studies of alkaline subhalides M_2X (M = Li, Na, K, Rb; X = F, Cl, Br, I), *J. Molec. Structure: Theochem.*, 1997, 390, 1–3, 121–126. https://doi.org/10.1016/S0166-1280(96)04766-5
30. P. Hebant, G. Picard, Electrochemical investigations of the liquid lithium/(LiCl–KCl eutectic melt) interface: Chronopotentiometric and electrochemical impedance spectroscopy measurements, *Electrochim. Acta*, 1998, 43, 14–15, 2071–2081. https://doi.org/10.1016/S0013-4686(97)10141-4
31. P. Hebant, G. Picard, Equilibrium reactions between molecular and ionic species in pure molten LiCl and in LiCl + MCl (M = Na, K, Rb) melts investigated by computational chemistry, *J. Molec. Structure: Theochem.*, 1995, 358, 1–3, 39–50. https://doi.org/10.1016/0166-1280(95)04359-4
32. E. Rehm, A. Boldyrev, P. V. R. Schleyer. Ab initio study of superalkalis. First ionization potentials and thermodynamic stability, *Inorg. Chem.*, 1992, 31, 23, 4834–4842. https://doi.org/10.1021/ic00049a022
33. G. L. Gutsev, A. I. Boldyrev, DVM-Xα calculations on the ionization potentials of MX_{k+1}^- complex anions and the electron affinities of MX_{k+1} "superhalogens", *Chem. Phys.*, 1981, 56, 3, 277–283. https://doi.org/10.1016/0301-0104(81)80150-4
34. S. Veličković, V. Djorđević, J. Cvetićanin, J. Djustebek, M. Veljković, O. Nešković, Ionization energies of Li_nX (n = 2,3; X = Cl, Br, I) molecules, *Rapid Commun. Mass Spectrom.*, 2006, 20, 20, 3151–3153. https://doi.org/10.1002/rcm.2712
35. J. Djustebek, S. Veličković, S. Jerosimić, M. Veljković, Mass spectrometric study of the structures and ionization potential of Li_nI (n=2, 4, 6) clusters, *J. Anal. Atom. Spectrom*, 2011, 26, 1641–1647. https://doi.org/10.1039/C1JA10078E
36. S. Veličković, J. Djustebek, F. Veljković, B. Radak, M. Veljković, Formation and ionization energies of small chlorine-doped lithium clusters by thermal ionization mass spectrometry, *Rapid Commun. Mass Spectrom.*, 2012, 26, 4, 443–448. https://doi.org/10.1002/rcm.6122

37. S. Veličković, J. Djustebek, F. Veljković, M. Veljković, Formation of positive cluster ions Li_nBr (n = 2–7) and ionization energies studied by thermal ionization mass spectrometry, *J. Mass Spectrom.*, 2012, 47, 5, 627–631. https://doi.org/10.1002/jms.3001
38. F. Veljković, J. Djustebek, M. Veljković, S. Veličković, A. Perić-Grujić, Production and ionization energies of K_nF (n=2–6) clusters by thermal ionization mass spectrometry, *Rapid Commun. Mass Spectrom.*, 2012, 26, 16,1–6. https://doi.org/10.1002/rcm.6284
39. F. Veljković, J. Djustebek, M. Veljković, A. Perić-Grujić, S. Veličković, Study of small chlorine-doped potassium clusters by thermal ionization mass spectrometry, *J. Mass Spectrom.*, 2012, 47, 11,1495–1499. https://doi.org/10.1002/jms.3076
40. M. Milovanović, S. Veličković, F. Veljković, S. Jerosimić, Structure and stability of small lithium-chloride $Li_nCl_m^{(0,1+)}$ (n < m, n = 1–6, m = 1–3) clusters, *Phys. Chem. Chem. Phys.*, 2017, 19, 45, 30481–30497. https://doi.org/10.1039/C7CP04181K
41. B. Milovanović, M. Milovanović, S. Veličković, F. Veljković, A. Perić-Grujić, S. Jerosimić, Theoretical and experimental investigation of geometry and stability of small potassium-iodide K_nI (n = 2–6) clusters, *Int. J. Quantum Chem.*, 2019, 119, 26009–26026. https://doi.org/10.1002/qua.26009
42. M. Mitić, M. Milovanović, F. Veljković, A. Perić-Grujić, S. Veličković, S. Jerosimić, Theoretical and experimental study of small potassium-bromide $K_nBr^{(0,1+)}$ (n =2–6) and $K_nBr_{n-1}^{(0,1+)}$ (n = 3–5) clusters, *J. Alloy Compd.*, 2020, 835, 155301–155310. https://doi.org/10.1016/j.jallcom.2020.155301
43. A. C. Reber, S. N. Khanna, A. W. J. Castleman, superatom compounds, clusters, and assemblies: Ultra alkali motifs and architectures, *J. Am. Chem. Soc.*, 2007, 129, 33, 10189–10194. https://doi.org/10.1021/ja071647n
44. S. A. Claridge, A. W. Castelman, S. N. Khanna, C. B. Murray, A. Sen, P. S. Weiss, Cluster-assembled materials, *ACS Nano.*, 3, 2009, 3, 2, 244–255. https://doi.org/10.1021/nn800820e
45. C. Paduani, Magnetic hyperalkali species of Gd-based clusters, *J. Phys. Chem. A*, 2018, 122, 22, 5037–5042. https://doi.org/10.1021/acs.jpca.8b02775
46. X. Zhang, Y. Wang, H. Wang, A. Lim, G. Gantefoer, H. K. Bowen, J. U. Reveles, S. N. Khanna, On the existence of designer magnetic superatoms, *J. Am. Chem. Soc*, 2013, 135, 12, 4856–4861. https://doi.org/10.1021/ja400830z
47. Y. Li, D. Wu, Z. R. Li, Compounds of superatom clusters: Preferred structures and significant nonlinear optical properties of the BLi_6-X (X = F, LiF_2, BeF_3, BF_4) motifs, *Inorg. Chem.*, 2008, 47, 21, 9773–9778. https://doi.org/10.1021/ic800184z
48. H. Park, G. Meloni, Reduction of carbon dioxide with a superalkali, *Dalton Trans.*, 2017, 46, 35, 11942–11949. https://doi.org/10.1039/C7DT02331F
49. A. K. Srivastava, Reduction of nitrogen oxides (NO_x) by superalkalis, *Chem. Phys. Lett.*, 2018, 695, 205–210. https://doi.org/10.1016/j.cplett.2018.02.029
50. H. Park, G. Meloni, Activation of dinitrogen with a superalkali species, Li_3F_2, *Chem. Phys. Chem.*, 2018, 19, 256–260. https://doi.org/10.1002/cphc.201800089
51. D. Roy, A. N. Vazquez, P. V. R. Schleyer, Modeling dinitrogen activation by lithium: A mechanistic investigation of the cleavage of N_2 by stepwise insertion into small lithium clusters, *J. Am. Chem. Soc.*, 2009, 131, 13045–13053. https://doi.org/10.1021/ja902980j
52. K. Wang, Z. Liu, X. Wang, X. Cui, Enhancement of hydrogen binding affinity with low ionization energy Li_2F coating on C_{60} to improve hydrogen storage capacity, *Int. J. Hydrogen Energy*, 2014, 39, 28, 15639–15645. https://doi.org/10.1016/j.ijhydene.2014.07.132

53. C. Binns, Nanoclusters deposited on surfaces, *Surf. Sci. Rep.*, 2001, 44, 1–2, 1–49. https://doi.org/10.1016/S0167-5729(01)00015-2
54. C. E. Rechsteiner, R. P. Buck, L. Pedersen, Experimental and theoretical studies on M_2X^+ (M=Li, Na; X=F, Cl), *J. Chem. Phys.*, 1976, 65, 1659. https://doi.org/10.1063/1.433310
55. F. W. Röllgen, "Principles of field desorption mass spectrometry (review)": Ion formation from organic solids. *Springer Ser. Chem. Phys.*, 1983, 25, 2–13. https://doi.org/10.1007/978-3-642-87148-1_1
56. C. E. Rechsteiner, T. L. Youngless, M. M. Bursey, R.-P. Buck, Field desorption of cluster ions: A method of testing the mechanism of ion formation, *Int. J. Mass. Spectrom. Ion Phys.*, 1978, 28, 4, 401–407. https://doi.org/10.1016/0020-7381(78)80082-5
57. X. B. Cox, R. W. Linton, M. M. Bursey, Formation of small cluster ions from alkali halides in SIMS, *Int. J. Mass. Spectrom.*, 1984, 55, 3, 281–290. https://doi.org/10.1016/0168-1176(84)87091-3
58. J. E. Campana, T. M. Barlak, R. J. Colton, J. J. DeCorpo, J. R. Wyatt, B. I. Dunlap, Effect of cluster surface energies on secondary-ion-intensity distributions from ionic crystals, *Phys. Rev. Lett.*, 1981, 47, 15, 1046–1049. https://doi.org/10.1103/PhysRevLett.47.1046
59. J. M. Miller, R. Theberge, Fast atom bombardment mass spectrometry of alkali metal fluorides and chlorides; a comparison of the solid salt and glycerol matrix spectra, *Organic Mass Spectrom.*, 1985, 20, 10, 600–605. https://doi.org/10.1002/oms.1210201003
60. M. A. Baldwin, C. J. Proctor, J. Amster, F. W. Mclaffertx, The behaviour of caesium iodide cluster ions produced by fast-atom bombardment, *Int. J. Mass Spectrom.*, 1983, 54, 1–2, 97–107. https://doi.org/10.1016/0168-1176(83)85009-5
61. F. A. Fernandez-Lima, M. A. C. Nascimento, E. F. da Silveira, Alkali halide clusters produced by fast ion impact, *Nucl. Instrum. Meth. B*, 2012, 273, 102–104. https://doi.org/10.1016/j.nimb.2011.07.050
62. A. P. Bruins, Mechanistic aspects of electrospray ionization, *J. Chromatogr. A*, 1998, 794, 1–2, 345–357. https://doi.org/10.1016/S0021-9673(97)01110-2
63. S. Banerjee, S. Mazumdar, Electrospray ionization mass spectrometry: A technique to access the information beyond the molecular weight of the analyte, *Int. J. Environ. Anal. Chem.*, 2012, 2012, 1–40. https://doi.org/10.1155/2012/282574
64. A. T. Blades, M. Peschke, U. H. Verkerk, P. Kebarle, Hydration energies in the gas phase of select $(MX)_mM^+$ ions, where $M^+ = Na^+, K^+, Rb^+, Cs^+, NH_4^+$ and $X^- = F^-, Cl^-, Br^-, I^-, NO_2^-, NO_3^-$, observed magic numbers of $(MX)_mM^+$ ions and their possible significance, *J. Am. Chem. Soc.*, 2004, 126, 38, 11995–12003. https://doi.org/10.1021/ja030663r
65. A. Wakisaka, Nucleation in alkali metal chloride solution observed at the cluster level, *Faraday Discuss.*, 2007, 136, 299–308. https://doi.org/10.1039/B615977J
66. C. Hao, R. E. March, T. R. Croley, J. C. Smith, S. P. Rafferty, Electrospray ionization tandem mass spectrometric study of salt cluster ions: Part 1 – Investigations of alkali metal chloride and sodium salt cluster ions, *J. Mass Spectrom.*, 2001, 36, 79–96. https://doi.org/10.1002/jms.107
67. G. Wang, R. B. Cole, Solvation energy and gas-phase stability influences on alkali metal cluster ion formation in electrospray ionization mass spectrometry, *Anal. Chem.*, 1998, 70, 873–881. https://doi.org/10.1021/ac970919+
68. G. Wang, R. B. Cole, Charged residue versus ion evaporation for formation of alkali metal halide cluster ions in ESI, *Anal. Chim. Acta*, 2000, 406, 1, 53–65. https://doi.org/10.1016/S0003-2670(99)00599-1

69. www.crawfordscientific.com/chromatography-blog/post/the-beauty-of-the-quadrupole-mass-analyser
70. www.utsc.utoronto.ca/~traceslab/PDFs/MassSpec_QuadsInfo.pdf
71. R.E.March,Anintroductiontoquadrupoleiontrapmassspectrometry,*J.MassSpectrom.*, 1997, 32, 351–369. https://doi.org/10.1002/(SICI)1096-9888(199704)32:4<351::AID-JMS512>3.0.CO;2-Y
72. B. Jost, B. Schueler, F. R. Krueger, Ion formation from alkali halide solids by high power pulsed laser irradiation, *Z. Naturforsch.*, 1982, 37, 7, 18–2. https://doi.org/10.1515/zna-1982-0106
73. F. Hillenkamp, M. Karas, *MALDI MS – A Practical Guide to Instrumentation, Methods and Applications*, Wiley – VCH GmbH & Co. KGaA, Weinheim, 2001, 3, 1–28. ISBN: 978-3-527-61046-4
74. J. E. Campana, T. M. Barlak, R. J. Colton, J. J. DeCorpo, J. R. Wyatt, B. I. Dunlap, Effect of cluster surface energies on secondary-ion-intensity distributions from ionic crystals, *Phys. Rev. Lett.*, 1981, 47, 15, 1046–1049. https://doi.org/10.1103/PhysRevLett.47.1046
75. R. Pflaum, P. Pfau, K. Sattler, E. Recknagel, Electron impact studies on sodium halide microclusters, *Surf. Sci.*, 1985, 156, 1,165–172. https://doi.org/10.1016/0039-6028(85)90570-9
76. C. W. S. Conover, Y. A. Yang, L. A. Bloomfield, Laser vaporization of solids into an inert gas: A measure of high-temperature cluster stability, *Phys. Rev. B*, 1988, 38, 5, 3517–3520. https://doi.org/10.1103/PhysRevB.38.3517
77. P. Labastie, J. M. L'Hermite, Ph. Poncharal, Spectral signatures and metallization sequences of alkali-halide clusters, *Z. Phys. Chem. Bd.*, 1998, 203, 15–35. https://doi.org/10.1524/zpch.1998.203.Part_1_2.015
78. E. C. Honea, M. L. Homer, P. Labastie, R. L. Whetten, Localization of an excess electron in sodium halide clusters, *Phys. Rev. Lett.*, 1989, 63, 4, 394–397. https://doi.org/10.1103/PhysRevLett.63.394
79. P. Labastie, J.-M. L'Hermite, P. Poncharal, M. Sence, Two-photon ionization of alkali-halide clusters spectroscopy of excess-electron excited states, *J. Chem. Phys.*, 1995, 103, 15, 6362–6367. https://doi.org/10.1063/1.470417
80. G. Rajagopal, R. N. Barnett, U. Landman, Metallization of ionic clusters, *Phys. Rev. Lett.*, 1991, 67, 6, 727–730. https://doi.org/10.1103/PhysRevLett.67.727
81. P. Jena, S. N. Khanna, K. Rao, *Physics and Chemistry of Finite Systems: From Clusters to Crystals*, Vol. I, ed. by U. Landman, R. N. Barnett, C. L. Cleveland, G. Rajagopal, Kluwer Academic Publishers, Dordrecht/Boston/London, Published in Cooperation with NATO Scientific Affairs Division, 1992, 165.
82. Y. J. Twu, C. W. S. Conover, Y. A. Yang, L. A. Bloomfield, Alkali-halide cluster ions produced by laser vaporization of solids, *Phys. Rev. B*, 1990, 42, 8, 5306–5316. https://doi.org/10.1103/PhysRevB.42.5306
83. P. Lievens, P. Thoen, S. Bouckaert, W. Bouwen, E. Vandeweert, F. Vanhoutte, H. Weidele, R. E. Silverans, Threshold photoionization behaviour of $(Li_2O)Li_n$ clusters produced by a laser vaporization source, *Z. Phys. D*, 1997, 42, 231–235. https://doi.org/10.1007/s004600050359
84. P. Lievens, P. Thoen, S. Bouckaert, W. Bouwen, F. Vanhoutte, H. Weidele, R. E. Silverans, A. Navarro-Vazquez, P. V. R. Schleyer, Ionization potentials of Li_nO ($2 \leqslant n \leqslant 70$) clusters: Experiment and theory, *J. Chem. Phys.*, 1999, 110, 21, 10316–10329. https://doi.org/10.1063/1.478965
85. F. Despa, W. Bouwen, F. Vanhoutte, P. Lievens, R. E. Silverans, The influence of O and C doping on the ionization potentials of Li-clusters, *Eur. Phys. J. D.*, 2000, 11, 403–411. https://doi.org/10.1007/s100530070069

86. E. de Hoffmann, V. Stroobant, *Mass Spectrometry Principles and Applications*, 3rd Edition, John Wiley & Sons Ltd, The Atrium, Southern Gate, Chichester, West Sussex.
87. O. W. Richardson, On the positive ionization produced by hot platinum in air at low pressures, *Phill. Mag.*, 1903, 6, 80–96. http://dx.doi.org/10.1080/14786440309462993
88. K. N. Kingdon, J. Langmuir, Thermionic phenomena due to alkali vapors, *Phys. Rev.*, 1923, 21, 380–382. https://doi.org/10.1103/PhysRev.21.366
89. H. Ives, Thermionic phenomena due to alkali vapors, *Phys. Rev.*, 1923, 21, 385, https://doi.org/10.1103/PhysRev.21.366
90. I. Langmuir, K. Kingdon, Thermionic effects caused by vapours of alkali metals, *Proc. R. Soc. Lond. A*, 1925, 107, 61. https://doi.org/10.1098/rspa.1925.0005
91. M. Saha, On the physical properties of elements at high temperatures, *Phill. Mag.*, 1923, 46, 534–543. http://dx.doi.org/10.1080/14786442308634276
92. J. Taylor, I. Langmuir, The evaporation of atoms, ions and electrons from caesium films on tungsten, *Phys. Rev.*, 1933, 44, 423. https://doi.org/10.1103/PhysRev.44.423
93. N. D. Morgulis, On the quantum-mechanical theory of ionization and neutralization on a metal surface, *Zh. Eksp. Teor. Fiz.*, 1934, 4, 7, 684–689.
94. P. P. Sutton, J. E. Mayer, A direct experimental determination of electron affinities, the electron affinity of iodine, *J. Chem. Phys.*, 1935, 3, 20. https://doi.org/10.1063/1.1749548
95. E. Y. Zandberg, N. I. Iosov, *Poverkhnosmaya lonizatsiya (Surface Ionization)*, Nauka, Moscow, 1969.
96. M. Studier, E. Sloth, L. Moore, The chemistry of uranium in surface ionization sources, *J. Phys. Chem.*, 1962, 66, 1, 133–134. https://doi.org/10.1021/j100807a029
97. Y. Kawaia, M. Nomura, Y. Fujii, T. Suzuki, Calcium ion generation from calcium iodide by surface ionization in mass spectrometry, *Int. J. Mass Spectrom.*, 1999, 193, 1, 29–34. https://doi.org/10.1016/S1387-3806(99)00134-7
98. Y. Kawaia, M. Nomura, H. Murata, T. Suzuki, Y. Fujii, Surface ionization of alkaline-earth iodides in double-filament system, *Int. J. Mass Spectrom.*, 2001, 26, 1–2,1–5. https://doi.org/10.1016/S1387-3806(00)00351-1
99. R. Arevalo Jr, Z. Ni, R. M. Danell, Mass spectrometry and planetary exploration: A brief review and future projection, *J. Mass Spectrom.*, 2020, 55, e4454. https://doi.org/10.1002/jms.4454
100. Y. K. Xiao, H. Z. Wei, W. G. Liu, Q. Z. Wang, Y. M. Zhou, Y. H. Wang, H. Lu, Emission of M_2X^+ cluster ions in thermal ionization mass spectrometry in the presence of graphite, *Fresenius J. Anal. Chem.*, 2001, 371, 1098–1103. https://doi.org/10.1007/s002160101086
101. Y. K. Xiao, C. C. Zhang, High precision isotopic measurement of chlorine by thermal ionization mass spectrometry of the Cs_2Cl^+ ion, *Im. J. Mass Spectrom. Ion. Processes*, 1992, 116, 3, 183–192. https://doi.org/10.1016/0168-1176(92)80040-8
102. Y. K. Xiao, Y. M. Zhou, W. G. Liu, Precise Measurement of Chlorine Isotopes Based on Cs_2Cl_2 by Thermal Ionization Mass Spectrometry, *Anal. Lett.*, 1995, 28, 7, 1295–1304. https://doi.org/10.1080/00032719508000346
103. Y. K. Xiao, H. Lu, C. C. Zhang, Q. Wang, H. Wei, A. Sun, W. Liu, Major factors affecting the isotopic measurement of chlorine based on the Cs_2Cl^+ ion by thermal ionization mass spectrometry, *Anal. Chem.*, 2002, 74, 11, 2458–2464. https://doi.org/10.1021/ac0107352
104. C. Volpe, A. J. Spivack, Stable chlorine isotopic composition of marine aerosol particles in the western Atlantic Ocean, *Geophys. Res. Lett.*, 1994, 21, 1161–1164. https://doi.org/10.1029/94GL01164

105. D. A. Banks, R. Green, R. A. Cliff, B. W. D. Yardley, Chlorine isotopes in fluid inclusions: Determination of the origins of salinity in magmatic fluid, *Geochim. Cosmochim. Acta*, 2000, 64, 10, 1785–1789. https://doi.org/10.1016/S0016-7037(99)00407-X
106. A. J. Magenheim, A. J. Spivack, C. Volpe, B. Ransom, Precise determination of stable chlorine isotopic ratios in low-concentration natural samples, *Geohim. Cosmochim. Acta*, 1994, 58, 14, 3117–3121. https://doi.org/10.1016/0016-7037(94)90183-X
107. B. Ransom, A. I. Spivack, M. Kastner, Stable Cl isotopes in subduction-zone pore waters: Implications for fluid-rock reactions and the cycling of chlorine, *Geology*, 1995, 23, 8, 715–718. https://doi.org/10.1130/0091-7613(1995)023<0715:SCIISZ>2.3.CO;2
108. O. M. Nesković, M. V. Veljković, S. R. Veličković, A. J. Đerić, N. R. Miljević, D. D. Golobočanin, Precise measurement of chlorine isotopes by thermal ionization mass spectrometry, *Nukleonika*, 2002, 47, 85–87. ISSN: 0029–5922
109. J. Đustebek, M. Veljković, S. Veličković, Major factors affecting the emission of dilithium-fluoride cluster ion in thermal ionization mass spectrometry, *Dig. J. Nanomater. Bios.*, 2013, 8, 359–366. www.chalcogen.ro/359_DUSTEBEK.pdf
110. S. Veličković, V. Koteski, J. Belošević Čavor, V. Đorđević, J. Cvetićanin, J. Đustebek, M. Veljković, O. Nešković, Experimental and theoretical investigation of new hypervalent molecules Li_nF (n=2–4), *Chem. Phys. Lett.*, 2007, 448, 4–6, 151–155. https://doi.org/10.1016/j.cplett.2007.09.082
111. S. Veličković, V. Đorđević, J. Cvetićanin, J. Đustebek, M. Veljković, O. Nešković, Ionization energies of the non-stoichiometric Li_nF_{n-1} (n=3, 4, 6) clusters, *Vacuum*, 2009, 83, 2, 378–380. https://doi.org/10.1016/j.vacuum.2008.05.026
112. K. Yokoyama, N. Haketa, H. Tanaka, K. Furukawa, H. Kudo, Ionization energies of hyperlithiated Li_2F molecule and Li_nF_{n-1} (n=3,4) clusters, *Chem. Phys. Lett.*, 2000, 330, 339–346, https://doi.org/10.1016/S0009-2614(00)01109-X
113. M. Knudsen, Die Gesetze der Molekularströmung und der inneren Reibungsströmung der Gase durch Röhren, *Annalen der Physik.*, 1909, 28, 75–130. http://dx.doi.org/10.1002/andp.19093330106
114. N. I. Ionov, Ionisation of KI, NaI and CsCl molecules by electrons. *Dokl. Akad. Nauk. SSSR.*, 1948, 59, 467–469.
115. E. H. Copland, N. S. Jacobson, Thermodynamic activity measurements with knudsen cell mass spectrometry, *Electrochem. Soc. Interface*, 2001, 10, 28–31. https://doi.org/10.1149/2.F05012IF
116. R. T. Grimley, J. A. Forsman, *Characterization of High Temperature Vapors by Angular Distribution Mass Spectrometry*, Proceedings of the 10th Materials Research Symposium Held at the National Bureau of Standards, Gaithersburg, MD, September 18–22, 1978, 211.
117. C. H. Wu, Experimental investigation of a stable lithium cluster: The thermochemical study of the molecule Li_4, *J. Phys. Chem.*, 1983, 87, 9, 1534–1540. https://doi.org/10.1021/j100232a017
118. C. H. Wu, Thermochemical properties of the lithium clusters Li_5, *J. Chem. Phys.*, 1989, 91, 546–551. https://doi.org/10.1063/1.457491
119. J. Berkowitz, H. A. Tasman, W. A. Chupka, Double-oven experiments with lithium halide vapors, *J. Chem. Phys.*, 1962, 36, 8, 2170–2179. https://doi.org/10.1063/1.1732848
120. S. Veličković, X. Kongo, *"Superalkali" Clusters, Production, Potential Application Like Energy Storage Materials*, 8th International Conference on Renewable Electrical Power Sources, Belgrade, 2020, 16, 15–22. ISBN 978-86-85535-06-2

121. J. Đustebek, M. Milovanović, S. Jerosimić, M. Veljković, S. Veličković, Theoretical and experimental study of the non-stoichiometric Li_nI (n = 3 and 5) clusters, *Chem. Phys. Letter*, 2013, 556, 380–385. https://doi.org/10.1016/j.cplett.2012.11.086
122. J. M. A. Frazão, J. M. C. Lourenço, M. Áurea Cunha, M. F. Laranjeira, J. Los, A. M. C. Moutinho, The sulfur reaction in small ionized carbonyl sulfide clusters, *J. Chem. Phys.*, 1996, 104, 8393–8404. https://doi.org/10.1063/1.471589
123. M. Veljković, O. Nešković, L. Petkovska, K. Zmbov, On autoionizing states in Xe-electron impact study, *J. Serb. Chem. Soc.*, 1990, 55, 3, 153–161.
124. J. Dustebek, S. Veličković, F. Veljković, M. Veljković, Production of heterogeneous superalkali clusters LinF (n=2–6) by Knudsen cell mass spectrometry, *Dig. J. Nanomater. Bios.*, 2012, 7, 1365–1372. https://chalcogen.ro/1365_Dustebek.pdf

8 Superalkalis in the Design of Strong Bases and Superbases

Harshita Srivastava and Ambrish Kumar Srivastava

8.1 INTRODUCTION

A fundamental attribute that would be crucial for comprehending many chemical and biological processes is the propensity of a compound to either receive or release a proton. The phenomena of basicity helped scientists to comprehend the connections between structure and reactivity across time. A key aspect of determining the reactivity of a compound is how basic it is in the gas phase. Basicity is the protons-accepting capacity of a molecule or an ion. The tendency of neutral bases to spread their positive charge upon the protonation determines its basicity [1]. Strong bases not only act as electron donors but also as acceptors of protons (H^+ ions) having large proton affinity (PA). They can deprotonate weak acids and form aqueous sluggish and foamy solutions. Some of the popular inorganic bases that completely dissociate into cation and hydroxide (OH^-) in water are alkali and alkaline metal hydroxides [2] such as LiOH, NaOH, $Ca(OH)_2$, *etc.* The PA of these strong bases is larger than that of weak bases such as $Al(OH)_3$, NH_4OH, *etc.* Superbases are molecules with larger PA than these strong bases [3]. As per Caubere [4], the word "superbase" should refer to bases formed by the reaction of two (or maybe more) bases, resulting in new basic species with inherent new features. Examples of superbases include lithium diisopropylamide [$LiN(CH(CH_3)_2)_2$] and several organic superbases such as phosphazenes, phosphanes, amidines and guanidines. In 2015, Srivastava and Misra [5] proposed for the first time a family of even stronger bases than alkali hydroxide using the species whose ionization energy (IE) is lower than alkali atoms, known as superalkalis.

Superalkalis were proposed by Gutsev and Boldyrev in 1982 [6]. According to them, superalkali is a hypervalent cluster having lower IE than alkali atoms. They proposed a generalized formula for superalkalis, XM_{k+1}, where k is the formal valence of the electronegative element (X). Several theoretical studies have focused on such exotic species over the past 40 years including traditional mononuclear superalkalis [7] to binuclear superalkali groups [8,9], and eventually to polynuclear superalkalis [10,11]. The first superalkali cluster Li_3O^+ in a gas phase

DOI: 10.1201/9781003384205-8

was experimentally identified by Wu [12] during the mass spectrometric investigation of lithium oxide in early 1976. Other experimental works on superalkalis include XLi_2 (X = F, Cl, Br and I) [13,14], and OM_3 (M = Li, Na and K) [15,16]. There are several applications of superalkalis have been explored in the last decade including the activation of small molecules such as CO_2, nitrogen oxides [17–20], and N_2 [21,22]. This led to the exploration of new species such as nonmetallic superalkali cations [23], aromatic superalkali species [24,25], organic superalkalis [26,27], *etc.*

In this chapter, we demonstrate the role of superalkalis in the design of superbases. Srivastava and Misra [5] designed MOH compounds for M = FLi_2, OLi_3, NLi_4 and compared their basicity with LiOH. Subsequently, they reported hydroxides of FLi_n [28] and Li_n [29] for n = 2–5. This motivated the studies on the exploration of neutral bases by other researchers [30,31]. Anionic bases generally contain more basicity than those of neutral species. Tian et al [32] reported lithium monoxide anion (LiO^-) as the strongest ionic base, with a basicity greater than that of methyl anion (CH_3^-) known previously. Srivastava and Misra extended the idea to design the strongest anionic base to date using superalkali.

8.2 COMPUTATIONAL DETAILS

The results discussed in this chapter were obtained by quantum chemical calculations Gaussian 09 software [33] using an appropriate method and basis set as mentioned in subsequent sections as mentioned during the discussion of results. The equilibrium geometries of MOH and their protonated species (MOH_2^+) were confirmed by real vibrational harmonic frequencies. The natural atomic charges were calculated by using the natural bond orbital (NBO) method [34]. To examine the basicity, PA along with the gas phase basicity (GB) is calculated considering the fictitious gas phase protonation equation as MOH + H^+ → MOH_2^+.

The calculations for the value of PA and GB are carried out using equations (1) and (2) below:

$$PA = \Delta H(\mathrm{MOH}) + \Delta H(\mathrm{H}^+) - \Delta H(\mathrm{MOH}_2^+) \tag{8.1}$$

$$GB = \Delta G(\mathrm{MOH}) + \Delta G(\mathrm{H}^+) - \Delta G(\mathrm{MOH}_2^+) \tag{8.2}$$

where Δ *H* and Δ *G* are defined as thermal enthalpy and Gibb's free energy changes, respectively, calculated at 293.15 K. Δ $H(H^+)$ = +6.1923 kJ/mol and Δ $G(H^+)$ = -26.2755 kJ/mol were taken from published works [35]. The ionic dissociation energy of MOH is obtained by equation (3) in which *E*(..) represents the total electronic energy of the respective species including zero-point energy.

$$\Delta E = E\left(\mathrm{M}^+\right) + E\left(\mathrm{OH}^-\right) - E\left(\mathrm{MOH}\right) \tag{8.3}$$

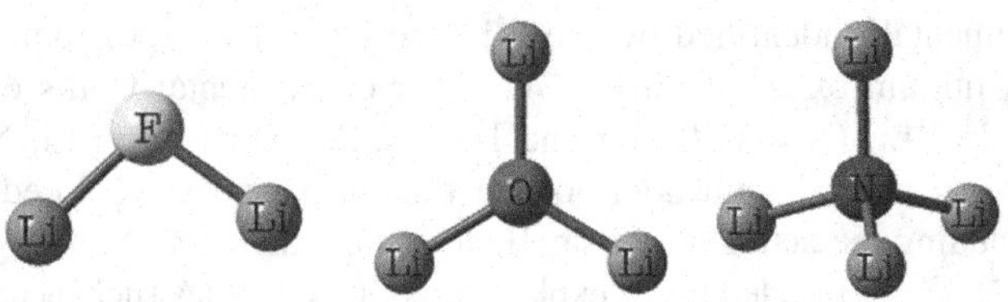

FIGURE 8.1 Equilibrium structures of superalkalis from Ref. [5] with the permission of Royal Society of Chemistry, Copyright 2015.

8.3 RESULTS AND DISCUSSION

Srivastava and Misra [5] reported the hydroxides of typical superalkalis such as FLi_2, OLi_3 and NLi_4 using the MP2 [36] method in 2015. These superalkalis are displayed in Figure 8.1 as proposed by Gutsev and Boldyrev [6]. The IE of these superalkalis follows the decreasing trend, *i.e.*, as 3.85, 3.46 and 3.23 eV for FLi_2, OLi_3 and NLi_4, respectively, lower than those of alkali atoms [37].

Subsequently, they designed their hydroxides, MOH for M = FLi_2, OLi_3 and NLi_4 and compared their properties with the LiOH, a strong base. The equilibrium structures of MOH species are displayed in Figure 8.2. They found that the superalkalis interact through two Li atoms with the OH group in the lowest energy structures of MOH. The bond lengths of superalkali moieties are found to be elongated in MOH due to the charge transfer. However, the OH bond length remains constant (0.95 Å) in all the species including LiOH. Like LiOH, the interaction between superalkalis and the OH group is essentially ionic. This is confirmed by partial charge analysis in which all superalkalis have a positive charge close to unity in their hydroxides [5], just like Li in LiOH. From Figure 8.2, it is also evident that the superalkalis retain their identity in their hydroxides.

To calculate PA and GB values, the authors considered the protonation of MOH and studied MOH_2^+ whose structures are also displayed in Figure 8.2 The PA and GB values calculated by equations (1) and (2) for MOH species are included in Table 8.1. LiOH has PA and GB values of 1,000 and 986 kJ/mol, which was consistent with the values in the literature [38]. One can see that the values of PA and GB of both OLi_3OH and NLi_4OH are found to be greater than LiOH, except for the FLi_2OH molecule. The computed PA and GB values for FLi_2OH are somewhat lower than LiOH but still permit it to be significantly basic. For instance, the PA of $Al(OH)_3$, a weak base is calculated to be 811 kJ/mol using the same method. The authors assumed that the existence of a highly electronegative F atom, whose electron affinity is substantially higher than those of O and N atoms, might be the reason for the reduced PA and GB values of FLi_2OH. Based on the results of OLi_3OH and NLi_4OH, they proposed that superalkali hydroxides may become superbases.

The ionic dissociation energy (ΔE) of MOH (M = FLi_2, OLi_3 and NLi_4) is found to be lower than that of LiOH and gets progressively lowered from $FLi_2 \rightarrow OLi_3 \rightarrow NLi_4$. This trend of decreasing ΔE is found to be consistent with

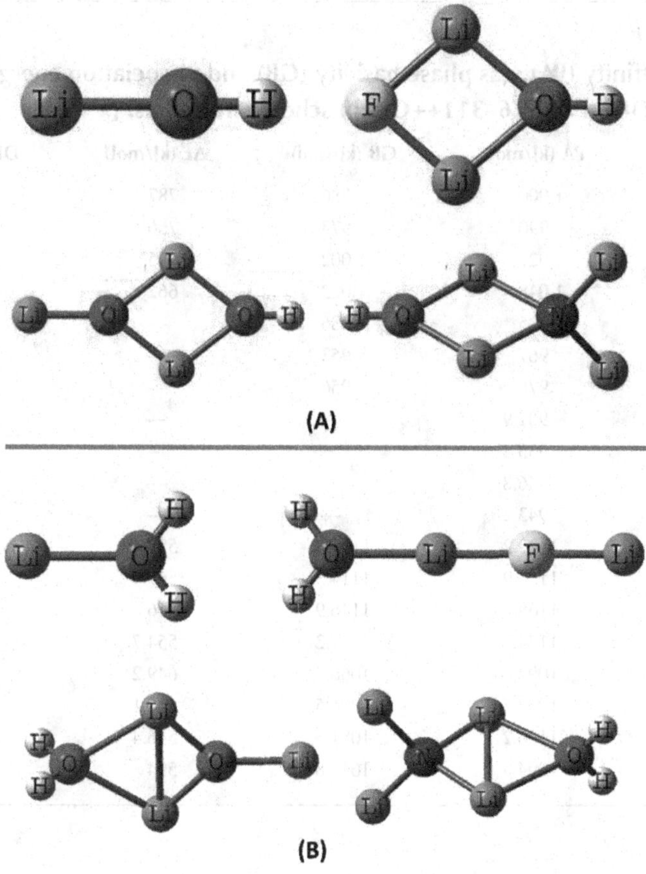

FIGURE 8.2 Equilibrium structures of MOH bases (a) and MOH_2 $^+$protonated bases (b) (M = FLi_2, OLi_3 and NLi_4) from Ref. [5] with the permission of Royal Society of Chemistry, Copyright 2015.

the decrease in the IE of M. Furthermore, the authors confirmed the stability of MOH species against dissociation to LiOH and stable molecules, namely LiF, Li_2O and Li_3N for M = FLi_2, OLi_3 and NLi_4, respectively. These dissociation energy (DE) values are also listed in Table 8.1, which ensures their stability and the possibility to realize the MOH species in the gas phase.

Subsequently, Srivastava and Misra [28] reported FLi_n superalkalis by attaching a hydroxyl group for n = 2–5 using the same MP2 method. These superalkali clusters were studied by Velickovic et al. [39] and Ivanvic et al. [40]. The MP2 computed IEs of FLi_n clusters were in good agreement with experimental results obtained by Djustebek et al. [41]. Schleyer et al. [7] reported that the Li atoms in FLi_2 create a cage of positively charged ions (Li_2^+). The authors noticed, however, that the Li_2^+ molecule is also contained for FLi_3 as well as FLi_4. The lowest

TABLE 8.1
Proton affinity (PA), gas phase basicity (GB) and dissociation energies (ΔE and DE) at MP2/6–311++G(d,p) scheme from Refs. [5,28,29,30].

System	PA (kJ/mol)	GB (kJ/mol)	ΔE (kJ/mol)	DE (kJ/mol)
LiOH	1,000	986	787	—
FLi_2OH	990	973	766	264.0
OLi_3OH	1,027	1,002	695	274.0
NLi_4OH	1,048	1,022	662	259.0
FLi_3OH	905	886	—	212.2
FLi_4OH	967	953	—	172.7
FLi_5OH	976	956	—	127.3
Li_2OH	901.9	—	—	142.7
Li_3OH	935.4	—	—	183.3
Li_4OH	946.8	—	—	250.8
Li_5OH	943.2	—	—	220.9
KOH	1129.2	1104.8	589	—
FK_2OH	1126.9	1114.3	567.9	203.2
OK_3OH	1168.4	1146.9	516	197.1
NK_4OH	1134.7	1117.2	554.7	164.6
NaOH	1092.6	1066.4	649.2	—
FNa_2OH	1073.5	1056.5	645.1	244.6
ONa_3OH	1106.2	1081.5	546.4	246.2
NNa_4OH	1094.9	1079.8	594	220.7

energy structures of the FLi_nOH are shown in Figure 8.3. Like MOH species (Figure 8.2), One can see that the two Li atoms in the FLi_n species interact with the OH group to form FLi_nOH. These species have Li-O bond lengths between 1.796 and 1.801 Å, however, the O-H bond length is 0.957 Å. It was noticed that in FLi_2OH, both Li atoms acquired a charge of 0.52*e* value; thus, forming Li_2^+ moiety, like in FLi_2. The ionic interaction between the Li_2^+-OH and Li_2^+–F and also the covalency between the Li atoms of the Li_2^+ moiety ensures the stability of FLi_2OH [28]. To verify that stability, the authors calculated the dissociation (DE) against the elimination of LiOH, which is also listed in Table 8.1. It is clear from the mentioned data that all DE values are positive, making FLi_nOH species stable for $n = 2–5$ for LiOH elimination. Therefore, it is energetically viable to realize these FLi_nOH species by adding successive Li atoms to LiF and LiOH, at least in the gas phase.

The protonated structures of FLi_nOH are also displayed in Figure 8.3. The PA and GB values are calculated by equations (1) and (2). The structures of $FLi_3OH_2^+$ and $FLi_5OH_2^+$ closely resemble those of their parent systems, with the extra H bonded to OH. $FLi_2OH_2^+$ and $FLi_4OH_2^+$ may be considered as $FLi_2^+(H_2O)$ and

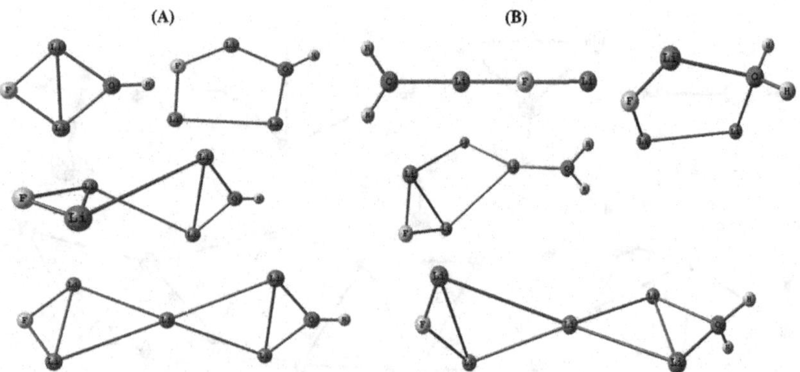

FIGURE 8.3 The optimized structures of the FLi_nOH species ($n = 2–5$) (a) along with its protonated forms (b) from Ref. [28] with the permission of Royal Society of Chemistry, Copyright 2015.

$FLi_4^+(H_2O)$ in which the positive charge is transferred to the FLi_2 and FLi_4 superalkali moieties during the protonation, respectively. From Table 8.1, the computed values PA and GB of FLi_nOH are less than those of LiOH by 50–100 kJ/mol. However, these values are large enough to support the strong basicity of these compounds.

Srivastava and Misra continued to explore the basicity of Li-based compounds. They used [29] small Li_n clusters and studied Li_nOH compounds for $n = 2–5$. It was found that the IE of these Li_n clusters is lower than that of the Li atom [42,43] and, therefore, such species can be treated as superalkali as well. The equilibrium geometries of Li_nOH species are given in Figure 8.4. In comparison to pure Li_n clusters and LiOH molecule, Li_nOH (n =2–5) has longer bonds as expected because of the charge transport from several Li atoms to the OH group. Li_2OH and Li_4OH are the planar molecules whereas there exist two conformers for Li_3OH and Li_5OH. For Li_3OH, the planar structure has lower energy over the nonplanar structure by 0.14 eV, unlike Li_5OH whose planar structure has higher energy by 0.19 eV.

Figure 8.4 also depicts the structures of $Li_nOH_2^+$, protonated bases of Li_nOH. Table 8.1 includes the computed PA values for the Li_nOH. It should be noted that the PA of Li_nOH is 50–100 kJ/mol lower than those of LiOH, comparable to those found for FLi_nOH species [5]. It was seen that adding Li atom to LiOH, causes the PA of Li_2OH and Li_3OH to drop around ~900 kJ/mol. With further addition, however, the PA gradually increases for Li_4OH and Li_5OH, becoming 950 kJ/mol. Therefore, the hypothesis that hydroxides of small Li_n clusters are strong bases in the gas phase offers a promising route for developing new inorganic bases.

Following the work of Srivastava and Misra [5] on typical superalkali based bases, Pandey [30] reported the superalkali hydroxides, $XM_{n+1}OH$ for M = Na and K. The IE of alkali atoms follows the trend IE(Li) > IE(Na) > IE(K). Likewise, the IE of XM_{n+1} superalkalis becomes in the order IE(XLi_{n+1}) > IE(XNa_{n+1}) >

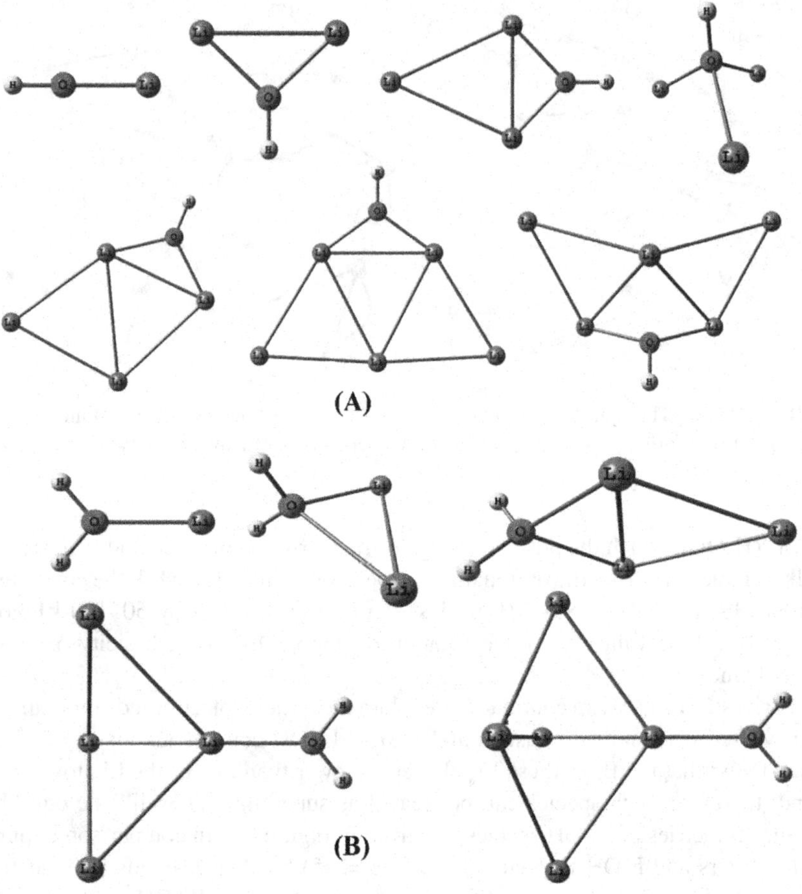

FIGURE 8.4 Equilibrium structure of Li_nOH ($n = 1–5$) species neutral (a) along with its protonated complexes (b) Ref. [29] with the permission of Wiley Periodicals, Copyright 2016.

$IE(XK_{n+1})$. The author divided these hydroxides into two sets, namely Na-based species and K-based species. The equilibrium structures of these species are displayed in Figure 8.5. One can see that these species have a close resemblance with those of Li-based species (see Figure 8.2). The partial charge on the K atom of the FK_2OH, OK_3OH and NK_4OH has been calculated to be 0.97, 0.95 and 0.90 *e*, respectively. This decreasing trend is caused by the increased X-K and O-K bond lengths from FK_2OH to NK_4OH species. The optimized $XNa_{n+1}OH$ species also share geometrical characteristics with the $XK_{n+1}OH$. However, the partial charge on Na is slightly diminished due to its lower electropositivity as compared to K. The ionic dissociation energy (ΔE) of the $XM_{n+1}OH$. calculated by equation (3), have been listed in Table 8.1. One can see that ΔE of Na-and K-based hydroxides species follows the same trend. For example, the ΔE values (in kJ/mol) are in the

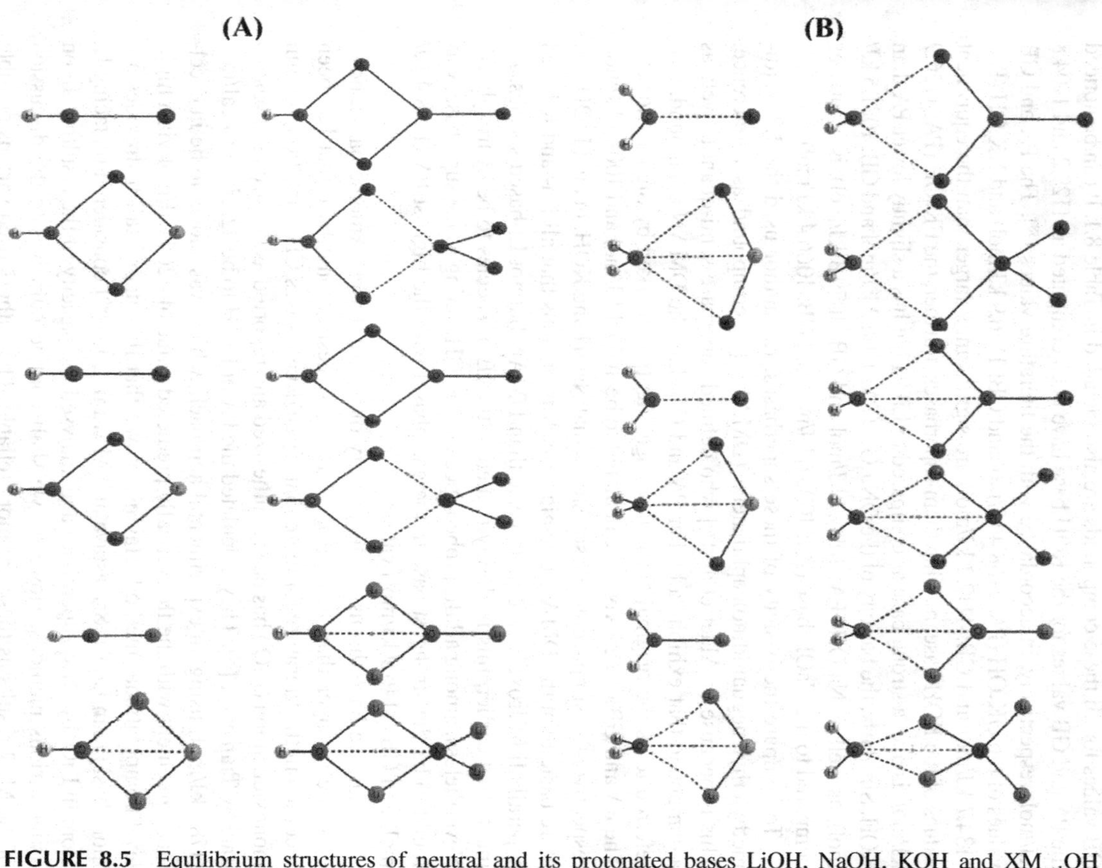

FIGURE 8.5 Equilibrium structures of neutral and its protonated bases LiOH, NaOH, KOH and $XM_{n+1}OH$ (XM_{n+1} = (FK_2, OK_3 and NK_4), (FNa_2, ONa_3 and NNa_4), and (FLi_2, OLi_3 and NLi_4) from Ref. [30].

order KOH (589) > FK_2OH (567.9) > NK_4OH (554.7) > OK_3OH (516) for such K-based species. This is in contrast to Li-based species explored earlier in which ΔE of OLi_3OH is greater than that of NLi_4OH.

To obtain the PA and GB parameters of $XM_{n+1}OH$, the author studied their protonated species ($XM_{n+1}OH_2^+$) also displayed in Figure 8.5. The calculated PA and GB values for all these compounds are also included in Table 8.1. It can be noticed the PA and GB values for the KOH base have been computed as 1129.2 and 1104.8 kJ/mol, respectively, in accordance with the literature values [38]. The PA and GB values of the OK_3OH (PA: 1168.4 kJ/mol and GB: 1146.9 kJ/mol) and NK_4OH (PA: 1134.7 kJ/mol and GB: 1117.2 kJ/mol) are significantly larger than the equivalent values of the KOH base, making them superbases. Further, the FK_2OH (PA: 1126.9 kJ/mol) is also a strong base as compared to LiOH but has a slightly lower PA than KOH. Similarly, the basicity of the ONa_3OH (PA: 1106.2 kJ/mol and GB: 1081.5 kJ/mol), as well as Na_4OH (PA: 1094.9 kJ/mol and GB: 1079.8 kJ/mol), is greater as compared to the NaOH base (PA: 1092.6 kJ/mol and GB: 1066.4 kJ/mol).

To compare the basicity of these superbases, the author used the 1,8-bis(dimethylethyleneguanidino)naphthalene (DMAN) [44] compound as a reference. This is because of Alder et al. [45] who defined superbase semi-quantitatively as a compound that exhibits a higher PA and GB than the DMAN compound, also referred to as the original "proton sponge" (as trademarked by Sigma-Aldrich). The PA and GB of DMAN are calculated to be 1027.4 kJ/mol and 1001.7 kJ/mol, respectively. A simple comparison shows that NaOH and KOH, but not LiOH, are more basic than the DMAN organic base. It is obvious that all Na- and K-based superalkali hydroxides are more basic than DMAN but not Li-based species. As mentioned earlier, this is closely related to the IE of species to be hydroxylated. Nevertheless, among all the probed species, OK_3OH, acting as the superbase, was found to be the strongest base in the gas phase) with the highest PA (1168.4 kJ/mol) and GB (1146.9 kJ/mol) values.

So far, we have discussed the basicity of neutral bases. Anionic bases, however, are stronger than neutral ones. The strongest anionic base that has been recorded in the literature for more than three decades was CH_3^-. In 2008, lithium monoxide anion (LiO^-) has been synthesized and reported as the strongest base to date by Tian et al. [32]. They calculated the PA of LiOH to be 425.7 ± 6.1 kcal/mol (1782 kJ/mol) using an experimental approach, which was shown to be in excellent agreement with the theoretically predicted value of 426 kcal/mol via a high-level computational method. Having known that the superalkali hydroxides are more basic than LiOH, Srivastava and Misra in 2016 [46] considered replacing Li atom in LiO^- by OLi_3 superalkali and analysed the basicity of the resulting anion using various methods. However, we shall confine ourselves to the discussion of the MP2 results as these are more reliable. The authors tried various possible structures in which O interacts with OLi_3 superalkali to form OLi_3O^- anion and performed calculations in singlet as well as triplet states. The equilibrium structures along with their ground-state spin multiplicity are displayed in Figure 8.6.

Table 8.2 lists the relative energy of these isomers. One can see that the ground-state structure of OLi_3O^- is a singlet in which O interacts with two Li atoms, see

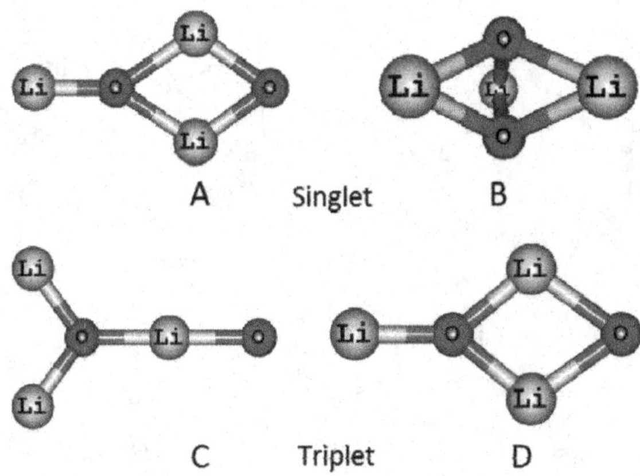

FIGURE 8.6 Equilibrium structures of OLi_3O^- anion and its isomers at MP2 level from Ref. [46] the permission of Elsevier, Copyright 2016.

TABLE 8. 2

Relative energy (RE), proton affinity (PA) and Gibbs' free energy change for reaction (ΔG_R) of OLi_3O› at MP2/6–311++G(d,p) basis set. Ref. [46].

OLi_3O^-	RE (eV)	PA (kJ/mol)	ΔG_R (kcal/mol)
A	0	1,803	–52.0
B	1.3	1,928	–19.3
C	0.7	1,866	–38.2
D	0.2	1,813	–49.6

conformer **A** in Figure 8.6. The authors considered a protonated base OLi_3OH to determine the PA of OLi_3O^- and its isomers, which is also listed in Table 8.2. It should be noted that at the MP2 level, the PA of ground-state OLi_3O^- anion is 1803 kJ/mol, which is higher than the PA of LiO› (1785 kJ/mol). This implies that $OLi_3\bar{O}$ is more basic than LiO›. Further, the PA of the $OLi_3\bar{O}$ isomers rises and reaches a maximum of 1928 kJ/mol for conformer **D**. This can be expected due to the fact that the PA of $OLi_3\bar{O}$ isomers rises when their energy relative to the ground state structure increases. Thus, the increased stability of the $OLi_3\bar{O}$ anion supports a lower PA value. The thermodynamic stability of $OLi_3\bar{O}$ anions and its conformers has been verified by considering a fictitious gas phase reaction $Li_2O + LiO^- \rightarrow OLi_3O^-$ and calculated the Gibbs' free energy change for this reaction (ΔG_R) also listed in Table 8.2. It should be noted that all ΔG_R values remain negative, *i.e.*, $\Delta G_R < 0$. This suggests that at room temperature, the above reaction is exothermic and spontaneous. This not only suggests the stability of OLi_3O› but also indicates the possible route of synthesis of this strongest anionic base.

In 2017, Wingfough and Meloni [31] investigated neutral and anionic species by the addition of an O atom and one or two OH groups in binuclear Li_3F_2

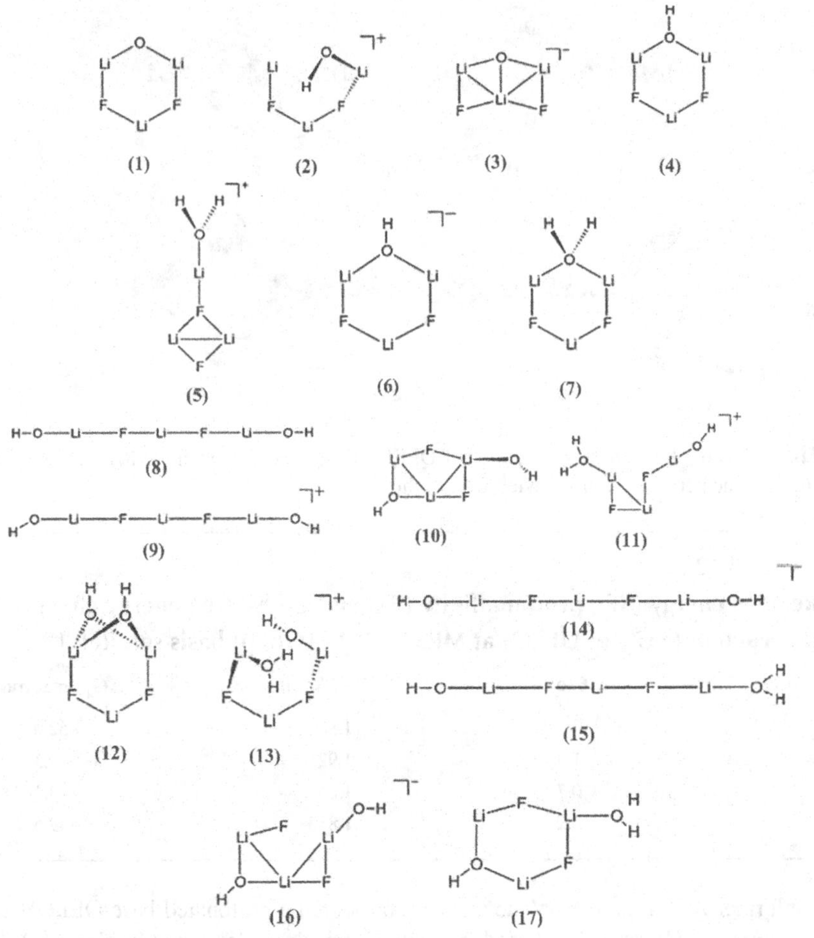

FIGURE 8.7 The optimized neutral bases and their protonated forms as well as the anionic bases and their protonated form for Li_3F_2O, $Li_3F_2(OH)$ and three structural isomers for $Li_3F_2(OH)_2$ from Ref. [31] the permission of Royal Society of Chemistry, Copyright 2018.

superalkali using the CBS-QB3 [47] method. The structures of the resulting monoxide, hydroxide and dihydroxide species and their protonated complexes are displayed in Figure 8.7. The corresponding PA and GB are listed in Table 8.3.

One can see that Li_3F_2O (1) converges to a planar ring with equal bond lengths on each side of the oxygen whose protonation leads to $Li_3F_2OH^+$ (2) such that its GB and PA become 897.2 and 893.7 kJ/mol. According to the authors, (1) does not qualify to become a superbase due to its lower PA and GB values than those of DMAN (Proton Sponge). Its anion (3), which can be protonated to Li_3F_2OH (4), has PA and GB values of 1639.9 and 1645.2 kJ mol^{-1}. Li_3F_2OH (4) is also planar, which assumes a "fin"-like structure (5) upon protonation. The PA and GB of (4), 952.9, and 957.2 kJ/mol are also lower than that of DMAN by around 60 kJ/mol.

TABLE 8.3
Calculated proton affinities, gas-phase basicities and the symmetries for each of the studied systems using the CBS-QB3 composite method. Ref. [31]

System	Isomers	Spin	PA (kJ/mol^{-1})	GB (kJ/mol^{-1})
Li_3F_2O	(1)	Doublet	893.7	897.2
$[Li_3F_2O]^-$	(3)	Singlet	1640.0	1645.3
Li_3F_2OH	(4)	Singlet	952.9	957.2
$[Li_3F_2OH]^-$	(6)	Doublet	1331.5	1331.5
$Li_3F_2(OH)_2$	(8)	Doublet	1201.1	1201.1
	(10)	Doublet	992.0	992.0
	(12)	Doublet	926.3	926.3
$[Li_3F_2(OH)_2]^-$	(14)	Singlet	1338.2	1338.2
	(16)	Singlet	1488.2	1488.2

The anion of hydroxide, $Li_3F_2OH^-$ (6) is planar like its parent neutral that retains its ring structure (7) with marginal changes in the bond lengths upon protonation. This anionic base (6) was predicted to have a GB of 1331.5 kJ/mol and a PA of 1325.8 kJ/mol, respectively. The authors next considered its dihydroxide $Li_3F_2(OH)_2$ that apparently possesses three isomers designated by (8), (10) and (12) in Figure 8.7. The first linear structure (8) remains quasi-linear with the increase in the Li-O bond lengths upon protonation (9). This isomer has a GB of 1201.1 kJ/mol and a PA of 1183.6 kJ/mol, respectively. The second isomer (10) contains one hydroxide group inside the ring, while the second hydroxide has linked to the structure as a "tail." The cage enclosing the inner oxygen is ruptured upon protonation of the tail oxygen (11). The third isomer (12) has both hydroxide groups attached to the same terminal Li atoms, which are elongated upon protonation (13). The PA and GB values of (10) and (12) are calculated to be below 1,000 kJ/mol (see Table 8.3).

There were two structures for the anionic form of dihydroxide. The first isomer (15) is similar (9) and found to be almost linear whereas the second isomer (17) is derived from (11). Their protonation leads to an increase in the Li-O bond length and a marginal decrease in the Li-F bond length as in (16) and (18), respectively. The PA for both anionic structures (15) and (17) are 1308.7 and 1486.7 kJ/mol whereas their GB values are 1328.2 and 1488.2 kJ/mol, respectively. The authors concluded, based on the trend for these neutral bases, that the proton affinity increases as the number of hydroxide groups increases. However, the pattern for such an anionic base does not seem to depend on the quantity of hydroxide group or oxygen atoms. They inferred that the basicity of anionic species is governed by the parent neutral species such that species with lower electron affinities lead to greater basicity of anions.

8.4 CONCLUSION AND PERSPECTIVES

Based on the results of quantum chemical calculations, we conclude that the superalkalis can be successfully used in the design of superbases. The alkali hydroxides are the strongest inorganic neutral bases most frequently used in the gas phase. Using typical superalkalis, it was noticed that the hydroxides of OLi_3 and NLi_4 superalkalis are more basic than LiOH. The basicity of such hydroxides can be further increased by using Na- and K-based superalkalis such that their PA exceeds that of DMAN, a reference base compound. The gas phase basicity of FLi_nOH and Li_nOH ($n = 2$–5), however, is lower than that of LiOH by 50–100 kJ/mol. Anionic bases are stronger than neutral bases due to their high PA values with CH_{3-} and LiO^- being the strongest anionic bases. Using OLi_3 superalkali, OLi_3O^- has been proposed to be the strongest anionic base hitherto. The design and synthesis of novel superbases using superalkalis may open the path for their practical uses. More research is needed in this direction to accomplish the goal of achieving the strongest neutral base. These accomplishments will show that the applications for newly designed superbases are boundless. This will, in turn, spark more interest in the building of a variety of superbases using this novel strategy.

ACKNOWLEDGMENTS

A.K.S. acknowledges the Science & Engineering Research Board (SERB) for approval of a project under the State University Research Excellence (SURE) scheme.

REFERENCES

1. Leito, I., Koppel, I. A., Koppel, I., Kaupmees, K., Tshepelevitsh, S., & Saame, J. (2015). Basicity limits of neutral organic superbases. *Angewandte Chemie International Edition*, 54(32), 9262–9265.
2. Searles, S. K., Dzidic, I., & Kebarle, P. (1969). Proton affinities of the alkali hydroxides. *Journal of the American Chemical Society*, 91(10), 2810–2811.
3. Schlosser, M. (1988). Superbases for organic synthesis. *Pure and Applied Chemistry*, 60(11), 1627–1634.
4. Caubere, P. (1993). Unimetal super bases. *Chemical Reviews*, 93(6), 2317–2334.
5. Srivastava, A. K., & Misra, N. (2015). Superalkali-hydroxides as strong bases and superbases. *New Journal of Chemistry*, 39(9), 6787–6790.
6. Gutsev, G. L., & Boldyrev, A. I. (1982). DVM Xα calculations on the electronic structure of "superalkali" cations. *Chemical Physics Letters*, 92(3), 262–266.
7. Rehm, E., Boldyrev, A. I., & Schleyer, P. V. R. (1992). Ab initio study of superalkalis. First ionization potentials and thermodynamic stability. *Inorganic Chemistry*, 31(23), 4834–4842.
8. Tong, J., Li, Y., Wu, D., Li, Z. R., & Huang, X. R. (2009). Low ionization potentials of binuclear superalkali B 2 Li 11. *The Journal of Chemical Physics*, 131(16), 164307.
9. Tong, J., Li, Y., Wu, D., Li, Z. R., & Huang, X. R. (2011). Ab initio investigation on a new class of binuclear superalkali cations M2Li2 k+ 1+(F2Li3+, O2Li5+, N2Li7+, and C2Li9+). *The Journal of Physical Chemistry A*, 115(10), 2041–2046.
10. Tong, J., Wu, Z., Li, Y., & Wu, D. (2013). Prediction and characterization of novel polynuclear superalkali cations. *Dalton Transactions*, 42(2), 577–584.

11. Tong, J., Li, Y., Wu, D., & Wu, Z. J. (2012). Theoretical study on polynuclear superalkali cations with various functional groups as the central core. *Inorganic Chemistry*, 51(11), 6081–6088.
12. Wu, C. H. (1976). Thermochemical properties of gaseous Li2 and Li3. *The Journal of Chemical Physics*, 65(8), 3181–3186.
13. Yokoyama, K., Haketa, N., Tanaka, H., Furukawa, K., & Kudo, H. (2000). Ionization energies of hyperlithiated Li2F molecule and LinFn– 1 (n= 3, 4) clusters. *Chemical Physics Letters*, 330(3–4), 339–346.
14. Veličković, S. R., Veljković, F. M., Perić-Grujić, A. A., Radak, B. B., & Veljković, M. V. (2011). Ionization energies of K2X (X= F, Cl, Br, I) clusters. *Rapid Communications in Mass Spectrometry*, 25(16), 2327–2332.
15. Dao, P. D., Peterson, K. I., & Castleman Jr, A. W. (1984). The photoionization of oxidized metal clusters. *The Journal of Chemical Physics*, 80(1), 563–564.
16. Hampe, O., Koretsky, G. M., Gegenheimer, M., Huber, C., Kappes, M. M., & Gauss, J. (1997). On the ground and electronically excited states of Na 3 O: Theory and experiment. *The Journal of Chemical Physics*, 107(18), 7085–7095
17. Ma, J. F., Ma, F., Zhou, Z. J., & Liu, Y. T. (2016). Theoretical investigation of boron-doped lithium clusters, BLi n (n= 3–6), activating CO 2: An example of the carboxylation of C–H bonds. *RSC Advances*, 6(87), 84042–84049.
18. Park, H., & Meloni, G. (2017). Reduction of carbon dioxide with a superalkali. *Dalton Transactions*, 46(35), 11942–11949.
19. Srivastava, A. K. (2018). Single-and double-electron reductions of CO2 by using superalkalis: An ab initio study. *International Journal of Quantum Chemistry*, 118(14), e25598.
20. Srivastava, A. K. (2018). Reduction of nitrogen oxides (NOx) by superalkalis. *Chemical Physics Letters*, 695, 205–210.
21. Park, H., & Meloni, G. (2018). Activation of dinitrogen with a superalkali species, Li3F2. *ChemPhysChem*, 19(3), 256–260.
22. Riyaz, M., & Goel, N. (2018). Computational design of boron doped lithium (BLin) cluster-based catalyst for N2 fixation. *Computational and Theoretical Chemistry*, 1130, 107–112.
23. Srivastava, A. K. (2019). Design of the $N_nH_{3n+1}^+$ series of "non-metallic" superalkali cations. *New Journal of Chemistry*, 43(12), 4959–4964.
24. Srivastava, A. K. (2020). MC6Li6 (M= Li, Na and K): A new series of aromatic superalkalis. *Molecular Physics*, 118(16), e1730991.
25. Sun, W. M., Li, Y., Wu, D., & Li, Z. R. (2013). Designing aromatic superatoms. *The Journal of Physical Chemistry C*, 117(46), 24618–24624.
26. Srivastava, A. K. (2018). Organic superalkalis with closed-shell structure and aromaticity. *Molecular Physics*, 116(12), 1642–1649.
27. Parida, R., Reddy, G. N., Ganguly, A., Roymahapatra, G., Chakraborty, A., & Giri, S. (2018). On the making of aromatic organometallic superalkali complexes. *Chemical Communications*, 54(31), 3903–3906.
28. Srivastava, A. K., & Misra, N. (2015). Ab initio investigations on the gas phase basicity and nonlinear optical properties of FLI n OH species (n= 2–5). *RSC Advances*, 5(91), 74206–74211.
29. Srivastava, A. K., & Misra, N. (2016). Structures and basicity of Li_NOH (N= 1– 5) species. *International Journal of Quantum Chemistry*, 116(7), 524–528.
30. Pandey, S. K. (2021). Novel and polynuclear K-and Na-based superalkali hydroxides as superbases better than Li-related species and their enhanced properties: An ab initio exploration. *ACS Omega*, 6(46), 31077–31092.

31. Winfough, M., & Meloni, G. (2018). Ab initio analysis on potential superbases of several hyperlithiated species: Li 3 F 2 O and Li 3 F 2 OH n (n= 1, 2). *Dalton Transactions*, 47(1), 159–168.
32. Tian, Z., Chan, B., Sullivan, M. B., Radom, L., & Kass, S. R. (2008). Lithium monoxide anion: A ground-state triplet with the strongest base to date. *Proceedings of the National Academy of Sciences*, 105(22), 7647–7651.
33. Frisch, M. J., Trucks, G. W., Schlegel, H. B., Scuseria, G. E., Robb, M. A., Cheeseman, J. R., Scalmani, G., Barone, V., Mennucci, B., Petersson, G. A., Nakatsuji, H., Caricato, M., Li, X., Hratchian, H. P., Izmaylov, A. F., Bloino, J., Zheng, G., Sonnenberg, J. L., Hada, M., Ehara, M., Toyota, K., Fukuda, R., Hasegawa, J., Ishida, M., Nakajima, T., Honda, Y., Kitao, O., Nakai, H., Vreven, T., Montgomery Jr. J. A., Peralta, J. E., Ogliaro, F., Bearpark, M., Heyd, J. J., Brothers, E., Kudin, K. N., Staroverov, V. N., Keith, T., Kobayashi, R., Normand, J., Raghavachari, K., Rendell, A., Burant, J. C., Iyengar, S. S., Tomasi, J., Cossi, M., Rega, N., Millam, J. M., Klene, M., Knox, J. E., Cross, J. B., Bakken, V., Adamo, C., Jaramillo, J., Gomperts, R., Stratmann, R. E., Yazyev, O., Austin, A. J., Cammi, R., Pomelli, C., Ochterski, J. W., Martin, R. L., Morokuma, K., Zakrzewski, V. G., Voth, G. A., Salvador, P., Dannenberg, J. J., Dapprich, S., Daniels, A. D., Farkas, O., Foresman, J. B., Ortiz, J. V., Cioslowski, J., & Fox, D. J. *Gaussian 09, Revision C02*; Gaussian, Inc., Wallingford, CT, 2009.
34. Reed, A. E., Weinstock, R. B., & Weinhold, F. (1985). Natural population analysis. *The Journal of Chemical Physics*, 83(2), 735–746.
35. Topol, I. A., Tawa, G. J., Burt, S. K., & Rashin, A. A. (1997). Calculation of absolute and relative acidities of substituted imidazoles in aqueous solvent. *The Journal of Physical Chemistry A*, 101(51), 10075–10081.
36. Møller, C., & Plesset, M. S. (1934). Note on an approximation treatment for many-electron systems. *Physical Review*, 46(7), 618.
37. Lias, S. G., Bartmess, J. E., Liebman, J. F., Holmes, J. L., Levin, R. D., & Mallard, W. G. (1988). Gas-phase ion and neutral thermochemistry. *Journal of Physical and Chemical Reference Data*, 17, 1–861.
38. Hunter, E. P., & Lias, S. G. (1998). Evaluated gas phase basicities and proton affinities of molecules: An update. *Journal of Physical and Chemical Reference Data*, 27(3), 413–656.
39. Veličković, S. R., Koteski, V. J., Čavor, J. N. B., Djordjević, V. R., Cvetićanin, J. M., Djustebek, J. B., Veljković, M. V., & Nešković, O. M. (2007). Experimental and theoretical investigation of new hypervalent molecules LinF (n= 2–4). *Chemical Physics Letters*, 448(4–6), 151–155.
40. Ivanic, J., Marsden, C. J., & Hassett, D. M. (1993). Novel structural principles in poly-lithium chemistry: Predicted structures and stabilities of XLi 3, XLi 5 (X= F, Cl), YLi 6 (Y= O, S), SLi 8 and SLi 10. *Journal of the Chemical Society, Chemical Communications*, 10, 822–825.
41. Đustebek, J., Veličković, S. R., Veljković, F. M., & Veljkovićvinča, M. (2012). Production of heterogeneous superalkali clusters Li_nF (n= 2–6) by knudsen-cell mass spectrometry. *Digest Journal of Nanomaterials & Biostructures*, 7(4), 1365–1372.
42. Gardet, G., Rogemond, F., & Chermette, H. (1996). Density functional theory study of some structural and energetic properties of small lithium clusters. *The Journal of Chemical Physics*, 105(22), 9933–9947.
43. Knickelbein, M. B. (1999). Reactions of transition metal clusters with small molecules. *Annual Review of Physical Chemistry*, 50(1), 79–115.

44. Raab, V., Harms, K., Sundermeyer, J., Kovačević, B., & Maksić, Z. B. (2003). 1, 8-Bis (dimethylethyleneguanidino) naphthalene: Tailoring the basicity of bisguanidine "proton sponges" by experiment and theory. *The Journal of Organic Chemistry*, 68(23), 8790–8797.
45. Alder, R. W., Bowman, P. S., Steele, W. R. S., & Winterman, D. R. (1968). The remarkable basicity of 1, 8-bis (dimethylamino) naphthalene. *Chemical Communications (London)*, 13, 723–724.
46. Srivastava, A. K., & Misra, N. (2016). OLi3O– anion: Designing the strongest base to date using OLi3 superalkali. *Chemical Physics Letters*, 648, 152–155.
47. Montgomery Jr, J. A., Frisch, M. J., Ochterski, J. W., & Petersson, G. A. (1999). A complete basis set model chemistry. VI: Use of density functional geometries and frequencies. *The Journal of Chemical Physics*, 110(6), 2822–2827.

9 Ab Initio Study on Complexes of Superalkali Li_nF_{n-1} (n = 2 – 4) Clusters with $Li@C_{60}$ and C_{60} Fullerenes

Milan Milovanović

9.1 INTRODUCTION

The C_{60} fullerene is the most investigated nano-carbon material, and a huge amount of literature has accumulated on it [1–4]. This molecule was discovered by Kroto et al. [5] in 1985. Following that, the method for producing macroscopic quantities of the C_{60} in pure solid form was propose [6], which has opened the possibility for many applications, such as the usage of C_{60} in solar cells [7,8] or transistors [9,10]. It is well-known that the exterior surface of the fullerenes may be subjected to numerous chemical reactions. Moreover, fullerenes have the unique feature of a hollow interior, which enables them to entrap atoms, ions or small molecules and clusters inside the carbon cage [11–13]. Those species are known as endohedral fullerenes, and the notation $A@C_n$ has been widely adopted, where A is the encapsulated species and n is the number of carbons in the fullerene cage [14]. The flexibility of electronic and chemical properties gained by encapsulation of different species, without altering the shape of the fullerene cage, facilitate endofullerenes practical use in many fields [15].

Alkali metal cations trapped inside fullerene cages comprise a distinct class of endohedral complexes [16]. The first representative of this class, namely $Li^+@C_{60}$, was successfully synthesized, in the form of its salt using counter anions such as $[SbCl_6^-]$ and $[PF_6^-]$ [17–19]. Since then, this new family of metal-ion endohedral C_{60} has received considerable attention as a new kind of functional material. Specifically, the unique features of $Li^+@C_{60}$, such as the high ionic conductivity in solution, optical properties, and enhanced chemical reactivity have been reported [20,21]. X-ray diffraction experiments revealed that the lithium cation is displaced from the cage centre [17,19]. The original HF/DZP prediction [22] and more recent

 DOI: 10.1201/9781003384205-9

theoretical prediction [23–25] are in agreement with experimental findings. One important features of $Li^+@C_{60}$ is the stabilization of the orbital energy levels of C_{60} due to the presence of the Li^+ ion, that causes an enhanced electron-accepting character as compared to that of pristine C_{60} [15]. Matsuo, Okada and Ueno published a comprehensive but still concise review on computational studies of $Li@C_{60}$ in the book Endohedral Lithium-containing Fullerenes [26]. It includes molecular dynamics simulations of mechanism of encapsulating the Li atom into the C_{60} cage, theoretical prediction [7] Li NMR spectra, absorption due to the motion of the inner Li^+, the interactions of $Li^+@C_{60}$ with nucleobases, corannulene, H_3^-, halogens, *etc.* Recently, the reactivity of the ion (Li^+, Li_2F^+, Cl^-, LiF_2^-) encapsulating systems was compared to that of the C_{60} fullerene for the Diels–Alder cycloaddition reaction, by means of DFT calculations. The enhancement in reactivity was found for cation-encapsulating complexes [27]. The surface interaction of C_{60} with superalkali clusters FLi_2, OLi_3 and NLi_4 was studied by Srivastava [28]. Also, properties of endofullerenes doped with the same superalkali clusters were theoretically investigated [29]. Sikorska showed on the basis of *ab initio* calculations that AlF_4, MgF_3 and LiF_2 superhalogens, as extremely strong oxidizing agents, are capable of ionizing the fullerene molecule and forming strongly bound radical cation salts [30].

Our intention is to examine the interaction of pristine C_{60} and endohedral $Li@C_{60}$ with superalkalis (SA), and possible formation of stable supramolecular complexes, by means of *ab initio* methods. According to the initial definition, superalkalis are species characterized by their lower IE than those of alkali metal atoms [31]. A typical representative of superalkalis are small clusters consisting of excess metal atoms, as for example heterogeneous clusters of lithium with halogens. Those clusters of type, Li_nX_m (X = F, Cl, Br, I and $n > m$) have been experimentally obtained and theoretically examined in detail [32–38]. Furthermore, those clusters can be classified as hyperlithiated species, because they have at least one excess electron, which originates from lithium atoms, and is not transferred to halogen atoms. Upon ionization, this electron can be more easily detached than others. In the present study, we focus on the superalkali lithium fluoride clusters, Li_nF_{n-1}; n = 2–4. They were generated by laser ablation of a mixture of lithium fluoride and nitride crystals in a supersonic beam and ionization energies determined by the photoionization experiment were 3.78 ± 0.2 eV for Li_2F, 4.32 ± 0.2 eV for Li_3F_2, and 4.30 ± 0.2 eV for Li_4F_3 [39]. Later, Li_nF_{n-1}, n = 3, 4, 6 were detected in a cluster beam generated by a mixture of lithium fluoride, lithium iodide and bismuth trifluoride. The ionization energies were determined by thermal ionization using a triple rhenium filament impregnated with fullerene C_{60} [35]. The results were in accord with previous studies. It was concluded that the presence of the fullerene C_{60} is necessary for producing those clusters.

The objectives of the present chapter are threefold: first to find structures of supramolecular complexes, consisting of endohedral fullerenes $Li@C_{60}$ and superalkali lithium fluoride clusters Li_nF_{n-1}; n = 2–4 second to consider the stability of such supramolecular complexes and compare it to complexes made of pristine C_{60} and superalkalis; and finally to describe the interaction between fragments of complexes using population analysis, theory of atoms in molecules and local energy decomposition analysis.

9.2 METHODS

All calculations were carried out using the ORCA program package [40,41]. Optimizations and frequency calculations were done with RIJCOSX-B3LYP-D3/def2-svp+ def2/J (abbreviated with B3LYP-D3/svp). There were no imaginary frequencies for any of the presented structures. The B3LYP-D3 stands for a popular B3LYP functional [42,43] in conjunction with the atom-pairwise dispersion correction of Grimme [44]. The RIJCOSX is a "chain of spheres" (COSX) approximation combined with the resolution of identity (RI) algorithm for the Coulomb term [45,46]. In the latest version of ORCA, this approximation is used by default with hybrid functionals, because it offers large speed-ups with almost no loss in accuracy of results. For the use of RIJCOSX approximations, two basis sets must be specified. We used the def2-svp, double-zeta valence basis set of the Karlsruhe group [47], and auxiliary Coulomb fitting def2/J basis set [48]. To obtain accurate results, we chose TightSCF convergence and large integration grids (DefGrid3 option).

For the optimized structures single point energy calculations were performed using the RIJCOSX-B3LYP-D3 method with def2-tzvp basis set (abbreviated with B3LYP-D3/tzvp) and RIJCOSX-DLPNO-CCSD/def2-tzvp+def2-tzvp/C+def2/J (abbreviated with DLPNO-CCSD/tzvp) method. The DLPNO-CCSD is a domain-based local pair natural orbital (DLPNO) [49] coupled cluster method with single- and double-excitations (CCSD). In correlation calculations, def2-tzvp/C is an auxiliary basis set. All single-point energies were corrected for zero-point vibrational energy (ZPVE) obtained at the B3LYP-D3/svp level of theory.

To gain insights into the nature of intermolecular interactions between encapsulated lithium, C_{60} cage and superalkalis, Local Energy Decomposition (LED) analysis [50–53] was performed. In this approach, within the DLPNO-CCSD framework, interaction energy is decomposed into physically meaningful contributions. As a further means of understanding the nature of interactions between superalkalis and fullerenes, Bader's quantum theory of atoms in molecules (QTAIM) [54] was utilized. Also, the atom dipole moment corrected Hirshfeld (ADCH) [55] partial charges were calculated. The ADCH charge was used in achieving favourable results in studying the interactions between metal elements and carbon nanotubes [56]. Topology and population analysis were carried out using B3LYP-D3/tzvp wave functions in the Multiwfn program [57]. Finally, visualization of structures was done in Chemcraft [58] and VMD [59] graphical software.

9.3 RESULTS AND DISCUSSION

9.3.1 Endohedral Li@C_{60} Fullerene

Structures of neutral Li@C_{60} and cation [Li^+@C_{60}] endohedral fullerene, calculated at B3LYP-D3/svp level, are shown in Figure 9.1. The most significant

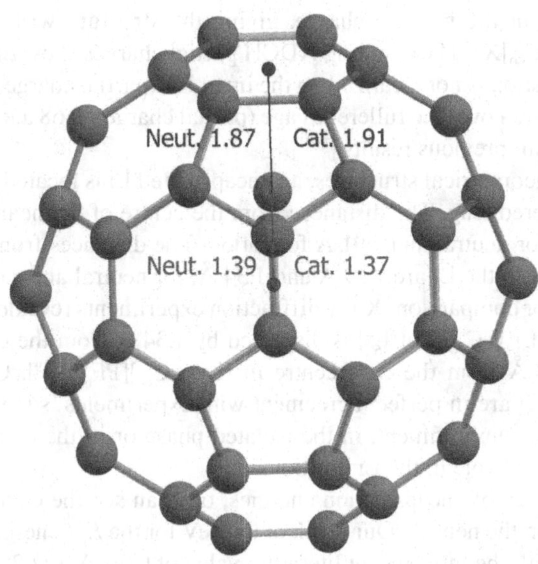

FIGURE 9.1 Structure of neutral $Li@C_{60}$ and cation $[Li^+@C_{60}]$. The blue dot is placed in the centre of the nearest ring, and the red dot is placed in the centre of carbon cage.

TABLE 9.1
Properties of neutral $Li@C_{60}$ and cation $[Li^+@C_{60}]$ endohedral fullerenes.

	DLPNO-CCSD/tzvp	B3LYP-D3/tzvp	
Species	E_{enc} (eV)	q(Li)(a.u.)	$q(C_{60})$(a.u.)
$Li@C_{60}$	0.78	0.36	−0.36
$[Li^+@C_{60}]$	0.99	0.32	0.68

properties of both species are given in Table 9.1. The encapsulating energy, E_{enc}, is energy required to split up $[Li@C_{60}]^{0,+}$ into fragments, $Li^{0,+}$ and fullerene cage C_{60}. The E_{enc} are obtained as differences between the sum of electronic energies of Li and C_{60} and electronic energy of endohedral fullerene, calculated with DLPNO-CCSD/tzvp method with ZPVE corrections. The ADCH charges for C_{60} cage result from summing the partial charges of all carbon atoms.

To avoid confusion, we will draw a distinction in meaning between "$Li@C_{60}$" and "$Li^+@C_{60}$." The neutral $Li@C_{60}$ is typical endohedral metallofullerene, which undergoes electron transfer from the inner Li atom to the C_{60} cage, giving an electronic state of $Li^+@C_{60}^-$ [60]. This is clearly viewed from partial charges of neutral $Li@C_{60}$ in Table 9.1, +0.36 a.u. on Li and −0.36 a.u. on C_{60}. The negative charges on the fullerene cages are effectively cancelled by internal ion-pair formation. On the other hand, lithium-ion-containing fullerene, $Li^+@C_{60}$, includes a lithium-ion in a neutral C_{60} cage. It requires a counter anion (X^-) outside the fullerene cage for balancing out the positive charge, giving the structure written as an ionic formula $[Li^+@C_{60}]X^-$. The resulting ADCH partial charges show that the positive charge of the cation is not localized on the inner Li^+ (partial charge +0.32 a.u.) but widely delocalized over the fullerene cage (partial charge +0.68 a.u.). This finding is consistent with previous results [19,61].

Regarding geometrical structures, an encapsulated Li is located in the vicinity of a six-membered ring. The distances from the centre of the nearest ring to the Li are 1.87 Å for neutral and 1.91 Å for cation. The distances from the centre of the carbon cage to the Li are 1.39 Å and 1.34 Å, for neutral and cationic species, respectively. For comparison, X-ray diffraction experiments revealed that the lithium cation in $[Li^+@C_{60}][SbCl_6^-]$ is displaced by 1.34 Å from the cage centre [17] and 1.40±0.01 Å from the cage centre in $[Li^+@C_{60}][PF_6^-]$ [19]. Our results for cation $[Li^+@C_{60}]$ are in perfect agreement with experiments, suggesting that the difference in the environment, in the isolated phase or in the condensed phase, plays only a minor role in the Li position.

By comparison of encapsulation energies, one can see the cationic species is more stable than the neutral. Our result of 0.99 eV for the E_{enc} energy of $[Li^+@C_{60}]$ is consistent with the most recent literature value of 0.96 eV (22.2 kcal/mol) [25].

9.3.2 Structures of SA-Li@C_{60} and SA-C_{60} Complexes

The equilibrium structures with corresponding parameters of all complexes are presented in Figure 9.2. The distances from the centre of the carbon cage to Li (Li-centre) and the shortest distances from superalkali to the carbon cage (SA-cage) are listed in Table 9.2. As can be seen, the type of superalkalis outside the cage (including Li atom) only slightly affects the position of Li inside the cage. The Li is placed at 1.43 Å, 1.45 Å, 1.44 Å and 1.46 Å from centre of cage, in neutral complexes with Li, Li_2F, Li_3F_2 and Li_4F_3, respectively. The distances from the centre of the six-membered ring to the Li are in the range of 1.83 to 1.86 Å for neutrals. Also, in all cationic complexes, the position of the encapsulated Li atom is practically the same, regardless of species outside the cage. However, the encapsulated Li atom is closer to the centre of the cage in cationic than in neutral complexes, for 0.02 Å, 0.07 Å, 0.06 Å and 0.09 Å in the case of Li, Li_2F, Li_3F_2 and Li_4F_3, respectively.

Turning now to the position of superalkalis (SA), one can see that in all complexes superalkalis are located away from the centre of the six-membered ring nearby the encapsulated lithium. The structure with exohedral Li above the six-membered ring of Li@C_{60} is a transition state. The lowest energy location of the Li atom is above the five-membered ring in this case. However, Li is located above the six-membered ring in the complex of Li and C_{60}. The distances between the superalkali and the carbon cage, including ring centre and neighbouring carbon atoms are shown in Figure 9.2. The shortest distances are given in Table 9.2. By taking these values into account, we can see that superalkalis, are closer to the carbon cage in neutral complexes with endohedral fullerenes Li@C_{60} than

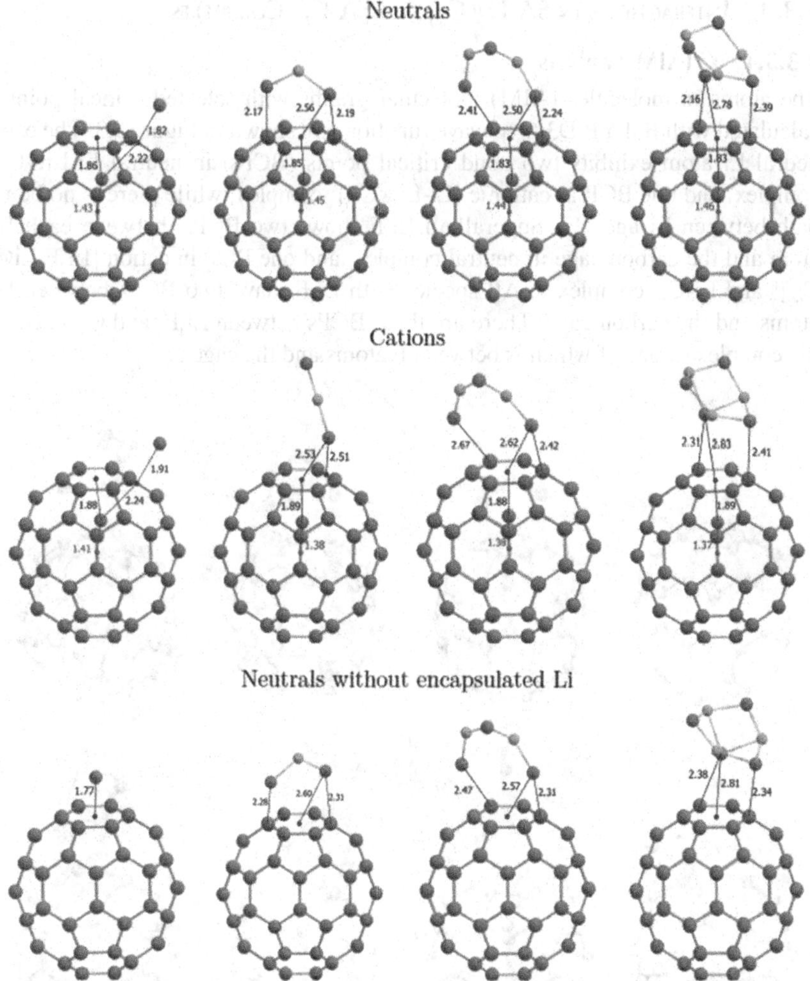

FIGURE 9.2 Optimized structures of superalkalis-Li@C_{60} and superalkalis-C_{60} complexes obtained at B3LYP-D3/svp level of theory. All distances are in Å.

in complexes with C_{60}. The shortening of distances upon encapsulation of Li is 0.11 Å, 0.06 Å and 0.18 Å for Li_2F, Li_3F_2 and Li_4F_3, respectively. The observed decrease in separation of complex constituents could be attributed to the favourable interaction of superalkali and the Li@C_{60} with respect to the interaction of superalkalis and the C_{60}. Further, ionization of SA-Li@C_{60} complexes increases the separation of SA and cage. Hence, the distances in cationic complexes are larger than in neutral SA-Li@C_{60}, as well as in SA-C_{60} complexes (except for Li_4F_3-C_{60}).

9.3.3 Interactions in SA-Li@C_{60} and SA-C_{60} Complexes

9.3.3.1 QTAIM Analysis

The atoms in molecules (AIM) molecular graphs with selected critical points, calculated with B3LYP-D3/tzvp wave function, are shown in Figure 9.3. The exohedral Li atom exhibits two bond critical points (BCPs) in neutral Li-Li@C_{60} complex, and one BCP in cationic [Li-Li@C_{60}]$^+$ complex, while there is no bond path between Li and C_{60}. Superalkali Li_2F shows two BCPs, between each Li atom and the carbon cage in neutral complex, and one BCP in cation [Li_2F-Li@C_{60}]$^+$ and Li-C_{60} complexes. All species with Li_3F_2 have two BCPs between Li atoms and the carbon cage. There are three BCPs between Li_4F_3 and C_{60} cage in the complexes, one of which is between F atoms and the cage.

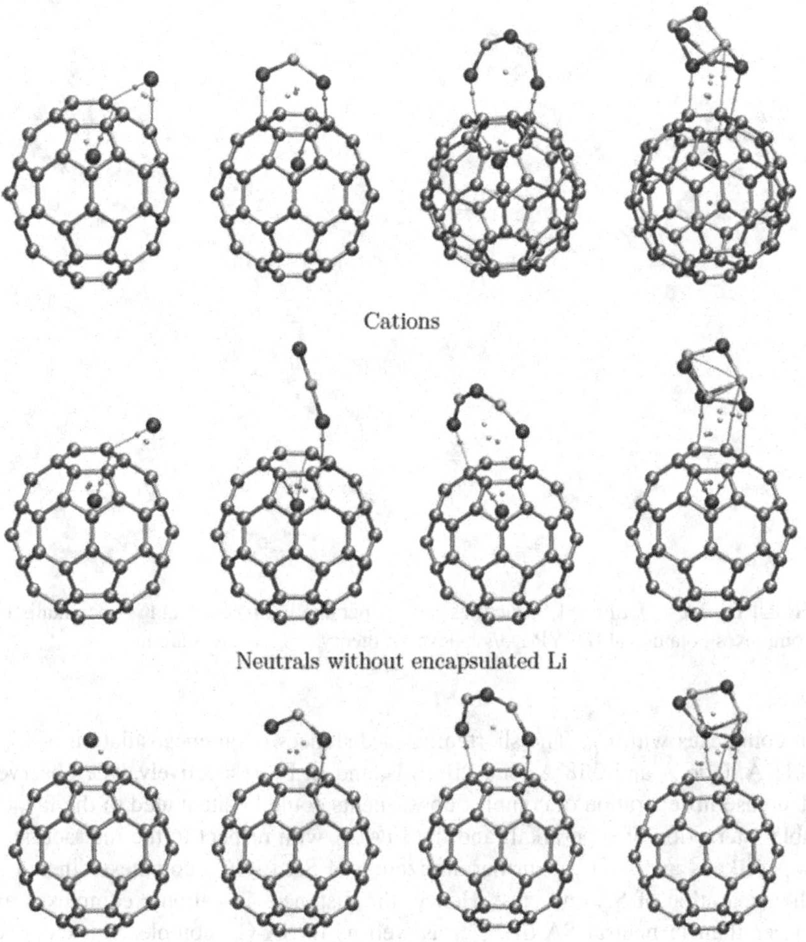

FIGURE 9.3 Atoms in molecules (AIM) molecular graphs.

TABLE 9.2
Structural and QTAIM properties of complexes of lithium atom and superalkalis with endohedral Li@C_{60} and empty C_{60} fullerene.

	Distances in Å		QTAIM
Species	**Li-centre**	**SA-cage**	ρ_{BCP} **(SA-C_{60}) in a.u.**
Li-Li@C_{60}	1.43	1.82	0.021, 0.022
Li_2F-Li@C_{60}	1.45	2.17	0.021, 0.022
Li_3F_2-Li@C_{60}	1.44	2.24	0.020, 0.022
Li_4F_3-Li@C_{60}	1.46	2.16	0.011, 0.015, 0.022
[Li-Li@C_{60}]$^+$	1.41	1.91	0.018
[Li_2F-Li@C_{60}]$^+$	1.38	2.51	0.021
[Li_3F_2-Li@C_{60}]$^+$	1.38	2.42	0.016, 0.016
[Li_4F_3-Li@C_{60}]$^+$	1.37	2.31	0.010, 0.012, 0.015
Li-C_{60}		1.77	no BCP
Li_2F-C_{60}		2.28	0.019
Li_3F_2-C_{60}		2.31	0.019, 0.019
Li_4F_3-C_{60}		2.34	0.010, 0.014, 0.019

The calculated electron density (ρ_{BCP}) at bond critical points is given in the last column of Table 9.2. The ρ_{BCP} of BCPs, between superalkalis and the C_{60} cage is the range 0.010–0.022 a.u. The highest values are obtained for neutral SA-Li@C_{60} complexes, which contributes to the conclusion that the interaction of SA and Li@C_{60} is stronger than SA and C_{60}. In all cases, BCPs show positive Laplacian values ($^2\rho_{BCP}$ in the range 0.03–0.13 a.u.). The small values of the electron density and the positive values of the Laplacian at the BCPs indicate that SA-cage interactions are dominated by the contraction of charge away from the BCPs. Consequentially, they correspond to ionic closed shell interactions.

9.3.3.2 Ionization Energies

The AIE is computed as the difference in E_{el}(DLPNO- CCSD/tzvp)+ ZPVE(B3LYP-D3/svp) of a pair of a relaxed cationic and neutral species. The values for superalkalis, fullerene C_{60}, endohedral Li@C_{60} fullerene, and complexes of endohedral fullerene with superalkalis, are presented in Table 9.3. The calculated ionization energies for lithium atoms and the smallest superalkali, Li_2F, match those experimentally observed [35,39]. The theoretical results for Li_3F_2 and Li_4F_3 are about 0.6 eV and 0.8 eV lower, respectively, than the experimental values.

A good agreement between theoretical (7.74 eV) and experimental (7.59 eV) [63] AIE values is found for C_{60}. Also, electron affinity (EA = 2.51 eV) for C_{60} obtained with DLPNO-CCSD/tzvp method agrees well with the experiments (2.6835 ± 0.0006 eV) [64]. The value EA = 2.73 eV is obtained for both C_{60} and Li@C_{60} with B3LYP-D3/tzvp method. The ionization potential of Li@C_{60}, roughly

TABLE 9.3
Ionization energies.

Species	B3LYP-D3	DLPNO-CCSD	Experimental
Li	5.63	5.37	5.3915[a]
Li_2F	4.18	3.77	3.78 ± 0.2[b]
Li_3F_2	4.02	3.70	4.32 ± 0.2[b]
Li_4F_3	3.87	3.52	4.30 ± 0.2[b]
C_{60}	7.41	7.74	7.59 ± 0.02[c]
Li@C_{60}	5.87	5.17	
Li-Li@C_{60}	5.58	5.89	
Li_2F-Li@C_{60}	5.79	5.82	
Li_3F_2-Li@C_{60}	5.63	5.80	
Li_4F_3-Li@C_{60}	5.15	5.24	

[a] Ref. [62]. [b] Ref. [39]. [c] Ref. [63].

estimated from gas phase experiments, was 6.5 eV [65]. This is consistent with the prediction from ultraviolet photoelectron spectra that ionization potential of Li@C_{60} is 1.1 eV smaller than that of C_{60} [66]. Our calculated AIE for Li@C_{60} is 5.17 eV at DLPNO-CCSD/tzvp level, which is considerably lower than those estimations. However, the calculated difference in AIE for C_{60} and Li@C_{60} is 2.57 eV, which is very close to the electron affinity of C_{60}. A possible explanation can be obtained by examining the ADCH charge in Table 9.1: upon ionization an electron is detached from the carbon cage; energy resolved in the process of electron transfer from the Li to carbon cage in Li@C_{60} is comparable to the electron affinity of C_{60}; this energy is invested in lowering the AIE of the Li@C_{60} with respect to the empty C_{60}.

As far as the SA-Li@C_{60} complexes are concerned, the values in Table 9.3 show that ionization energies of Li-Li@C_{60}, Li_2F-Li@C_{60} and Li_3F_2-Li@C_{60} are very close to each other, 5.89 eV, 5.82 eV and 5.80 eV, respectively. The lowest value of 5.24 eV was obtained for Li_4F_3-Li@C_{60}. Considering the ionization energy as a parameter of complex stability, we conclude that complexes containing Li_4F_3 superalkali are the least stable. Based on ADCH charges in Table 9.4, we see that all neutral species have C_{60} cage with partial charge of approximately –1, and in all cation species C_{60} cage is almost neutral. Thus, an electron is removed from the carbon cage upon ionization, and as a consequence the AIE value does not depend significantly on the species outside the fullerene cage.

9.3.3.3 ADCH Partial Charges

The ADCH partial charges of neutral and charged SA-Li@C_{60} and neutral SA-C_{60} complexes are given in Table 9.4. For all calculations the B3LYP-D3/tzvp wave functions were used.

TABLE 9.4
The atom dipole moment corrected Hirshfeld (ADCH) partial charges (a.u.) of SA-Li@C_{60} and SA-C_{60} complexes.

	Partial charges			DE(eV)
Species	*q*(SA)	*q*(Enc.Li)	*q*(C_{60})	
Li@C_{60}	—	0.36	−0.36	
Li-Li@C_{60}	0.68	0.39	−1.07	1.77
Li_2F-Li@C_{60}	0.67	0.38	−1.06	2.77
Li_3F_2-Li@C_{60}	0.67	0.36	−1.03	2.67
Li_4F_3-Li@C_{60}	0.53	0.37	−0.90	2.42
Li^+@C_{60}	—	0.32	0.68	
[Li-Li@C_{60}]$^+$	0.75	0.36	0.11	1.26
[Li_2F-Li@C_{60}]$^+$	0.87	0.37	0.24	0.72
[Li_3F_2-Li@C_{60}]$^+$	0.74	0.34	0.08	0.57
[Li_4F_3-Li@C_{60}]$^+$	0.63	0.36	0.00	0.67
Li-C_{60}	0.64		−0.64	
Li_2F-C_{60}	0.66		−0.66	1.78
Li_3F_2-C_{60}	0.64		−0.64	1.75
Li_4F_3-C_{60}	0.52		−0.52	1.47

The superalkalis have an extra electron, so they must transfer it to achieve stability. As expected, the partial charges of superalkalis (as well as Li) in the SA-C_{60} complexes are positive (0.52–0.66 a.u.), while the carbon cages are negatively charged. Accordingly, those complexes are stabilized by electron transfer from SA to the carbon cage, making them ionic in nature. Thus, we should represent them as $SA^+C_{60}{}^-$, and refer to them as supersalts made of superalkali cation and fullerene anion.

Now we discuss the SA-Li@C_{60} complexes. As previously stated, in neutral Li@C_{60}, the encapsulated Li atom is partial positive (–0.36 a.u.) while the carbon cage is partial negatively charged (–0.36 a.u.). Despite this, the values of charges in Table 9.4 indicate that electron transfer occurs efficiently from SA to the carbon cage. The partial charge of the Li atom inside the cage remains constant and is almost equal to the value in the separate Li@C_{60}. The partial charges of SA in complexes with endohedral fullerene and pristine fullerene are practically the same. A negative charge on the C_{60} cage increases from –0.36 a.u in Li@C_{60} to approximately –1 a.u. in the complexes. Thus, the SA- Li@C_{60} complexes are also ionic. They exhibit electron transfer from superalkalis outside the cage as well as from encapsulated Li to the carbon cage, and the most descriptive representation is SA^+[Li^+@C_{60}]$^-$. In this case, we have supersalts made of superalkali cations and endohedral fullerene anions. This is consisted with viewpoint of Li@C_{60} as an example of a superatom [60,67].

As was mentioned in the previous section, upon ionization of SA-Li@C_{60} complexes, an electron is removed from the carbon cage and in cationic species, C_{60} cage becomes almost neutral. The partial charge of encapsulated Li is unchanged, while the partial charges of SA are slightly increased, in compassion to neutral complexes. The appropriate representation of cationic endohedral complexes is $SA^+[Li^+@C_{60}]$.

9.3.3.4 Dissociation Energies

The quantity that yields useful information on complex stability is dissociation energy (DE) for the following reactions of fragmentation:

$$\text{SA-Li@C}_{60} \rightarrow \text{SA} + \text{Li@C}_{60}$$
$$[\text{SA-Li@C}_{60}]^+ \rightarrow \text{SA}^+ + \text{Li@C}_{60}$$
$$\text{SA-C}_{60} \rightarrow \text{SA} + \text{C}_{60}$$

DEs were calculated as the difference of electronic energy (with ZPVE correction) for products and reactants. The results of DLPNO-CCSD/tzvp calculation are presented in the last column of Table 9.4. In general, positive DE values indicate that all complexes are thermodynamically stable toward fragmentation. The DEs for reactions where cationic complexes are fragmented into SA and charged $[Li@C_{60}]^+$ are 1.06 eV, 2.12 eV, 2.05 eV and 2.35 eV for complexes with Li, Li_2F, Li_3F_2 and Li_4F_3, respectively. Comparing those values to the corresponding values in Table 9.4, we can see that cationic complexes are preferably fragmented into superalkali cation (Li_2F^+, $Li_3F_2^+$ and $Li_4F_3^+$) and neutral Li@C_{60}. In any case, due to electron deficiency cationic complexes are less stable than neutral species. Let us emphasize that the DE values for neutral endohedral complexes are higher than for complexes with pristine C_{60}. The differences are 1.0 eV, 0.92 eV and 0.95 eV for complexes with Li_2F, Li_3F_2 and Li_4F_3, respectively. Therefore, the presence of Li inside the fullerene leads to increased dissociation energy. Physically speaking, we see that electron transfer from encapsulated lithium to the surface of the fullerene cage leads to favourable interactions between superalkalis and fullerene. Among the neutral endohedral complexes, the lowest DE = 1.77 eV has Li-Li@C_{60}. However, when considering complexes containing superalkalis, the DE decreases as follows: Li_2F, Li_3F_2 and Li_4F_3. The least stable are Li_4F_3-Li@C_{60}, where the amount of electron density withdrawn from Li_4F_3 to the carbon cage is the smallest, 0.53 a.u. Similarly, the electron transfer and DE value are lowest for the Li_4F_3-C_{60}.

9.3.3.5 LED Analysis

In LED analysis, within the DLPNO-CCSD framework, interaction energy is decomposed into the following contributions:

$$\begin{aligned}\Delta E_{int} &= \Delta E_{int}^{HF} + \Delta E_{int}^{C} = \\ &= \Delta E_{prep}^{HF} + E_{elstat} + E_{exch} + \Delta E_{non-disp}^{C} + E_{disp}^{C}\end{aligned}$$

where ΔE_{int}^{HF} is a component from the Hartree-Fock (HF) level of theory and ΔE_{int}^{C} is the correction due to electron correlation at CCSD level of theory. The HF term is decomposed into the repulsive electronic preparation contribution ΔE_{prep}^{HF}, and attractive electrostatic E_{elstat} and exchange contribution E_{exch}. The CCSD correlation interaction energy is decomposed into the nondispersive $\Delta E_{non-disp}^{C}$ and London E_{disp}^{C} dispersion contribution. Term ΔE_{prep}^{HF} is obtained from the difference in the reference (HF) energy of the fragments in the interacting complex and isolated relaxed fragments. Similarly, ΔE^{C} is the corresponding difference in correlation energy.

To determine the interaction energies, DLPNO-CCSD/tzvp calculations of isolated superalkalis, lithium atoms, fullerene C_{60}, as well as the entire complexes were performed. The LED was carried out for endohedral complexes divided into three fragments (superalkali, lithium atom and C_{60}) and for complexes with fullerene divided into two fragments (superalkali and C_{60}). Furthermore, DLPNO-CCSD electronic energies, were used for calculation of the quantity ΔE, using following formula:

$$\Delta E = E_{int}(\text{complex}) - [-\text{DE}(\text{complex}) - E_{enc}(\text{Li@C}_{60})]$$

This definition of ΔE includes the difference of interaction energy in complex obtained with LED (negative quantity) and the energy released when all isolated constituents come to complex (positive quantity), obtained as the sum of negative DE and E_{enc} for endohedral complexes, as well as negative DE for complexes with pristine C_{60}. A larger ΔE implies a lower stability of the complex. This is because the energy released for formation of complex from its isolated constituents, is not fully converted into attractive interaction energy between constituents in complex.

The results of LED analysis and ΔE values are given in Table 9.5. The E_{int} for endohedral complexes includes interactions between superalkali and the carbon cage, superalkali and Li atom inside the cage, and between Li and cage, while for SA-C_{60} complexes E_{int} includes interaction between superalkali and the carbon cage. The net interaction in all complexes is attractive since E_{int} is negative for all complexes. This is in accord with the thermodynamic stability of complexes, previously deduced from the positive values of DE. We see that the most significant attractive contribution to the stability of complexes comes from the E_{elstat}. Although E_{elstat} includes repulsive part from the interaction of positively charged superalkali and encapsulated lithium atom in endohedral complexes, this repulsion is surpassed by attractive interactions between highly negative carbon cage with both superalkali and lithium (Table 9.6). The next attractive contribution is the nondispersive $\Delta E_{non\text{-}disp}^{C}$ term from the CCSD correlation interaction energy, and then exchange E_{exch} and dispersion E_{disp}^{C} contributions. As far as interacting energy at Hartree-Fock level ($\Delta E_{prep}^{HF} + E_{elstat} + E_{exch}$) is concerned, it yields attractive interaction only for neutral SA-Li@C_{60} complexes.

TABLE 9.5
Interaction energies and their components obtained using LED analysis at DLPNO-CCSD/tzvp level. All values are in eV units.

	Components of interaction energies						
Species	E_{int}	ΔE^{HF}_{prep}	E_{elstat}	E_{exch}	$\Delta E^{C}_{non\text{-}disp}$	E^{C}_{disp}	ΔE
Li-Li@C_{60}	−2.51	19.30	−19.10	−0.33	−2.23	−0.15	0.04
Li_2F- Li@C_{60}	−3.53	18.64	−19.44	−0.45	−2.00	−0.28	0.02
Li_3F_2- Li@C_{60}	−3.41	18.14	−18.67	−0.45	−2.15	−0.29	0.04
Li_4F_3- Li@C_{60}	−3.29	17.82	−17.94	−0.55	−2.21	−0.41	−0.09
[Li-Li@C_{60}]$^+$	−1.76	13.62	−12.78	−0.28	−2.19	−0.12	0.28
[Li_2F-Li@C_{60}]$^+$	−1.23	12.02	−10.84	−0.25	−2.03	−0.13	0.27
[Li_3F_2-Li@C_{60}]$^+$	−1.06	13.43	−11.64	−0.43	−2.16	−0.26	0.29
[Li_4F_3-Li@C_{60}]$^+$	−1.15	13.13	−11.38	−0.48	−2.09	−0.33	0.32
Li_2F-C_{60}	−1.48	8.59	−7.93	−0.30	−1.66	−0.17	0.30
Li_3F_2-C_{60}	−1.42	8.64	−7.67	−0.32	−1.91	−0.16	0.33
Li_4F_3-C_{60}	−1.25	8.30	−6.85	−0.41	−2.02	−0.27	0.22

TABLE 9.6
Electrostatic and exchange interaction energies between fragments. All values are in eV units.

	E_{elstat}			E_{exch}	
Species	SA-C_{60}	SA-Li	Li-C_{60}	SA-C_{60}	Li-C_{60}
Li-Li@C_{60}	−11.67	3.62	−11.05	−0.20	−0.13
Li_2F-Li@C_{60}	−11.99	3.75	−11.21	−0.32	−0.13
Li_3F_2-Li@C_{60}	−11.25	3.70	−11.12	−0.32	−0.13
Li_4F_3-Li@C_{60}	−9.98	3.13	−11.08	−0.41	−0.13
[Li-Li@C_{60}]$^+$	−8.43	3.52	−7.88	−0.16	−0.12
[Li_2F-Li@C_{60}]$^+$	−5.88	2.89	−7.85	−0.13	−0.12
[Li_3F_2-Li@C_{60}]$^+$	−7.09	3.32	−7.87	−0.31	−0.12
[Li_4F_3-Li@C_{60}]$^+$	−6.40	2.85	−7.83	−0.36	−0.12
Li_2F-C_{60}	−7.93	—	—	−0.30	—
Li_3F_2-C_{60}	−7.67	—	—	−0.32	—
Li_4F_3-C_{60}	−6.85	—	—	−0.41	—

Considering ΔE values, it is noticeable that they are almost independent of the type of superalkali that participates in the complex. Among the neutral endohedral complexes, $\Delta E \le 0.04$ eV, whereas among the cationic endohedral complexes and the complexes with pristine fullerene ΔE is in the range 0.22–0.33 eV. As previously explained, small values of ΔE favour the stability of SA-Li@C_{60} over cationic and SA-C_{60} complexes. We can conclude that ΔE may be a useful quantity for comparing the stability of different classes of complexes.

The individual contribution from SA-C_{60}, SA-Li and Li-C_{60} interactions to the E_{elstat} and E_{exch} are presented in Table 9.6. We can see that the interaction of Li-C_{60} is unaffected by superalkali, as E_{elstat}(Li-C_{60}) and E_{exch}(Li-C_{60}) remain practically constant with the change of SA, in both neutral and cationic complexes.

9.4 CONCLUSION

In the present chapter, we have reported structures and properties of supramolecular complexes, consisting of lithium and superalkali lithium fluoride clusters Li_nF_{n-1}, n = 2–4 from one side, and endohedral fullerenes Li@C_{60} and pristine C_{60} on the other side. To the extent of our knowledge the SA-Li@C_{60} complexes (SA = Li_2F, Li_3F_2, Li_4F_3), and SA- C_{60} complexes (SA = Li_3F_2, Li_4F_3) have not been described in the literature thus far. On the basis of B3LYP-D3 and DLPNO-CCSD calculations performed for those complexes, we conclude that they are thermodynamically stable and that the net interaction in all complexes is attractive. Those complexes are stabilized by electron transfer from SA to the carbon cage. Their ionic properties are also a result of electron transfer. Accordingly, they are referred to as supersalts made of superalkali cation and endohedral fullerene an-ion ($SA^+[Li^+@C_{60}]^-$) or superalkali cation and fullerene anion ($SA^+C_{60}^-$). Additionally, based on the QTAIM, we have shown that ionic closed shell interactions occur via Li-C bonds. Superalkali does not affect Li's position inside the cage, its partial charge, and interaction with the C_{60}. The significant conclusion is that encapsulation of lithium leads to favourable interactions between superalkalis and fullerene, making the endohedral complexes more stable than complexes with pristine fullerene. Moreover, the stability of complexes decreases when the transfer of electron density from SA to the fullerene cage is reduced.

ACKNOWLEDGMENTS

The financial support by Ministry of Education, Science and Technological Development of Republic of Serbia Contract number: 451–03–47/2023–01/200146.

REFERENCES

1. A. Hirsch, *The Chemistry of the Fullerenes*, Wiley, 2008.
2. R. Taylor (Ed.), *Fullerene Science and Technology*, 1996, 4, 1317–1318.
3. M. Dresselhaus, G. Dresselhaus and P. Eklund, *Science of Fullerenes and Carbon Nanotubes: Their Properties and Applications*, Elsevier Science, 1996.
4. K. M. Kadish and R. S. Ruoff, *Fullerenes: Chemistry, Physics, and Technology*, John Wiley & Sons, 2000.
5. H. W. Kroto, J. R. Heath, S. C. O'Brien, R. F. Curl and R. E. Smalley, *Nature*, 1985, 318, 162–163.
6. W. Krätschmer, L. D. Lamb, K. Fostiropoulos and D. R. Huffman, *Nature*, 1990, 347, 354–358.

7. G. Yu, J. Gao, J. C. Hummelen, F. Wudl and A. J. Heeger, *Science*, 1995, 270, 1789–1791.
8. G. Li, V. Shrotriya, J. Huang, Y. Yao, T. Moriarty, K. Emery and Y. Yang, *Nature Materials*, 2005, 4, 864–868.
9. A. Dodabalapur, H. Katz, L. Torsi and R. Haddon, *Science*, 1995, 269, 1560–1562.
10. E. Meijer, D. De Leeuw, S. Setayesh, E. Van Veenendaal, B.-H. Huisman, P. Blom, J. Hummelen, U. Scherf and T. Klapwijk, *Nature Materials*, 2003, 2, 678–682.
11. H. Shinohara, *Reports on Progress in Physics*, 2000, 63, 843.
12. T. Akasaka and S. Nagase, *Endofullerenes: A New Family of Carbon Clusters*, Springer Science & Business Media, 2002, vol. 3.
13. A. A. Popov, S. Yang and L. Dunsch, *Chemical Reviews*, 2013, 113, 5989–6113.
14. Y. Chai, T. Guo, C. Jin, R. E. Haufler, L. P. F. Chibante, J. Fure, L. Wang, J. M. Alford and R. E. Smalley, *The Journal of Physical Chemistry*, 1991, 95, 7564–7568.
15. Y. Yamada, A. V. Kuklin, S. Sato, F. Esaka, N. Sumi, C. Zhang, M. Sasaki, E. Kwon, Y. Kasama, P. V. Avramov and S. Sakai, *Carbon*, 2018, 133, 23–30.
16. J. Cioslowski, *Molecules*, 2023, 28, 1384.
17. S. Aoyagi, E. Nishibori, H. Sawa, K. Sugimoto, M. Takata, Y. Miyata, R. Kitaura, H. Shinohara, H. Okada, T. Sakai, Y. Ono, K. Kawachi, K. Yokoo, S. Ono, K. Omote, Y. Kasama, S. Ishikawa, T. Komuro and H. Tobita, *Nature Chemistry*, 2010, 2, 678–683.
18. H. Okada, T. Komuro, T. Sakai, Y. Matsuo, Y. Ono, K. Omote, K. Yokoo, K. Kawachi, Y. Kasama, S. Ono et al., *RSC Advances*, 2012, 2, 10624–10631.
19. S. Aoyagi, Y. Sado, E. Nishibori, H. Sawa, H. Okada, H. Tobita, Y. Kasama, R. Kitaura and H. Shinohara, *Angewandte Chemie International Edition*, 2012, 51, 3377–3381.
20. Y. Matsuo, H. Okada, M. Maruyama, H. Sato, H. Tobita, Y. Ono, K. Omote, K. Kawachi and Y. Kasama, *Organic Letters*, 2012, 14, 3784–3787.
21. H. Ueno, H. Kawakami, K. Nakagawa, H. Okada, N. Ikuma, S. Aoyagi, K. Kokubo, Y. Matsuo and T. Oshima, *Journal of the American Chemical Society*, 2014, 136, 11162–11167.
22. J. Cioslowski, *Spectroscopic and Computational Studies of Supramolecular Systems*, Ed. J. E. D. Davies, Kluwer, 1992.
23. H. Bai, H. Gao, W. Feng, Y. Zhao and Y. Wu, *Nanomaterials*, 2019, 9, 630.
24. Y. Noguchi, O. Sugino, H. Okada and Y. Matsuo, *Journal of Physical Chemistry C*, 2013, 117, 15362–15368.
25. A. Saroj, V. Ramanathan, B. Kumar Mishra, A. N. Panda and N. Sathyamurthy, *ChemPhysChem*, 2022, 23.
26. Y. Matsuo, H. Okada and H. Ueno, *Endohedral Lithium-Containing Fullerenes*, Springer, 2017, pp. 1–140.
27. G. George, A. J. Stasyuk and M. Solà, *Dalton Transactions*, 2022, 51, 203–210.
28. A. K. Srivastava, *Molecular Physics*, 2022, 120.
29. A. Kumar, A. K. Srivastava, G. Tiwari and N. Misra, *Atomic Clusters with Unusual Structure, Bonding and Reactivity*, Elsevier, 2023, pp. 173–183.
30. C. Sikorska, *Physical Chemistry Chemical Physics*, 2016, 18, 18739–18749.
31. G. Gutsev and A. Boldyrev, *Chemical Physics Letters*, 1982, 92, 262–266.
32. O. M. Nešković, M. V. Veljković, S. R. Veličković, L. T. Petkovska and A. A. Perić-Grujić, *Rapid Communications in Mass Spectrometry*, 2003, 17, 212–214.
33. S. Velicković, V. Djordjević, J. Cvetićanin, J. Djustebek, M. Veljković and O. Nesković, *Rapid Communications in Mass Spectrometry: RCM*, 2006, 20, 3151–3153.
34. S. R. Veličković, V. J. Koteski, J. N. Belošević Čavor, V. R. Djordjević, J. M. Cvetićanin, J. B. Djustebek, M. V. Veljković and O. M. Nešković, *Chemical Physics Letters*, 2007, 448, 151–155.

35. S. Veličković, V. Djordjević, J. Cvetićanin, J. Djustebek, M. Veljković and O. Nešković, *Vacuum*, 2008, 83, 378–380.
36. S. R. Veličković, J. B. Đustebek, F. M. Veljković and M. V. Veljković, *Journal of Mass Spectrometry*, 2012, 47, 627–631.
37. M. Z. Milovanović and S. V. Jerosimić, *International Journal of Quantum Chemistry*, 2014, 114, 192–208.
38. M. Milovanović, S. Veličković, F. Veljković and S. Jerosimić, *Physical Chemistry Chemical Physics*, 2017, 19, 30481–30497.
39. K. Yokoyama, N. Haketa, H. Tanaka, K. Furukawa and H. Kudo, *Chemical Physics Letters*, 2000, 330, 339–346.
40. F. Neese, *WIREs Computational Molecular Science*, 2012, 2, 73–78.
41. F. Neese, *WIREs Computational Molecular Science*, 2022, 12.
42. C. Lee, W. Yang and R. G. Parr, *Physical Review B*, 1988, 37, 785–789.
43. A. D. Becke, *The Journal of Chemical Physics*, 1993, 98, 5648–5652.
44. S. Grimme, J. Antony, S. Ehrlich and H. Krieg, *The Journal of Chemical Physics*, 2010, 132, 154104.
45. F. Neese, *Journal of Computational Chemistry*, 2003, 24, 1740–1747.
46. F. Neese, F. Wennmohs, A. Hansen and U. Becker, *Chemical Physics*, 2009, 356, 98–109.
47. F. Weigend and R. Ahlrichs, *Physical Chemistry Chemical Physics*, 2005, 7, 3297.
48. F. Weigend, *Physical Chemistry Chemical Physics*, 2006, 8, 1057.
49. C. Riplinger, P. Pinski, U. Becker, E. F. Valeev and F. Neese, *The Journal of Chemical Physics*, 2016, 144, 024109.
50. W. B. Schneider, G. Bistoni, M. Sparta, M. Saitow, C. Riplinger, A. A. Auer and F. Neese, *Journal of Chemical Theory and Computation*, 2016, 12, 4778–4792.
51. A. Altun, M. Saitow, F. Neese and G. Bistoni, *Journal of Chemical Theory and Computation*, 2019, 15, 1616–1632.
52. G. Bistoni, *WIREs Computational Molecular Science*, 2020, 10, 1–22.
53. A. Altun, R. Izsák and G. Bistoni, *International Journal of Quantum Chemistry*, 2021, 121, 1–13.
54. R. Bader, *Atoms in Molecules*, Claredon Press,1990.
55. T. Lu and F. Chen, *Journal of Theoretical and Computational Chemistry*, 2012, 11, 163–183.
56. F. Shojaie, *Computational and Theoretical Chemistry*, 2017, 1114, 55–64.
57. F. Tian Lu, *Journal of Computational Chemistry*, 2012, 33.
58. G. Andrienko, 2010, www.chemcraftprog.com.
59. W. Humphrey, A. Dalke and K. Schulten, *Journal of Molecular Graphics*, 1996, 14, 33–38.
60. Y. Matsuo, H. Okada and H. Ueno, *Endohedral Lithium-Containing Fullerenes*, Springer, 2017, pp. 105–115.
61. H. Ueno, K. Kokubo, Y. Nakamura, K. Ohkubo, N. Ikuma, H. Moriyama, S. Fukuzumi and T. Oshima, *Chemical Communications*, 2013, 49, 7376.
62. A. Kramida, *Atomic Data and Nuclear Data Tables*, 2010, 96, 586–644.
63. H. Steger, J. Holzapfel, A. Hielscher, W. Kamke and I. Hertel, *Chemical Physics Letters*, 1995, 234, 455–459.
64. D.-L. Huang, P. D. Dau, H.-T. Liu and L.-S. Wang, *The Journal of Chemical Physics*, 2014, 140, 224315.
65. F. Rohmund, A. Bulgakov, M. Hedén, A. Lassesson and E. Campbell, *Chemical Physics Letters*, 2000, 323, 173–179.
66. H. Yagi, N. Ogasawara, M. Zenki, T. Miyazaki and S. Hino, *Chemical Physics Letters*, 2016, 651, 124–126.
67. M. Feng, J. Zhao and H. Petek, *Science*, 2008, 320, 359–362.

10 A Computational Survey on Superalkali and Superhalogen Assisted Noble Gas Compounds

Ranajit Saha

10.1 INTRODUCTION

The modern periodic table has helium (He), neon (Ne), argon (Ar), krypton (Kr), xenon (Xe), radon (Rn) and oganesson (Og) in group 18, which are known as Noble gases (Ng) and are also termed as Inert gases, Rare gases and/or Aerogens [1]. Among these Ngs, Rn and Og are radioactive and less explored as compared to the others [2,3]. Over a few decades after the discovery of the first Ng, they were believed to be inert as no molecules containing Ngs were known. The outer electronic shell configurations of Ngs are ns^2np^6 (n = 2–6 for Ne-Rn, respectively), except for He ($1s^2$) and these filled valence orbitals are the reason for high ionisation potential (IP) [4] that means a huge amount of energy is required to ionise Ngs. However, ongoing from top to bottom of the Ng group, the IP value decreases (see Figure 10.1(I)) and the valence forces weaken, which suggests that it will be easier to knock out electron(s) from the outer orbital of heavier Ngs compared to that of the lighter congeners. Kossel predicted that fluorides of Kr and Xe could be formed [5]. In 1924, The valence of Ng atoms could be expanded up to eight which was claimed by Antropoff and in his words, "*one should not forget that as the valence number increases from one group to the next, the intensity of the valence forces decreases*" [6]. In 1933, based on the ionic radii of atoms, Linus Pauling predicted that the Xe atom can be coordinated with the oxygen atoms and can form "*Xenic acid, H_4XeO_6,*" which "*should form salts such as Ag_4XeO_6 and AgH_3XeO_6*" [7]. Moreover, Pauling further predicted the existence of krypton hexafluoride (KrF_6), xenon hexafluoride (XeF_6) and xenon octafluoride (XeF_8), the fluorinated compounds Kr and Xe.

Despite all these predictions, no success in the synthesis of Ng compounds was recorded. Several attempts to get fluorinated Ng compounds (XeF_6 and XeF_8) *via* electric discharge experiments were carried out by Prof. Yost and his student,

DOI: 10.1201/9781003384205-10

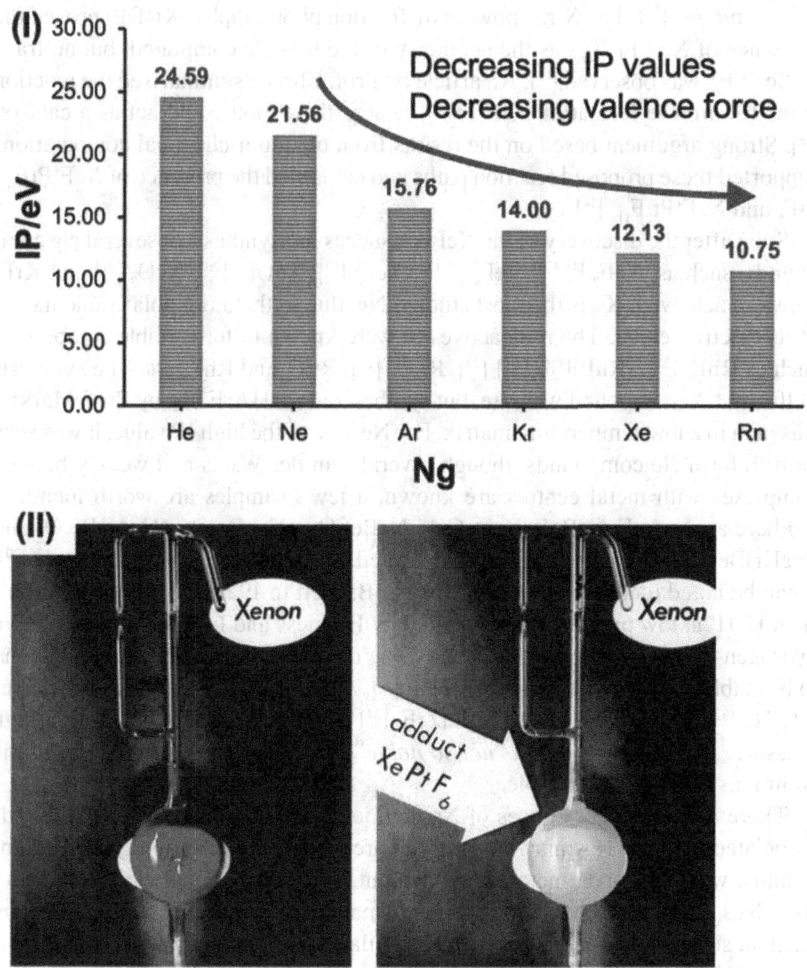

FIGURE 10.1 (I) The ionisation potential (IP) of the Ng atoms; (II) The Xe + PtF_6 reaction: left, the gases separated by a glass break-seal; right, the yellow reaction product formed on breaking the seal, from Ref. (10) the permission of Elsevier, Copyright 2000.

Kaye but they could not find success [8]. In 1962, Prof. Neil Bartlett synthesized $XePtF_6$ [9], an orange-yellow solid as shown in Figure 10.1 (II) [10], as the first Ng compound. The inspiration came from the synthesis of $O_2^+PtF_6^-$ [11] where he noticed that the IP values of O_2 (12.07 eV) and Xe (12.13 eV) are almost similar and so PtF_6 could be useful to synthesize the Xe complex and that concept produced the first Ng compound. Prof. Bartlet described the synthetic procedure of $XePtF_6$, as, "*When I broke the seal between the red PtF_6 gas and the colourless xenon gas, there was an immediate interaction, causing an orange-yellow solid* *to precipitate*" [12]. The X-ray powder diffraction photographs (XRDP) proved the existence of $XeF^+Pt_2F_{11}^-$, as the geometry of the first Xe compound, but no trace of Xe^+PtF_{6-} was observed [10]. An article by Prof. Christe summarised the reaction paths toward the formation $XeF^+Pt_2F_{11}$, where the F- ion could act as a catalyst [13]. Strong argument based on the results from quantum-chemical computations supported these proposed reaction paths and explained the presence of $XeF^+PtF_6^-$, PtF_5 and $XeF^+Pt_2F_{11}^-$ [14].

Soon after the discovery of the $XePtF_6$, successful synthesis of several Ng compounds, such as, XeF_2 [15,16], XeF_4 [17,18], XeF_6 [19], $XeOF_4$ [20], XeO_3 [21] and KrF_2 [22] were achieved. Xe is the most studied Ng, due to the more polarisable nature of its electron cloud. The radioactive Rn were known to form stable compounds such as RnF_2 [23], $[RnF][Sb_2F_{11}]$ [23], RnO_3 [24], RnF_4 and RnF_6 *etc.* The synthesis of the first Ar compound was late, but synthesized as, HArF [25], by Prof. Markku Räsänen in a low-temperature matrix. For Ne, due to the high IP value, it was very hard to form Ne compounds, though several Van der Waals and weakly bonded complexes with metal centres are known, a few examples are worth mentioning here as, NeAuF, NeBeS, $NeBeCO_3$, $NeBeSO_2$, $(Ne)_2Be_2O_2$, $(NeAr)Be_2O_2$ and $(NeKr)Be_2O_2$, were experimentally identified in a low-temperature matrix [26–29]. It can be noted that before the discovery of Bartlett in 1925, the presence of transient HeH^+ at low pressure was observed by Hogness and Lunn by bombarding a hydrogen-helium mixture [30]. Recently, the disodium helide (Na_2He) was found to be stable at high pressure (> 113 GPa) [31]. A few other He compounds, such as SiO_2He [32], $As_4O_6{\cdot}2He$ [33] and $HeCaZrF_6$ [34] *etc.* were known to be stable at high pressure. It could be said that "*not so noble*" [35] gases can form bonds with the commons of the periodic table.

There are two major classes of Ng compounds, where direct chemical bonds associated with the Ngs may be observed are *viz*., (i) noninsertion type Ng compounds, where an open metal site is present, which can accept electron density from Ngs. But, the presence of metal is not mandatory, while an electron-deficient element such as boron can perform the similar role. For simplicity, a general formula of NgMY can be considered in this chapter where the M and Y are an electropositive element and the counter anion, respectively, and the whole complex can be neutral, positive or negative and (ii) insertion type Ng compounds, with a general formula of XNgY, where Ng atoms are inserted within a X-Y bond, X and Y are H, electronegative atoms (such as halogens) or cluster of atoms (such as superhalogens) [36]. The X and Y can be the same or different as in NgF_2 or HNgF, respectively [15,25].

Several examples of noninsertion of Ng compounds can be given as, NgLiX (Ng = He; X = H, F) [37–39], NgBeY (Ng = He-Rn; Y = O, NH, NBO, NCN, CO_3, SO_4, CrO_4, HPO_4) [39–43], $NgBeCp^+$ (Ng = He-Rn; M = Be-Ba, Cp = η^5-C_5H_5) [44], NgMX (Ng = Ar-Rn, M = Cu, Ag, Au; X = Cl, CN) [45–49], NgMO (Ng = Ar-Rn, M = Cu, Ag, Au) [50], $Ng_2Be_2N_2$ (Ng = He-Rn), [51] $Ng_nM_3^+$ (M = Li, B, Cu, Ag, Au) [52–54], *etc.* The M in the noninsertion type of Ng compounds is the electrophilic centre where the Ng-atom get attached. An increase in the charge density (higher positive change and lower radius) on the M-centre showed an enhancement of the polarising ability of the M-atoms and increases the polarizability of Ng atoms, hence a stronger Ng-M bond is expected [39]. On the other hand, with an increase in the size of the Ng atoms, the polarizability of the Ng-atoms enhances and hence the strength of the Ng-M bond increases, which can be easily understood from a large number of known Xe compounds.

The degree of covalency for a particular Ng-M has been found to be increased for the heavier members of group 18. The energy decomposition analysis (EDA) helps to understand the origin of the nature of the interactions between the Ng and M. In Figure 10.2, two representatives, namely (a) $B_3Ar_3^+$ [53] and (b) $XeBeCp^+$ [44] have been presented which show two typical Ng-M bonding patterns. For both cases, the major contribution of the energies toward the Ng-M bond originated from the orbital interaction. The total orbital interactions were decomposed and found that the major stabilising came from the Ng→M σ-donation from the filled *p*-orbital of Ng-atom to the vacant orbital, here the lowest unoccupied molecular orbital (LUMO) of the second fragment. The later orbital contributions make the story more interesting. In Figure 10.2(I), the vacant orbitals on the Ngs can receive electron density from the occupied MOs that indicated Ng←M π-back donation, which is similar to the synergistic effect. In Figure 10.2(II), the minor orbital interactions came from the rest of the *p*-orbitals of the Ng atom to vacant orbitals on the other fragment and Ng→M π-donation. Hence, the major interactions could be similar, but other minor interactions can be different. Though all interactions led to the stabilisation of Ng-compounds.

In the insertion type of XNgY compounds, the Ng-atom is inserted within the X-Y bond. The X and Y can be the same atoms as in KrF_2, XeF_2, *etc.* or different atoms as in HArF, HKrF, *etc.* The bonding analysis showed the presence of 3c-4e bonds in XNgY compounds. The strength of the Ng-X and Ng-Y bonds is the major force behind the stability of such compounds. The XNgY compounds can be decomposed (i) into XY and Ng, known as two body (2B) dissociation path and (ii) into Ng, X and Y, known as three body (3B) dissociation path. A schematic potential energy surface showing the stability of the XNgY with respect to both dissociation paths is shown in Figure 10.3. In general, the 3B path is endergonic in nature, though it depends on the nature of X, Y and Ng. On the other hand, the 2B dissociation is exothermic in nature and proceeds toward the formation of stable XY and Ng. This 2B dissociation path goes *via* a bent transition state and the stability of XNgY depends on the energy required to achieve this transition state. Computational studies by Hu and co-workers have shown that the half-life of XNgY depends on the energy barrier height and a minimum energy barrier of 6, 13 and 21 $kcal{\cdot}mol^{-1}$ would have a half-life in the order of ~10^2 seconds at 100, 200 and 300 K, respectively [55].

Other than these two types of Ng compounds, Ngs encapsulated inside cages are known in literatures. Ng atoms such as, He, Ne, Ar, Kr and Xe inside C_{60} cage have been generated by experiments [56–58]. Several computational studies considering cages like C_{60} [59], C_{70} [60], $B_{12}N_{12}$ [61], $B_{16}N_{16}$ [61], $C_{20}H_{20}$ [62], cucurbit[n]uril (n = 6,7) [63,64], carbon nanotubes [65], B_{40}, [66] clathrate hydrate and its HF doped analogues [67], were studied to encapsulating Ng atoms inside the cage.

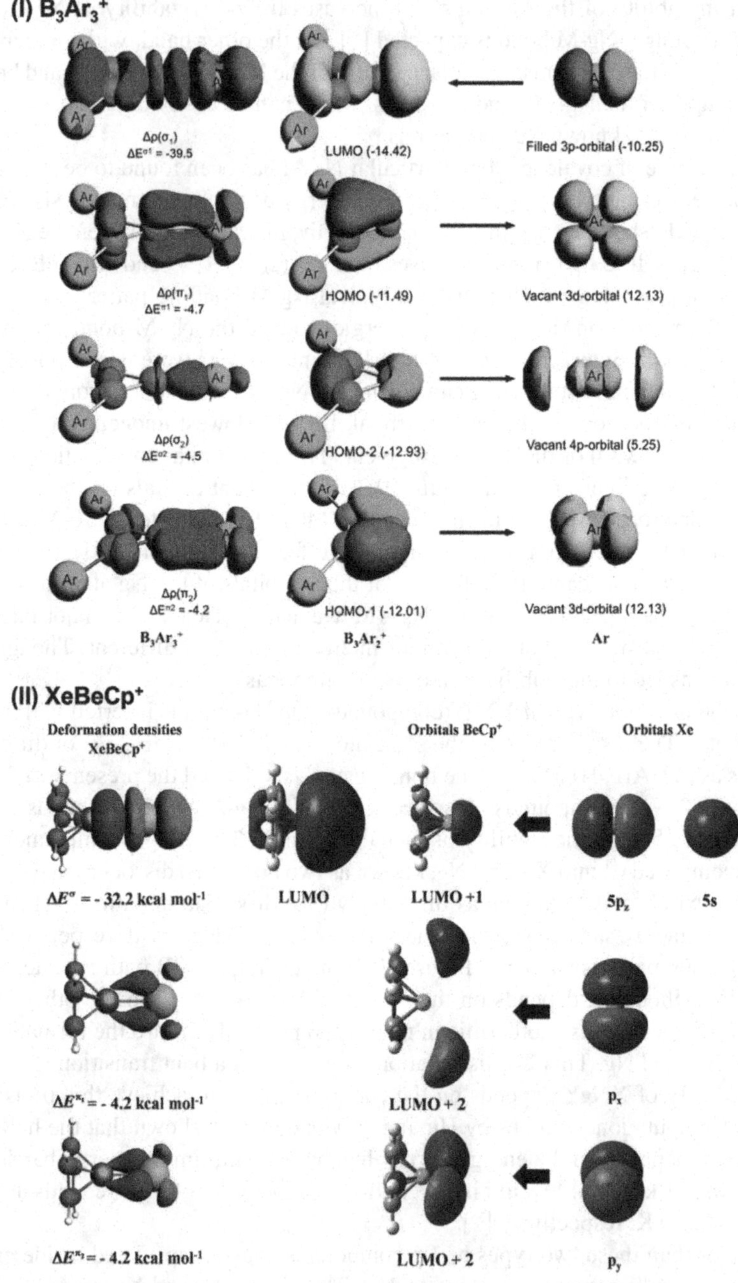

FIGURE 10.2 The donor-acceptor interactions in noninsertion type of Ng-compounds, *e.g.*, (I) $B_3Ar_3^+$; from Ref. [53] the permission of Royal Society of Chemistry, Copyright 2016 and (II) $XeBeCp^+$ (Cp = η-$C_5H_5^-$) from Ref. [44] the permission of American Chemical Society, Copyright 2017.

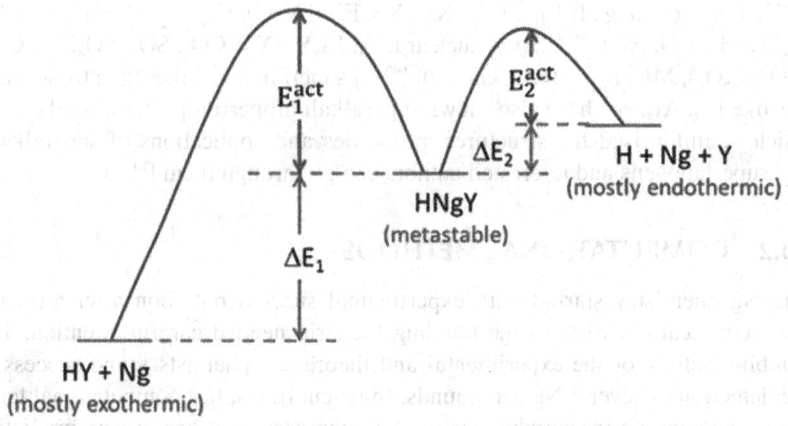

FIGURE 10.3 The schematic depiction of the 2B and 3B dissociation paths of XNgY from Ref. (76) the permission of American Chemical Society, Copyright 2014.

Superhalogens are clusters of atoms that have electron affinity (EA) higher than that of chlorine (Cl; EA = 3.62 eV). In 1981, the concept of superhalogen was proposed by Gutsev and Boldyrev by the formula MX_{n+1} where M, X and n are the central metal atom, halogen atoms and the maximum valence of M, respectively [68]. The extra electron added to MX_{n+1} to form MX_{n+1}^- is delocalized over all surrounding n+1 halogen atoms and is the reason for the extra stability and higher EA values of MX_{n+1} compared to the constituting halogen [69]. Some examples of this kind of superhalogens are MX_3 (M = Be, Mg, Ca; X = F, Cl, Br) [70,71], MX_4 (M = B, Al; X = F, Cl, Br) [72], MX_n clusters (M = Sc, Ti, V; X = F, Cl, Br; n = 1–7) [73] and many more to mention.

Calculations have shown that PtF_6 (EA = 7.09 eV) can be considered as a superhalogen and have high oxidising power that is the reason behind its success toward the first Ng compound [14]. On a similar route, superhalogens like BF_4 (EA = 6.86 eV) [74] and BO_2 (EA = 4.46 eV) [75] were shown to form metastable Ng compounds, HNgY, where Ng = Ar, Kr; Y = BF_4, BO_2 [76]. These Ng compounds act as Ng insertion compounds and are kinetically stabilised by an energy barrier. Other examples of this kind will be discussed in the later paragraphs.

Similarly, superalkali are those atomic clusters that have IP values lower than Caesium (IP = 3.89 eV) [77]. In 1982, Gutsev and Boldyrev used the term superalkali in their work on a group of atomic clusters having the formula, $M_{k+1}L$, where M and L are alkali and electronegative atoms and k is the valence of the L-atom [78]. Computationally, they have shown clusters of hypervalent electronegative atoms, *viz.*, F, O and N with alkali atoms, *viz.*, Li, Na and Cs have low IP values and are considered as superalkalis. Later on, extensive works on superalkali were reported and it is established that the territory of superalkali is beyond the above-mentioned formula $M_{k+1}L$ [79]. It is possible to categorise the superalkalis into three broad categories, mononuclear (*e.g.*, Li_2F, Li_3O, Li_4N, Li_5Si, Li_6B, *etc.*)

[80–84], binuclear (*e.g.*, $B_2Li_{11}^+$, $Li_{2k+1}X_2^+$ (X = F, O, N, C), $H_{2k+1}X_2^+$ (X = F, O, N, C), Li_7CO^+, Li_8CN, *etc.*) [85–89], polynuclear (*e.g.*, Li_3Y^+ (Y = CO_3, SO_3, SO_4, O_2, CO_4, C_2O_4, C_2O_6), $MLi_4F_6^+$ (M=Al, Ga, Sc)) [90–92] superalkalis. Other than these, clusters like Li_3, Al_3, *etc.* have also shown superalkali properties [93,94]. Several review articles summarised the structures, properties and applications of superalkalis and superhalogens and interested authors can go through them [79, 95].

10.2 COMPUTATIONAL METHODS

The Ng chemistry started with experimental success but soon after attracted the theoretical chemists as the bonding scenario needed careful attention. The combined study of the experimental and theoretical chemists found success in the detection of several Ng compounds. In recent times, fast computational techniques become an inseparable part of Ng chemistry as it has proven predictive ability for molecular, and periodic systems as well as high-pressure calculations. This chapter mainly focuses on the computational findings and discusses them in a summarised fashion.

The way into the Ng compounds starts with the geometry optimisation of the Ng compound and is the most crucial step in the computational section of Ng chemistry. Hence the choice of the computational methods as well as the basis sets are very critical. The aerogen interactions are considered to be weak in nature. A computational method such as coupled cluster with the inclusion of single and double substitutions and an estimate of connected triples (CCSD(T)) level is required for a correct description of the geometry and energies of the studied system [96]. But on the other hand, the CCSD(T) is computationally expensive so its usage has limitations. A fine alternative is the use of density functional theory (DFT) and Møller–Plesset perturbation theory truncated at the second order (MP2) methods. Benchmark studies considering DFT methods on Ng compounds have suggested some DFT methods that can compute the energetics and geometry of the Ng compounds [97–99]. The reference geometries and the energies were optimised at the CCSD(T)/aug-cc-pVTZ and CCSD(T)/CBS level of theories, respectively. The bond distances computed considering the following level of theories such as MPW1B95/6–311+G(2df,2pd), MPW1PW91/6–311+G(2df,2pd) and B3P86/6–311+G(2df,2pd), DSD-BLYP/6–311+G(2df,2pd) and DSD-BLYP/aug-cc-pVTZ methods performed well with a mean unsigned error (MUE) of 0.008–0.013 Å per bond. The computed bond energies at the MPW1B95/6–311+G(2df,2pd), BMK/aug-cc-pVTZ, B2GP-PLYP/aug-cc-pVTZ and DSD-BLYP/aug-cc-pVTZ level of theories worked well with a MUEs within the range of 2.0–2.3 kcal/mol per molecule. On the other hand, the semi-empirical MP2 method in conjunction with Aug-cc-pVDZ has shown an MUE value of 4.5 kcal/mol. One important part of the calculation was to take care of the relativistic effect present on the heavier Ng-atoms (such as Xe, Rn) and metal atoms (such as Ag, Au, Sr, Ba, Cd, Hg, *etc.*) using the quasi-relativistic effective core potentials (ECP) [100,101], or the zeroth-order regular approximation (ZORA) [102].

The strength of the Ng-M bond can be connected to the amount of electronic charge transfer from the Ng-atom to the electron-deficient binding site. Such parameters can be computed using natural bond orbital (NBO) [103], electron density [104], energy decomposition analysis (EDA) [105,106], *etc.* The NBO analysis is one of the most widely used computational techniques that find out the natural charge distribution on the atoms of the molecule and calculates the bond order between any pair of atoms known as the Wiberg bond index (WBI). The calculated natural charges on the atoms help to understand the direction of the charge flow within the system and the computed WBI values indicate the bond order between two atoms. In the case of the Ng compounds, the Ng-atoms get a positive charge on them and an increased positive charge indicates more polarization of the Ng atoms and hence a stronger Ng-M bond that is also reflected by the high WBI values.

Depending on the amount of the transfer of the electron density the nature of the Ng-M bond varies within a long range from weak van der Waals interaction to covalent type. The quantum theory of atoms in molecule (QTAIM) approach by Bader has been another great tool for the analysis of the Ng-M bonds that helps to calculate various electron density-based descriptors [107]. According to the QTAIM, a covalent bond has a negative value for the Laplacian of the electron density, $\nabla^2(\rho(r_c))$, at the bond critical point (BCP), which often failed for Ng compound cases as relatively weaker orbital involvement in Ng cases. Other examples involving F_2 and complexes of 3d-transition metals also showed positive $\nabla^2(\rho(r_c))$ values instead covalency of the bonds [108–111]. For such systems, the total energy density $H(r_c)$ are being considered and proved as a very useful descriptor. In such cases, $\nabla^2(\rho(r_c)) > 0$, $H(r_c) < 0$ suggests covalent bonding or the presence of partial covalent bonding therein, depending on the magnitude of the electron density ($\rho(r_c)$) [107,112].

For the analysis of the Ng-M bonds two fragments, one containing Ng-atom and the rest in another fragment are considered in the EDA [105,106]. In the EDA, the interaction energy (ΔE_{int}) between two fragments is decomposed into four energy terms, which can be written as follows,

$$\Delta E_{int} = \Delta E_{elstat} + \Delta E_{Pauli} + \Delta E_{orb} + \Delta E_{disp}$$

where the ΔE_{elstat} was computed classically by taking the two fragments at their optimized positions but considering the charge distribution was unperturbed on each fragment, by another one. The next one, ΔE_{Pauli}, appeared as the repulsive energy between electrons of the same spin and it was computed by using the Kohn-Sham determinant on the superimposed fragments to obey the Pauli principle by antisymmetrization and renormalization. The ΔE_{orb} originated from the mixing of orbitals, charge transfer and polarization between two fragments. Lastly, ΔE_{disp} represented the dispersion interaction between the two fragments.

10.3 SUPERHALOGEN-SUPPORTED NG COMPOUNDS

Computational work by Chakraborty et al. showed that neutral Xe-bound to transition metal fluorides with general formula MF_3 (where M = Ru, Os, Rh, Ir,

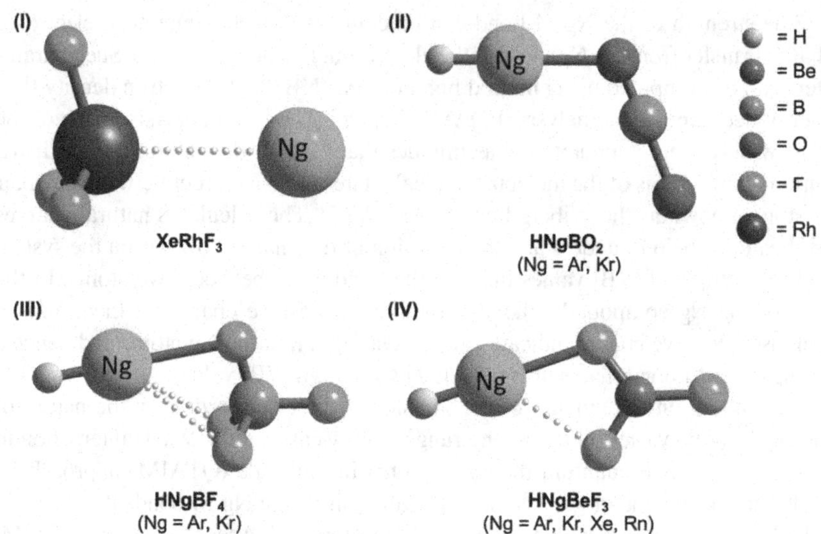

FIGURE 10.4 The geometry of superhalogen-supported Ng compounds; (I) $XeRhF_3$, (II) $HNgBO_2$, (III) $HNgBF_4$ and (IV) $HNgBeF_3$.

Pd, Pt, Ag and Au) are thermochemically viable as shown by density functional theory-based calculations at 298 K [113]. Among the Ng compounds studied, $XeRhF_3$ showed the strongest M-Xe interaction. The nature of the M-Xe interactions in $XeMF_3$ (shown in Figure 10.4 (I)) are partially covalent as evidenced by NBO, AIM, EDA analyses. The EDA results revealed that the major contribution toward the Xe-M interaction came from the orbital interactions therein.

The superhalogens, such as BO_2 and BF_4 showed their creditability to form stable Ng compounds with Ar and Kr.[76] Several dissociation paths have been considered to check the stability of HNgY (Ng = Ar, Kr; Y = BO_2, BF_4) and the stabilities of these superhalogen supported Ng compounds were compared with that of HNgF. The geometries of $HNgBO_2$ and $HNgBF_4$ (Ng = Ar and Kr) were shown in Figure 10.4 (II) and (III), respectively. Computations at MP2 and CCSD(T) level of theory showed these Ng-inserted compounds are kinetically stable and can be made viable. The NBO analysis confirmed that the HNg moiety carries a positive electronic charge that approaches +1 as the amount of charge transfer increases, which is reflected in the corresponding increase in the EA value of Y.

Work by Saha et al. showed that superhalogen, BeF_3, also follows the path of BO_2 and BF_4 and is able to form stable Ng compounds, $HNgBeF_3$ (Ng = Ar-Rn) (see Figure 10.4 (IV)) [114]. Computations using the DFT, MP2 and CCSD(T) methods in conjunction with Def2-QZVPD basis set showed that except for the 2B and 3B dissociations paths other dissociation channels were endothermic in nature at 298 K. The electronic charge distribution showed that $HNgBeF_3$ can be represented as n ion-pair $[HNg]^+[BeF_3]^-$. The calculated barrier height associated with the 2B dissociation path: $HNgBeF_3 \rightarrow Ng + HBeF_3$ were 1.0–13.9 kcal/mol for Ar to Rn analogues that suggested the Xe and Rn might be viable to synthesize and could have a half-life of ~10^2 seconds up to ~100 K. Further extensive study by Wu et al. have elaborated that other elements from groups 2 and 7, might form stable $HNgMX_3$ (Ng = Ar-Xe, M = Be-Ca, X = F-Br) [[115]] compounds. The stability order increased on going toward heavier Ng atoms as compared to the lighter analogues. The opposite trend was observed to be true for the halogen atoms and M atoms where Be and F showed maximum stability compared to the heavier counterparts as the barrier height follows the order as Be > Mg > Ca and F > Cl > Br, respectively, for a particular Ng atom. It was also established that the strength of the Ng···F interaction plays a decisive role in their stability. Polynuclear superhalogens could also form stable Ng-compounds, but the stability was found to be lower as compared to the similar mononuclear superhalogen analogue. The stability of the transition state associated with the 2B dissociation paths depended on the nature of Ng···F interactions, not on the number of superhalogen.

The work by Mayer et al. showed cage-like superhalogen formed Ar bound complex, $ArB_{12}(CN)_{11}^-$ (see Figure 5.I.a), which was stable at room temperature [[116]]. The authors considered cyanide-decorated boron cage, $B_{12}(CN)_{12}^{2-}$, from which one CN^- anion was removed and $B_{12}(CN)_{11}^-$ anion was formed. The EA value for $B_{12}(CN)_{11}^-$ anion was computed to be 8.49 eV, which supported its superhalogen property of it. The vacant B-side in $B_{12}(CN)_{11}^-$ became more positive as compared to the rest of B-atoms, which polarised the electron density of Ar atom and form a stable Ar-B bond in $ArB_{12}(CN)_{11}^-$. This mechanism is presented in Figure 5.II. The comparative study showed that the Ar-B bond in Ar-$B_{12}(CN)_{11}^-$ is stronger than the same in compounds like Ar-B, Ar-BF_2 and Ar-B_5O_7. Later, the computational study by Fang et al. at the DFT level of theory proved that several Ng compounds that were stabilised by superhalogens are stable and viable [[117]]. Boron clusters like $B_{12}X_{11}^-$ (X = H, CN and BO) and $B_{12}X_{10}$ (X = CN and BO) could bind to Ng atoms (Ng = Ne, Ar, Kr and Xe) and form one and two Ng bonded compounds like $B_{12}X_{11}(Ng)^-$ (X = H, CN and BO; Ng = Ne, Ar, Kr and Xe) and $B_{12}X_{10}(Ng)_2$ (X = CN and BO; Ng = Ne, Ar, Kr and Xe), respectively. The geometries of $NgB_{12}(CN)_{11}^-$ and $NgB_{12}(CN)_{10}$ were presented in Figure 5.I.a and Figure 5.I.b, respectively. The strength of the Ng-B bonds follows the ascending order as, Ne < Ar < Kr < Xe that seemed to be connected to the polarizability of the Ng-atoms. The study also showed that the Ng-B bond strength depends on increasing cluster size and the nature of the X, *i.e.*, the electron affinity of the terminal ligands, *e.g.*, CN showed the strongest binding among all of the X-ligands. Moreover, it was shown that the removal of two X ligands make the vacant B-centres more electrophilic that the Ng-B interactions became stronger. In another trial, they have also shown that small boron clusters cannot hold many electrons and hence weak Ng-B interactions were observed in Ng-compounds like $C_5BX_5(Ng)$ (X = H, F and CN; Ng = Ne and Ar) and $C_4B_2(CN)_4(Ng)_2$ (Ng = Ne and Ar) (see Figure 5.I.c).

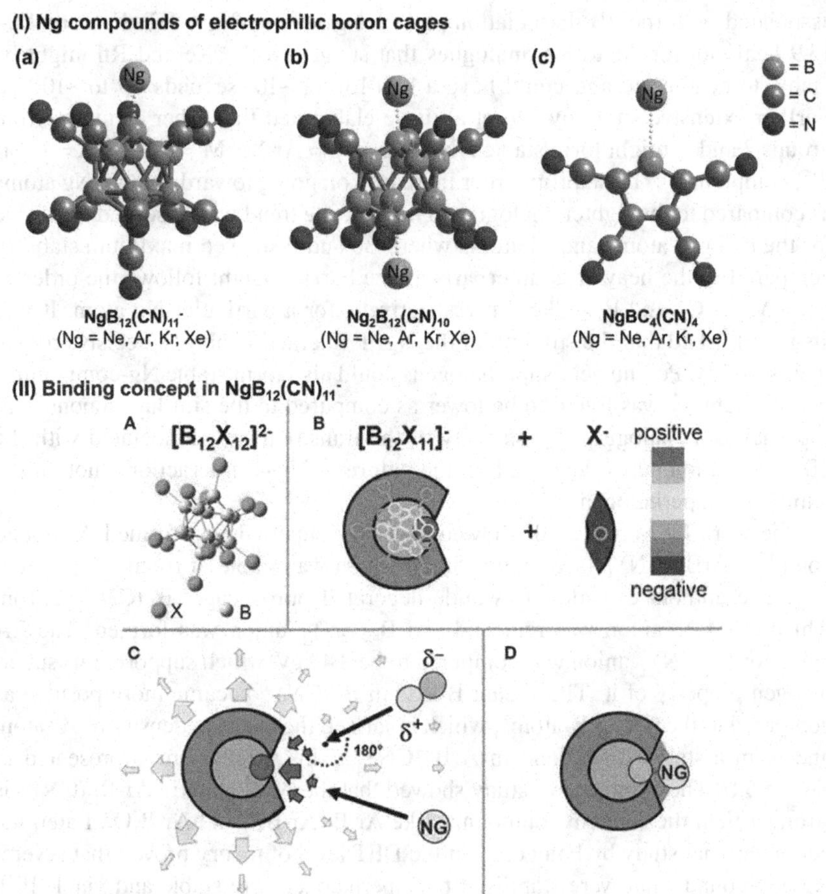

FIGURE 10.5 (I) The geometry of superhalogen-supported Ng compounds; (II) the binding between Ar and B centres, from Ref. [116] the permission of PNAS, Copyright 2017.

10.4 SUPERALKALI-SUPPORTED NG COMPOUNDS

The superalkalis were found to form complexes with atoms with the open metal sites presenting therein. An elaborate study by Pan et al. considered $O_2Li_5^+$ [118], the superalkali to bind Ng atoms [119,120]. The global minimum geometry of $O_2Li_5^+$ is in the D_{3h} point group of symmetry and the EA value was calculated to be 2.94 eV at OVGF/6–311G(3df)//MP2/6–311G(3df) level of theory.[118] Ng atoms (Ng = He, Ne, Ar, Kr, Xe) were considered for the study of Ng binding ability of $O_2Li_5^+$ (see Figure 10.6 (I-III)).[119,120] The five Li-atoms in $O_2Li_5^+$ were not equivalent, three of them were in equatorial (*eq.*) and the rest two were in axial (*ax.*) positions and showed different affinity toward Ng atoms. The *ax.* Ng···Li interactions were found to be stronger than that of the *eq.* Ng···Li interactions. Moreover, the

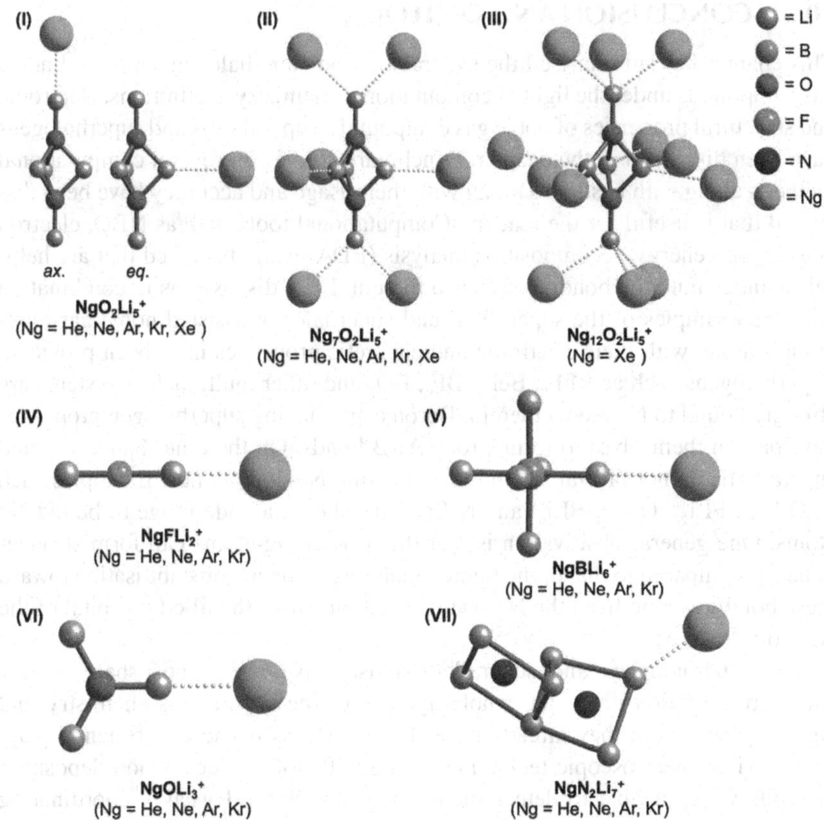

FIGURE 10.6 The geometries of Li-based superalkali-supported Ng compounds.

ax. Li atoms could bind to two Ng atoms simultaneously, proving that superalkali, $O_2Li_5^+$ can be bound to a total of seven Ng atoms (Ng = He, Ne, Ar, Kr) shown in Figure 10.6.II. For the case of Xe, due to the enhanced polarizability, the *ax.* and *eq.* Li-atoms interacted with three and two Xe atoms and a stable $Xe_{12}O_2Li_5^+$ cluster was predicted (see Figure 10.6.III). The EDA analysis on the Ng···$O_2Li_5^+$ clusters showed that polarisation and charge transfer were the major contributing factors toward the stabilisation of these complexes. The Ng-atoms (Ng = He, Ne, Ar, Kr) also bound to the one Li-atom in superalkalis, like FLi_2^+, OLi_3^+, BLi_6^+ and $N_2Li_7^+$ and the strength of the Ng···Li interaction followed the order Ar > Kr > Ne > He.

Li_3 cluster was considered as superalkali (IE = 4.08 ± 0.05 eV) and the removal of one electron from the neutral Li_3 cluster led to the formation of the stable Li_3^+ cation where the stability of Li_3^+ was explained by the jellium model [121]. The stable Li_3^+ cation formed stable Ng complexes (Ng = He, Ne, Ar) as shown by Chakraborty et al. and these compounds were able to accommodate a maximum of three Ng atoms via the Ng···Li interactions.[52]

10.5 CONCLUSION AND OUTLOOK

This chapter has summarised the superalkali and superhalogen supported noble gas compounds under the light of computational chemistry. Definitions, electronic and structural properties of noble gas compounds, superalkalis and superhalogens have been discussed in the chapter. Benchmark studies on several computational methods along with basis sets along with their usage and accuracy have been discussed that is useful for the readers. Computational tools such as NBO, electron density, and energy decomposition analysis (EDA) were discussed that are helpful to understand the bonding scenario therein. Brief discussions in combination with the examples of the superalkali and superhalogen assisted noble gas compounds along with their electronic and energetic properties have been provided. Superhalogens such as RhF_3, BeF_3, BF_4, BO_2 and other multinuclear systems are strongly bound to Ng atoms therein. Boron cages having superhalogen properties have proven themselves to form strong Ar-B bonds. On the other hand, lithiated superalkalis do not fall far behind in stabilising Ng-compounds. Examples such as $O_2Li_5^+$, FLi_2^+, OLi_3^+, BLi_6^+ and $N_2Li_7^+$ have shown a wide range to bound Ng atoms. One general observation is that the heavier Ng-atoms can form stronger bonds as compared to that of the lighter analogues. The major stabilisation toward these bonding came from the Ng→M σ-donation where the filled *p*-orbital of the Ng-atoms take part.

The superhalogens and superalkali assisted Ng-compounds share a very small cross-section under the whole dynasty of the superatoms chemistry and Ng-compounds, but they already have shown success in the experimental playground. The spectroscopic techniques along with soft molecular ion deposition methods were capable to detect such complexes. Several weakly coordinating counter-ions are being used in the soft-landing setups. The computational paths are also being followed to characterise the nature of bonding in these compounds. Although much progress has been made in understanding the properties and applications of superalkali and superhalogen assisted noble gas compounds, there are still many more Ng compounds with the help of superhalogen and superalkali clusters to make their way into this field of work.

ACKNOWLEDGEMENTS

The author gratefully acknowledges Institute for Chemical Reaction Design & Discovery (ICReDD), Hokkaido University, Japan. The author also acknowledges Prof. P. K. Chattaraj and Dr. Ambrish Kumar Srivastava for the invitation to contribute a chapter to the book entitled, "Superhalogens, Superalkalis and Supersalts: Bonding, Reactivity, Dynamics and Applications."

REFERENCES

1. Bauzá, A., and Frontera, A. 2015, *Angew. Chem. Int. Ed.* 54, 7340–7343.
2. EPA Facts About Radon. *United States Environmental Protection Agency*, pp. 1–3 (https://semspub.epa.gov/work/HQ/176336.pdf) last accessed 07 March 2023.
3. Koppenol, W. H. et al. 2016, *Pure Appl. Chem.* 88, 401–405.
4. Lide, D. R. 2003–2004, "Section 10: Atomic, Molecular, and Optical Physics." In *Ionization Potentials of Atoms and Atomic Ions: CRC Handbook of Chemistry and Physics* (84th edition). CRC Press, 10–178.
5. Kossel, W. 1916, *Ann. Phys.* 354, 229–362.
6. Antropoff, A. V. 1924, *Angew. Chem. Int. Ed.* 37, 695–696.
7. Pauling, L. 1933, *J. Am. Chem. Soc.* 55, 1895–1900.
8. Labinger, J. A. 2015, *Bull. Hist. Chem.* 40, 1, 29–36.
9. Bartlett, N. 1962, *Proc. Chem. Soc. Lond.* 1962, 218.
10. Graham, L. et al. 2000, *Coord. Chem. Rev.* 197, 321–334.
11. Bartlett, N., and Lohmann, D. 1962, *J. Chem. Soc.* 5253–5261.
12. Hargittai, I. 2009, *Neil Bartlett and the First Noble-Gas Compound.* Springer.
13. Christe, K. O. 2013, *Chem. Commun.* 49, 4588–4590.
14. Craciun, R. et al. 2010, *Inorg. Chem.* 49, 1056–1070.
15. Hoppe, R. et al. 1962, *Angew. Chem.* 74, 903.
16. Chernick, C. L. et al. 1962, *Science* 138, 136–138.
17. Claassen, H. H., Selig, H., and Malm, J. G. 1962, *J. Am. Chem. Soc.* 84, 3593.
18. Chernick, C. L., Malm, J. G., and Williamson, S. M. 1966, *Inorg. Synth.* 8, 254–258.
19. Chernick, C. L., Malm, J. G., and Williamson, S. M. 1966, *Inorg. Synth.* 8, 258–260.
20. Smith, D. F. 1963, *Science* 140, 899–900.
21. Templeton, D. H. et al. 1963, *J. Am. Chem. Soc.* 85, 817.
22. Grosse, A. V. et al. 1963, *Science* 139, 1047–1048.
23. Stein, L. 1970, *Science* 168, 362–364.
24. Holloway, J. H. and Hope, E. G. 1998, "Recent Advances in Noble-Gas Chemistry." In *Advances in Inorganic Chemistry* (Vol. 46). Academic Press, 51–100.
25. Khriachtchev, L. et al. 2000, *Nature* 406, 874–876.
26. Wang, Q., and Wang, X. 2013, *J. Phys. Chem. A* 117, 1508–1513.
27. Zhang, Q. et al. 2014, *J. Phys. Chem. A* 119, 2543–2552.
28. Yu, W. et al. 2016, *J. Phys. Chem. A* 120, 8590–8598.
29. Zhang, Q. et al. 2017, *Chem. Eur. J.* 23, 2035–2039.
30. Hogness, T. R., and Lunn, E. G. 1925, *Phys. Rev.* 26, 44–55.
31. Dong, X. et al. 2017, *Nat. Chem.* 9, 440–445.
32. Matsui, M., Sato, T., and Funamori, N. 2014, *Am. Min.* 99, 184–189.
33. Guńka, P. A. et al. 2015, *Cryst. Growth Des.* 15, 3740–3745.
34. Lloyd, A. J., et al. 2021, *Chem. Mater.* 33, 3132–3138.
35. Sanderson, K. 2008, *Nature* (www.nature.com/articles/news.2008.856), last accessed 22nd February 2023.
36. Saha, R. et al. 2019, *Molecules* 24, 2933.
37. Kaufman, J. J., and Sachs, L. M. 1969, *J. Chem. Phys.* 51, 2992–3005.
38. Barat, M. et al. 1973, *J. Phys. B: Atom. Mol. Phys.* 6, 2072–2087.
39. Frenking, G. et al. 1988, *J. Am. Chem. Soc.* 110, 8007–8016.
40. Thompson, C. A., and Andrews, L. 1994, *J. Am. Chem. Soc.* 116, 423–424.
41. Pan, S. et al. 2013, *J. Phys. Chem. A* 118, 487–494.
42. Saha, R. et al. 2015, *J. Phys. Chem. A* 119, 6746–6752.
43. Pan, S. et al. 2016, *RSC Adv.* 6, 92786–92794.

44. Saha, R., Pan, S., and Chattaraj, P.K. 2017, *J. Phys. Chem. A* 121, 3526–3539.
45. Evans, C. J., Lesarri, A., and Gerry, M. C. L. 2000, *J. Am. Chem. Soc.* 122, 6100–6105.
46. Reynard, L. M., Evans, C. J. M. and Gerry, C. L. 2001, *J. Mol. Spectrosc.* 206, 33–40.
47. Walker, N. R., Reynard, L. M., and Gerry, M. C. L. 2002, *J. Mol. Struct.* 612, 109–116.
48. Michaud, J. M., Cooke, S. A., and Gerry, M. C. L. 2004, *Inorg. Chem.* 43, 3871–3881.
49. Pan, S. et al. 2015, *J. Comp. Chem.* 36, 2168–2176.
50. Pan, S. et al. 2016, *Int. J. Quantum Chem.* 116, 1016–1024.
51. Pan, S. et al. 2014, *ChemPhysChem* 15, 2618–2625.
52. Chakraborty, A., Giri, S., and Chattaraj, P. K. 2010, *New J. Chem.* 34, 1936–1945.
53. Saha, R. et al. 2016, *RSC Adv.* 6, 78611–78620.
54. Pan, S. et al. 2016, *Phys. Chem. Chem. Phys.* 18, 11661–11676.
55. Li, T.-H. 2007, *Chem. Phys. Lett.* 434, 38–41.
56. Saunders M. et al. 1996, *Science* 271, 1693–1697.
57. Darzynkiewicz R. B., and Scuseria, G. E. 1997, *J. Phys. Chem. A* 101, 38, 7141–7144.
58. Khong, A. et al. 1998, *J. Am. Chem. Soc.* 120, 25, 6380–6383.
59. Khatua, M., Pan, S., and Chattaraj, P. K. 2014, *Chem. Phys. Lett.* 610–611, 351–356.
60. Gómez, S., and Restrepo, A. 2019, *Phys. Chem. Chem. Phys.* 21, 15815.
61. Khatua, M., Pan, S., and Chattaraj, P. K. 2014, *J. Chem. Phys.* 140, 164306.
62. Cerpa, E. et al. 2008, *Chem. Eur. J.* 14, 10232–10234.
63. Pan, S. et al. 2017, *Phys. Chem. Chem. Phys.* 19, 24448–24452.
64. Pan, S. et al. 2016, *J. Phys. Chem. C* 120, 13911–13921.
65. Chakraborty, D. and Chattaraj, P. K. 2015, *Chem. Phys. Lett.* 621, 29–34.
66. Pan, S. et al. 2018, *Phys. Chem. Chem. Phys.* 20, 1953–1963.
67. Mondal, S., and Chattaraj, P. K. 2014, *Phys. Chem. Chem. Phys.* 16, 17943–17954.
68. Gutsev, G. L., and Boldyrev, A. I. 1981, *Chem. Phys.* 56, 277–283.
69. Sun, W.-M., and Wu, D. 2019, *Chem. Eur. J.* 25, 9568–9579.
70. Anusiewicz, I., and Skurski, P. 2002, *Chem. Phys. Lett.* 358, 426.
71. Elliott, B. M. et al. 2005, *J. Phys. Chem. A* 109, 50, 11560–11567.
72. Sikorska, C. et al. 2008, *Inorg. Chem.* 47, 16, 7348–7354.
73. Pradhan, K. et al. 2010, *J. Chem. Phys.* 133, 144301.
74. Paduani, C. et al. 2011, *J. Phys. Chem. A* 115, 10237–10243.
75. Zhai, H. J. et al. 2007, *J. Phys. Chem. A* 111, 1030–1035.
76. Samanta, D. 2014, *J. Phys. Chem. Lett.* 5, 3151–3156.
77. Lide, D. R. 2008, *CRC Handbook of Chemistry and Physics*. CRC Press.
78. Gutsev, G. L., and Boldyrev, A. I. 1982, *Chem. Phys. Lett.* 92, 262–266.
79. Sun, W.-M., and Wu. D. 2019, *Chem. Eur. J.* 25, 9568–9579.
80. Yokoyama, K. et al. 2000, *Chem. Phys. Lett.* 330, 339–346.
81. Yokoyama, K., Tanaka, H., and Kudo, H. 2001, *J. Phys. Chem. A* 105, 4312–4315.
82. Rehm, E., Boldyrev, A. I., and Schleyer, P. V. R. 1992, *Inorg. Chem.* 31, 4834–4842.
83. Otten, A., and Meloni, G. 2018, *Chem. Phys. Lett.* 692, 214–223.
84. Li, Y. 2007, *J. Comput. Chem.* 28, 1677–1684.
85. Tong, J. 2009, *J. Chem. Phys.* 131, 164307.
86. Hou, N. et al. 2013, *Chem. Phys. Lett.* 575, 32–35.
87. Tong, J. et al. 2011, *J. Phys. Chem. A* 115, 2041–2046.
88. Xi, Y. J. et al. 2012, *Comput. Theor. Chem.* 994, 6–13.
89. Hou, D. et al. 2015, *J. Mol. Graphics Modell.* 59, 92–99.
90. Tong, J. et al. 2012, *Inorg. Chem.* 51, 6081–6088.
91. Tong, J. et al. 2013, *Dalton Trans.* 42, 577–584.
92. Liu, J.-Y. et al. 2018, *J. Cluster Sci.* 29, 853–860.
93. Dugourd, P. et al. 1992, *Chem. Phys. Lett.* 197, 433–437.
94. Zhao, T., Wang, Q., and Jena, P. 2017, *Nanoscale* 9, 4891–4897.

95. Pandey, S. K. et al. 2022, *Front. Chem.* 10, 1019166.
96. Raghavachari K. et al. 1989, *Chem. Phys. Lett.* 157, 479–483.
97. Lai, T.-Y. et al. 2011, *J. Chem. Phys.* 134, 244110.
98. Yu, L.-J. et al. 2020, *Chem. Phys.* 531, 110676.
99. Tsai, C.-C. et al. 2020, *Int. J. Quantum Chem.* 120, e26238.
100. Wadt, W. R., and Hay, P. J. 1985, *J. Chem. Phys.* 82, 284–310.
101. Wadt, W. R., and Hay, P. J. 1985, *J. Chem. Phys.* 82, 284–298.
102. Baerends, E. J., Gritsenko, O. V., and Snijders, J. G. 2000. "Relativistic Quantum Chemistry: The ZORA Approach." In *Reviews in Computational Chemistry*. Wiley-VCH, 15, 1–86.
103. Glendening, E.D., Landis, C.R., and Weinhold, F. 2013, *J. Comput. Chem.* 34, 1429–1437.
104. Bader, R.F. 1985, *Acc. Chem. Res.* 18, 9–15.
105. Michalak, A., Mitoraj, M., and Ziegler, T. J. 2008, *Phys. Chem. A* 112, 1933–1939.
106. Mitoraj, M.P., Michalak, A., and Ziegler, T. 2009, *J. Chem. Theory. Comput.* 5, 962–975.
107. Borocci, S., Grandinetti, F., Nunzi, F., and Sanna, N. 2020, *New J. Chem.* 44, 14536–14550.
108. Bader, R. F. W. 1990, *Atoms in Molecules: A Quantum Theory*. Oxford University Press.
109. Macchi, P., Proserpio, D. M., and Sironi, A. 1998, *J. Am. Chem. Soc.* 120, 13429–13435.
110. Macchi, P. et al. 1999, *J. Am. Chem. Soc.* 121, 10428–10429.
111. Novozhilova, I. V., Volkov, A. V., and Coppens, P. 2003, *J. Am. Chem. Soc.* 125, 1079–1087.
112. Cremer, D., and Kraka, E. 1984, *Angew. Chem., Int. Ed.* 23, 627–628.
113. Chakraborty, D., and Chattaraj, P. K. 2015, *J. Phys. Chem. A* 119, 3064–3074.
114. Saha, R., Mandal, B., and Chattaraj, P. K. 2018, *Int. J. Quantum Chem.* 118, e25499.
115. Wu, L.-Y. et al. 2019, *Phys. Chem. Chem. Phys.* 21, 19104–19114.
116. Mayer, M. et al. 2019, *Proc. Natl. Acad. Sci. U. S. A.* 116, 8167–8172.
117. Fang, H., Deepika, D., and Jena, P. 2021, *J. Chem. Phys.* 155, 014304.
118. Tong, J. et al. 2011, *J. Phys. Chem. A* 115, 2041–2046.
119. Pan, S. et al. 2013, *Chem. Eur. J.* 19, 2322–2329.
120. Pan, S. et al. 2013, *Comput. Theor. Chem.* 1021, 62–69.
121. Dugourd, P. et al. 1992, "Measurements of Lithium Clusters Ionization Potentials." In *Physics and Chemistry of Finite Systems: From Clusters to Crystals*. Springer, 555–560.

11 Superhalogens-Based Superacids

Jitendra Kumar Tripathi and Ambrish Kumar Srivastava

11.1 INTRODUCTION

Compounds with stronger acidity than 100% pure sulphuric acid (H_2SO_4) are referred to as superacids. The term "superacid" was first introduced by Hall and Conant in 1927 [1–3]. The gas phase acidity of compounds is evaluated in terms of deprotonation energies (ΔG_{acid}) of the corresponding conjugate bases. The powerful Brønsted acids are HCl, HNO_3 and H_2SO_4 having Gibbs deprotonation free energy (ΔG_{acid}) values of 328.1, 317.8 and 302.2 kcalmol^{-1}, respectively [4]. Koppel et al. [5] reported some superacids with free energies of deprotonation in the range of 249–270 kcalmol^{-1}. The stronger superacids can retain even smaller ΔG_{acid} values due to their easy emission of protons (H^+), For example, $HSbF_6$ (HF/SbF_5) is one of the known strongest mineral superacids because of its ΔG_{acid} value of 255.5 kcalmol^{-1} [6]. Basically, these systems are composed of Brønsted and Lewis acid components. In the $HSbF_6$ (HF/SbF_5) strongest mineral superacid, HF (Bronsted acid) releases H^+ (proton) and leaves F^- (fluoride) to interact with SbF_5 (Lewis acid) resulting in the SbF_6^- anion. Superacids can protonate extremely weak bases such as methane [7] easily due to their extremely high acidity. In addition, superacids and their derivatives are of great importance in organic synthesis and chemical applications, electrochemical technologies (fuel cells), electric double-layer capacitors, *etc.* [8–13].

The electronic stability of superacids can be attributed to their conjugate base anions. In past decades, it has been reported as superhalogen anions can work as superacids upon protonation. The term "superhalogen" was first introduced by Gutsev and Boldyrev [14] for the species with the general formula "MX_{k+1}" where 'M' is a central core element with valence 'k' and 'X' is an electronegative ligand such as Cl, F, O, *etc.* These are the radicals having higher electron affinities (EAs) than halogen or anions having higher vertical detachment energies (VDEs) than halides [15], which is limited to 3.6 eV (for Cl). They have various important applications, such as developing new materials for hydrogen storage [16], serving as electrolytes in Li-ion batteries [17–19] and contributing to organic superconductors [20], *etc.* Additionally, they can be used as building blocks for properties of other materials, due to their stable single negatively charged state [21]. Due to these applications, various superhalogen have been designed using different strategies [22–26]. Recent progress on the design and applications of superhalogens has

DOI: 10.1201/9781003384205-11

been reported by Srivastava [27]. In this chapter, we will provide a comprehensive account of the role of superhalogens in superacids.

11.2 METHODS

The superhalogen anions (X^-) and their protonated complexes (HX) discussed in this chapter were obtained by suitable quantum chemical methods (as mentioned in the next section) as implemented in the Gaussian 09 program [28]. The deprotonation process of the acidic species are given as following –

$$HX \rightarrow [X]^- + [H]^+$$

The gas phase acidity of the protonated species are calculated as follows:

$$\Delta G_{acid} = \Delta G(HX) + \Delta G(H^+) - \Delta G(X^-)$$

This ΔG is Gibbs' free energy, which includes zero-point correction and thermal enthalpy and entropy terms at 298.15 K. The value of $\Delta G(H^+) = -6.28$ kcalmol^{-1} [29] taken from the literature. The VDE of X^- are determined through the following formula

$$VDE = E[X]_{sp} - E[X^-]_{opt}$$

Where $E[X^-]_{opt}$ represents the overall energy of the optimized anion states, while $E[X]_{sp}$ denotes the single point energy of the corresponding neutrals at anionic geometries.

11.3 RESULTS AND DISCUSSION

11.3.1 SUPERACIDS BASED ON POLYNUCLEAR SUPERHALOGENS

Polynuclear superhalogens are generally designed by substituting the same superhalogens in the place of ligands in a typical superhalogen. Take $Al_nF_{3n+1}^-$ anions [30] as an example, which are designed by substituting AlF_4 moieties successively in the place of F in AlF_4^-, a typical superhalogen anion. Note that the VDE of $Al_nF_{3n+1}^-$ polynuclear superhalogen anions ranges from 9.79 to 12.40 eV, listed in Table 11.1. also Czapla and Skurski [31] protonated $Al_nF_{3n+1}^-$ anions and studied resulting HAl_nF_{3n+1} species obtained at the MP2 level [32] are displayed in Figure 11.1, in their lowest energy conformations.

They calculated ΔG_{acid} of HAl_nF_{3n+1} between 269.2 and 246.3 kcalmol^{-1} for n = 1 to 4 as listed in Table 11.1. It should be noticed that these values are less than those of H_2SO_4 (302.2 kcalmol^{-1}). Thus, the authors advised that the acidity of HAl_nF_{3n+1} in gas phase is more than that of the known powerful mineral acid. This makes them to behave as superacids. The rise in acidity can be attributed

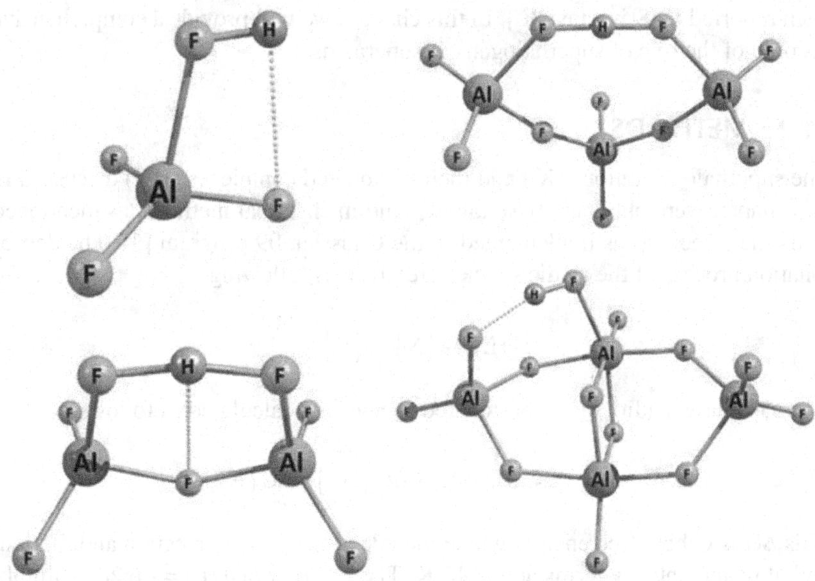

FIGURE 11.1 The structures of HAl_nF_{3n+1} for n = 1–4 from Ref. [31] with permission of Elsevier, Copyright 2015.

to the reduction in ΔG_{acid} values. Furthermore, it was observed that the ΔG_{acid} of HAl_4F_{13} (246.3 kcalmol^{-1}) is even less than that of $HSbF_6$ (255.5 kcalmol^{-1}). The authors also emphasized that while these calculations were conducted in the gas phase, the findings are likely applicable to liquids as well.

The preceding discussion establishes that the concept of polynuclear superhalogens serves as a powerful approach for design superacids. Subsequently, Czapla and Skurski [33] also utilized the B3LYP [34] method to report the HIn_xF_{3x+1}, HSn_xF_{4x+1} and HSb_xF_{5x+1} series of compounds. These were obtained through the protonation of $In_xF_{3x+1}^-$, $Sn_xF_{4x+1}^-$ and $Sb_xF_{5x+1}^-$ anions, respectively for x = 1–3. The VDE of these anionic systems are tabulated in Table 11.1. One can note that VDE values of anions ranges from 9.93 to 11.98 eV for $In_xF_{3x+1}^-$, from 10.25 to 12.56 eV for $Sn_xF_{4x+1}^-$, and from 10.15 to 13.27 eV for $Sb_xF_{5x+1}^-$, classifying them as polynuclear superhalogens. The evaluated ΔG_{acid} values (listed in Table 11.1) falls within the range 273.6–268.9 kcalmol^{-1} for HIn_xF_{3x+1} and 275.6–243.5 kcalmol^{-1} for HSn_xF_{4x+1}, establishing their superacidic nature. Notably, the HSb_xF_{5x+1} series exhibited remarkably lower ΔG_{acid} values of 260.5 to 230.3 kcalmol^{-1}, making them the strongest reported superacids to date. This result can be attributed to their exceptionally high VDEs, as will be explained later.

11.3.2 Superacids Based on Typical Superhalogens

Srivastava and Misra [35] explored the potential of protonating typical superhalogens to form superacids. They focused on MX_{k+1}^- superhalogen anions, including

TABLE 11.1
Vertical detachment energy (VDE) of superhalogen anions (X^-) from Ref. [14,30,36–39] and gas phase acidity (ΔG_{acid}) of their protonated complexes from Ref. [31,33,35,41–48] using different computational methods as discussed in the accompanying text.

Superhalogen anion		Protonated complex	
X^-	VDE (eV)	HX	ΔG_{acid} (kcalmol^{-1})
AlF_4^-	9.79	$HAlF_4$	269.2
$Al_2F_7^-$	11.36	HAl_2F_7	261.1
$Al_3F_{10}^-$	12.04	HAl_3F_{10}	257.4
$Al_4F_{13}^-$	12.40	HAl_4F_{13}	246.3
InF_4^-	9.93	$HInF_4$	273.6
In_2F_7	11.40	HIn_2F_7	268.8
In_3F_{10}–	11.98	HIn_3F_{10}	268.9
SnF_5–	10.25	$HSnF_5$	275.6
Sn_2F_9–	12.01	HSn_2F_9	250.5
Sn_3F_{13}–	12.56	HSn_3F_{13}	243.5
SbF_6–	10.15	$HSbF_6$	260.5
Sb_2F_{11}–	12.54	HSb_2F_{11}	244.7
$Sb_3F_{16}^-$	13.27	HSb_3F_{16}	230.3
LiF_2^-	5.82	$HLiF_2$	319.1
$LiCl_2^-$	5.90	$HLiCl_2$	283.6
BeF_3^-	6.99	$HBeF_3$	286.8
$BeCl_3^-$	—	$HBeCl_3$	272.4
BF_4^-	7.70	HBF_4	290.6
PF_6^-	8.49	HPF_6	281.0
HSO_4^-	5.12	H_2SO_4	302.8
AlH_4^-	4.89	$HAlH_4$	326.3
BO_2^-	4.89	HBO_2	326.3
ClO_4^-	5.44	$HClO_4$	297.3
BF_4^-	7.96	HBF_4	289.4
AlF_4^-	8.73	$HAlF_4$	266.0
$B(BF_4)_4^-$	10.59	$HB(BF_4)_4$	257.7
$B(AlF_4)_4^-$	10.70	$HB(AlF_4)_4$	287.7
$Al(BF_4)_4^-$	10.52	$HAl(BF_4)_4$	236.4
$Al(AlF_4)_4^-$	10.30	$HAl(AlF_4)_4$	253.7
BeF_3^-	7.63	$HBeF_3$	286.8
$BeCl_3^-$	6.17	$HBeCl_3$	272.4
$BH_4^-(B_nH_{3n+1})^-$	4.62	HBH_4	323.8
$B_2H_7^-$	5.71	HB_2H_7	292.7
$B_3H_{10}^-$	6.25	HB_3H_{10}	276.0
$B_4H_{13}^-$	7.07	HB_4H_{13}	265.3

TABLE 11.1 (*Continued*)
Vertical detachment energy (VDE) of superhalogen anions (X^-) from Ref. [14,30,36–39] and gas phase acidity (ΔG_{acid}) of their protonated complexes from Ref. [31,33,35,41–48] using different computational methods as discussed in the accompanying text.

Superhalogen anion		Protonated complex	
X^-	VDE (eV)	HX	ΔG_{acid} (kcalmol^{-1})
$B_5H_{16}^-$	7.61	HB_5H_{16}	258.8
NaF_2^-	6.97	$HNaF_2$	337.0
$NaCl_2^-$	5.81	$HNaCl_2$	292.8
KF_2^-	6.07	HKF_2	350.9
KCl_2^-	5.37	$HKCl_2$	302.5
MgF_3^-	8.79	$HMgF_3$	294.8
$MgCl_3^-$	6.68	$HMgCl_3$	270.8
CaF_3^-	8.62	$HCaF_3$	309.9
$CaCl_3^-$	6.37	$HCaCl_3$	274.0
BCl_4^-	6.02	$HBCl_4$	292.0
AlF_4^-	8.71	$HAlF_4$	269.2
$AlCl_4^-$	6.83	$HAlCl_4$	263.4
GaF_4^-	—	$HGaF_4$	274.4
$GaCl_4^-$	—	$HGaCl_4$	265.4
CF_5^-	6.83	HCF_5	366.6
CCl_5^-	—	$HCCl_5$	328.5
SiF_5^-	6.04	$HSiF_5$	300.0
$SiCl_5^-$	—	$HSiCl_5$	311.5
GeF_5^-	—	$HGeF_5$	285.6
$GeCl_5^-$	—	$HGeCl_5$	303.9
$[CB_2(CN)_3]^-$	5.36	$H/[B_2C(CN)_3]$	208.58
$[CB_2(NO_2)_3]^-$	6.13	$H/[B_2C(NO_2)_3]$	271.02
$[C_5(CN)_5]^-$	5.72	$H/[C_5(CN)_5]$	255.26
$[C_5(NO_2)_5]^-$	6.25	$H/[C_5(NO_2)_5]$	252.11
$[BC_5(CN)_6]^-$	5.98	$H/[BC_5(CN)_6]$	248.15
$[BC_5(NO_2)_6]^-$	6.44	$H/[BC_5(NO_2)_6]$	243.78
a-$[BC_9(CN)_8]^-$	5.91	H/a-$[BC_9(CN)_8]$	247.00
a-$[BC_9(NO_2)_8]^-$	6.33	H/a-$[BC_9(NO_2)_8]$	246.26
b-$[BC_9(CN)_8]^-$	6.07	H/b-$[BC_9(CN)_8]$	252.74
b-$[BC_9(NO_2)_8]^-$	6.49	H/b-$[BC_9(NO_2)_8]$	248.03

LiF_2^-, $LiCl_2^-$, BeF_3^-, $BeCl_3^-$, BF_4^- and PF_6^- and examined their protonated complexes (Figure 11.2) using the MP2 method. The VDE values of these anions have been previously reported [14,36–39], listed in Table 11.1. Their protonated complexes can be considered as two-component systems containing HF or HCl moiety. The

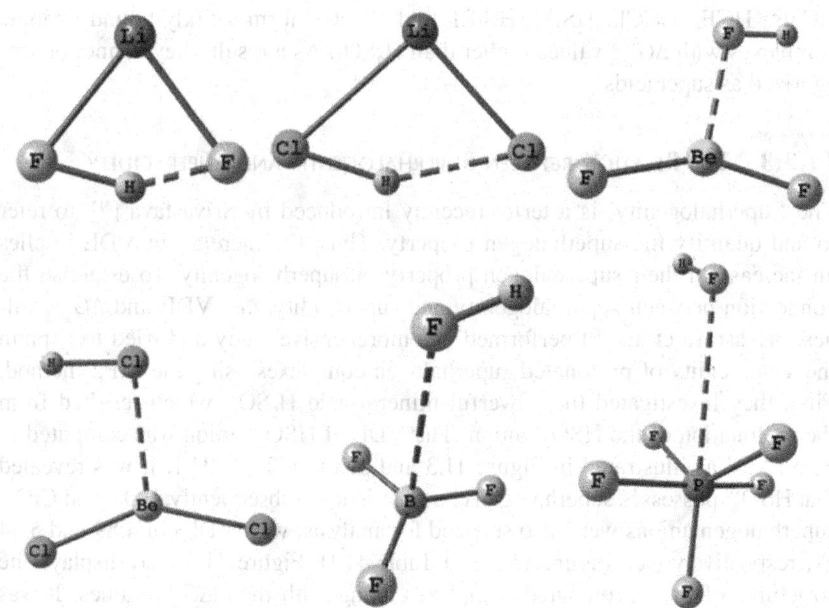

FIGURE 11.2 The structures of the protonated species HX for X = LiF_2, $LiCl_2$, BeF_3, $BeCl_3$, BF_4 and PF_6 optimized from Ref. [35] with permission of Elsevier, Copyright 2015.

researchers further investigated the stability of these complexes against dissociation to HF or HCl and confirmed their stability.

They calculated that the ΔG_{acid} values of these complexes as listed in Table 11.1. The values for all complexes fell within the range of 290.6–272.4 kcalmol^{-1}, with the exception of $HLiF_2$, which had a ΔG_{acid} value of 319.1 kcalmol^{-1} (Table. 11.1). Based on their ΔG_{acid} values, the gas phase acidities were found to be in the order of $HBeCl_3$ > HPF_6 > $HLiCl_2$ > $HBeF_3$ > HBF_4 > H_2SO_4 > $HLiF_2$ > HCl. Consequently, the authors concluded that all protonated species, with the exception of the $HLiF_2$ complex, exhibit superacidic behaviour. Despite this, the $HLiF_2$ complex remains more acidity than HCl. Later on, Czapla et al. [40] also looked into this aspect and re-examined the relation between the protonation of superhalogen anions and the occurrence of superacids using the MP2 method. They examined the process of protonating a more extensive range of MX_{k+1}^- superhalogen anions where M represents Li, Na, K, Be, Mg, Ca, B, Al, Ga, C, Si, Ge and X corresponds to F, Cl. The calculated ΔG_{acid} values for these protonated complexes ranged from 263.4–366.6 kcalmol^{-1}, as shown in Table 11.1. They found that all the protonated species as $HLiCl_2$, $HNaCl_2$, $HBeF_3$, $HBeCl_3$, $HMgF_3$, $HMgCl_3$, $HCaCl_3$, HBF_4, $HBCl_4$, $HAlF_4$, $HAlCl_4$, $HGaF_4$, $HGaCl_4$ and $HGeF_5$ contain HF or HCl moieties and exhibit lower ΔG_{acid} values compared to H_2SO_4. Consequently, these complexes can be classified as superacids, confirming the earlier hypothesis made by Srivastava and Misra [35]. In contrast, $HLiF_2$, $HNaF_2$, HKF_2, $HKCl_2$,

$HCaF_3$, HCF_5, $HCCl_5$, $HSiF_5$, $HSiCl_5$ and $HGeCl_5$ form weakly-bound or ionic complexes with ΔG_{acid} values higher than H_2SO_4. As a result, they cannot be categorized as superacids.

11.3.3 The Relation between Superhalogenity and Superacidity

The 'superhalogenity' is a term, recently introduced by Srivastava [41], to refer to and quantify the superhalogen property. Thus, the increase in VDE implies an increase in their superhalogen property or superhalogenity. To establish the connection between superhalogenity and superacidity, *i.e.*, VDE and ΔG_{acid} values, Srivastava et al. [42] performed a comprehensive study and tried to explain the superacidity of protonated superhalogen complexes using the MP2 method. First, they investigated the powerful mineral acid H_2SO_4, which resulted from the protonation of the HSO_4^- anion. The VDE of HSO_4^- anion was computed to be 5.12 eV as illustrated in Figure 11.3 and given in Table 11.1. It was revealed that HSO_4^- possesses superhalogen characteristics. Subsequently, BO_2^- and ClO_4^- superhalogen anions were also selected for analysis, with VDEs of 4.89 and 5.44 eV, respectively (see Figure 11.3 and Table 11.1). Figure 11.3 also displays the structures of their protonated complexes along with their ΔG_{acid} values. It was interesting to note that the ΔG_{acid} of HBO_2 is higher than that of H_2SO_4 whereas

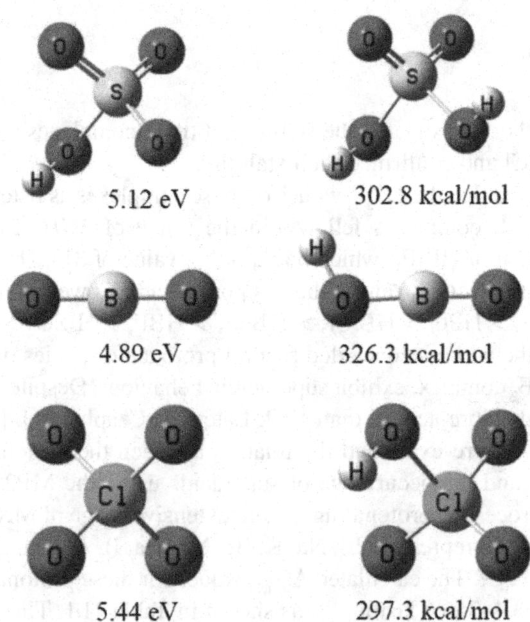

FIGURE 11.3 The structures of superhalogen anions (X^-) and their protonated species (HX) for X = HSO_4, BO_2 and ClO_4 from Ref. [42] with permission of Royal Society of Chemistry, Copyright 2016.

the VDE of BO_2^- is lower than that of HSO_4^-. On the contrary, ΔG_{acid} of $HClO_4$ is lower than that of H_2SO_4 whereas the VDE of ClO_4^- is higher than that of HSO_4^-. This suggests that lower ΔG_{acid} value, indicating higher gas-phase acidity of protonated species, relates to a higher VDE of the corresponding superhalogen anions. Hence, there is a direct relation between the superacidity of the protonated complex and the superhalogenity of the anions.

To further explain this, the authors considered $B_nH_{3n+1}^-$, polynuclear superhalogen anions for n from 1 to 5. The VDE of these anions shows a gradual rise with an increasing value of 'n', falling within the range of 4.62 to 7.61 eV as given in Table 11.1. The researchers formulated and examined their protonated complexes B_nH_{3n+2} as shown in Figure 11.4. It should be noted that the ΔG_{acid} of BH_5 should be higher than that of H_2SO_4 because of lower VDE of BH_4^- (4.62 eV) as compared to HSO_4^- (5.12 eV). The ΔG_{acid} value of BH_5, which amounted to 323.8 kcalmol^{-1} was demonstrably greater than that of H_2SO_4. On the contrary, the VDE of $B_2H_7^-$ (5.71 eV) is higher than that of HSO_4^- and consequently, the ΔG_{acid} of B_2H_8 (292.7 kcalmol^{-1}) was lower than that of H_2SO_4. The authors observed a gradual decrease in the ΔG_{acid} value of B_nH_{3n+2} species, going from 276.0 kcalmol^{-1} (for $n = 3$) to 258.8 kcal/mol (for $n = 5$). The authors proposed a statistical correlation between the VDEs of $B_nH_{3n+1}^-$ anions and the (gas-phase) acidity, represented as the inverse of ΔG_{acid} for the B_nH_{3n+2} series. This relationship is visually depicted in Figure 11.4. Essentially, the study establishes a clear and direct proportional connection between the gas-phase acidity of B_nH_{3n+2} species and superhalogenity of $B_nH_{3n+1}^-$ anions.

11.3.4 Superacids Based on Hyperhalogen Anions

Hyperhalogen anions [43] are similar to polynuclear superhalogens in which all ligands are replaced by other superhalogens. For instance, AlF_4^- superhalogen anion in which all F ligands are replaced by BF_4 superhalogens will result into $Al(BF_4)_4^-$ hyperhalogen anion whose VDE (10.52 eV) is higher than that of AlF_4^- (8.73 eV) as listed in Table 11.1. Having established that the gas phase acidity is related to the superhalogenity, Srivastava et al. [44] proposed the design of stronger superacids using hyperhalogen anions. They used $M(M'F_4)_4^-$ (M, M' = B, Al) polynuclear superhalogen/hyperhalogen anions with the VDEs > 10 eV and studied their protonated complexes at the ωB97xD level [45]. The geometries of $Al(BF_4)_4^-$ and $Al(AlF_4)_4^-$ hyperhalogen anionic systems and their protonated complexes are shown in Figure 11.5. The ΔG_{acid} values calculated for the protonated species $HB(BF_4)_4$, $HB(AlF_4)_4$, $HAl(AlF_4)_4$ and $HAl(BF_4)_4$ were 257.7, 287.7, 253.7 and 236.4 kcalmol^{-1}, respectively (see Table 11.1). Therefore, all the protonated complexes $HM(M'F_4)_4$ appear to be superacidic species with ΔG_{acid} lower than 300 kcalmol^{-1} and the order of gas-phase acidity is $HB(AlF_4)_4 < HB(BF_4)_4 < HAl(AlF_4)_4 < HAl(BF_4)_4$. Interestingly, $HAl(BF_4)_4$ possesses even higher acidity than known strongest superacid $HSbF_6$. These findings strongly indicate a promising approach for creating stronger superacids using the concept of superhalogens/hyperhalogens.

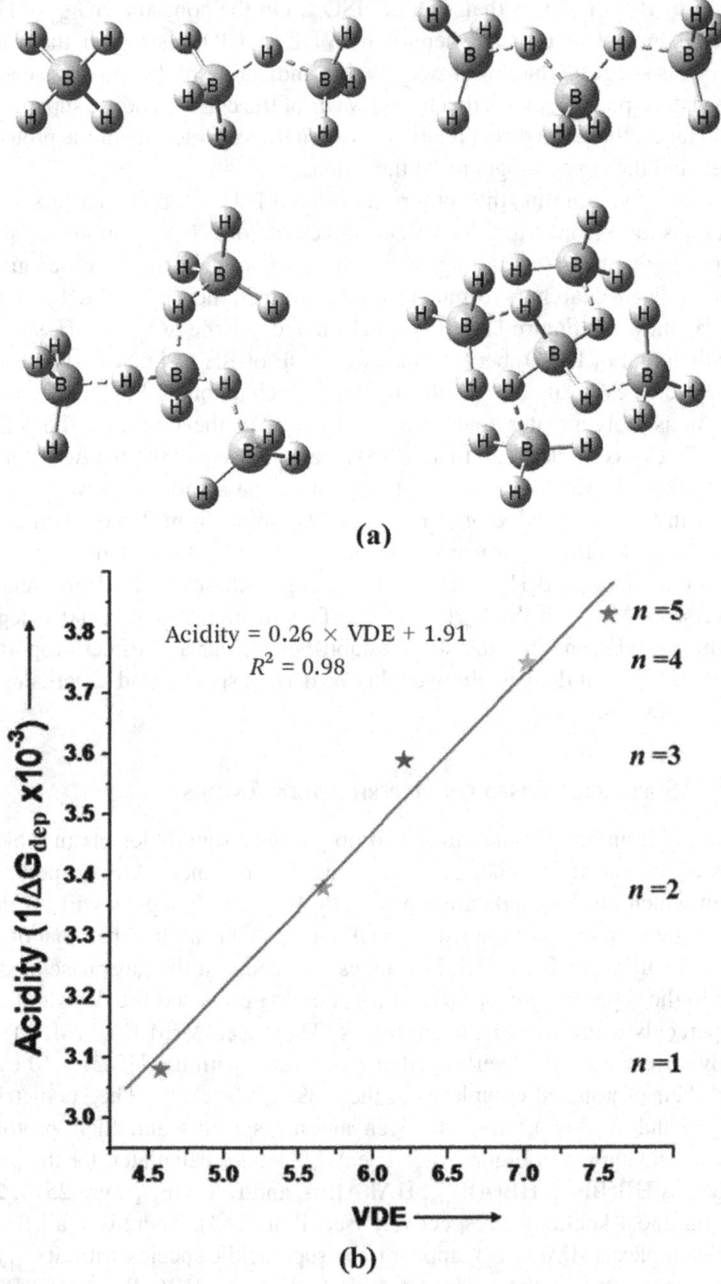

FIGURE 11.4 The structures of B_nH_{3n+2} complexes (a) and the relationship between the acidity of B_nH_{3n+2} and the vertical detachment energy (VDE) of $B_nH_{3n+1}^-$ anions (b) from Ref. [42] with permission from the Royal Society of Chemistry, Copyright 2016.

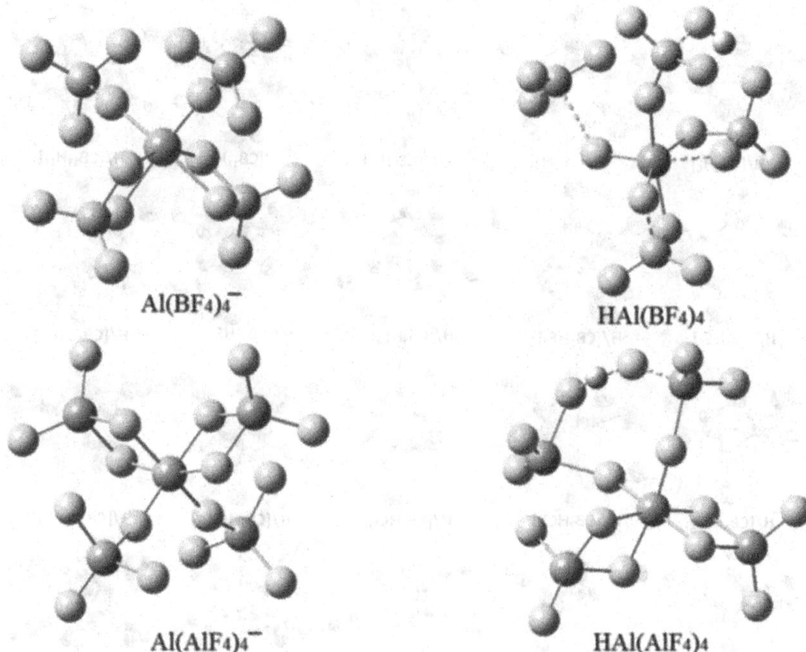

FIGURE 11.5 The structures of $Al(BF_4)_4^-$ and $Al(AlF_4)_4^-$ hyperhalogen anions, as well as their protonated complexes from Ref. [44] with permission of Elsevier, Copyright 2017.

11.3.5 Superacids Based on Other Superhalogen Anions

The superhalogen behaviour of some anions can be attributed to their increased stability following the Wade–Mingos rule [15]. According to this rule, a polyhedron with *n* vertices requires (n + 1) electrons for stability. For instance, borane ($B_{12}H_{12}^-$) becomes superhalogen anions due to its stability as per the Wade–Mingos rule. Replacing boron atoms from boranes results in carboranes. Zhao and co-workers [46] exploited carborane-based anions, $CB_nHX_n^-$ for X = H, F, Cl, CN; $n = 5, 7, 9, 11, 13$; which were designed by choosing borane $B_{n+1}H_{n+1}^-$ and replacing one of boron with carbon as well as *n* hydrogen atoms by X. Subsequently, they examined 63 complexes designed via protonation of these anions, as illustrated in Figure 11.6. The corresponding ΔG_{acid} values for these complexes are presented in Table 11.2. The authors found that most of these complexes become superacids having ΔG_{acid} lower than 300 kcalmol^{-1}, expect a few. For instance, the ΔG_{acid} of H/[$CB_{11}HF_{11}$], H/[$CB_{13}HF_{13}$], H/[$CB_{11}F_{12}$] and H/[$CB_{13}F_{14}$] were obtained in the range 215–220 kcalmol^{-1}. This study was further continued by Luo et al. [47] for superhalogen anions based on silaborane, $SiB_nH_mX_{n-m+1}^-$ for X = H, F, Cl, CN. They examined their protonated composites as $H/SiB_nH_mX_{n-m+1}$ for *n* from 5 to 13 and *m* from 0 to 1 in the gas phase as well as in solvents. The corresponding ΔG_{acid} values are also listed in Table 11.2, which suggests that all these complexes, excluding

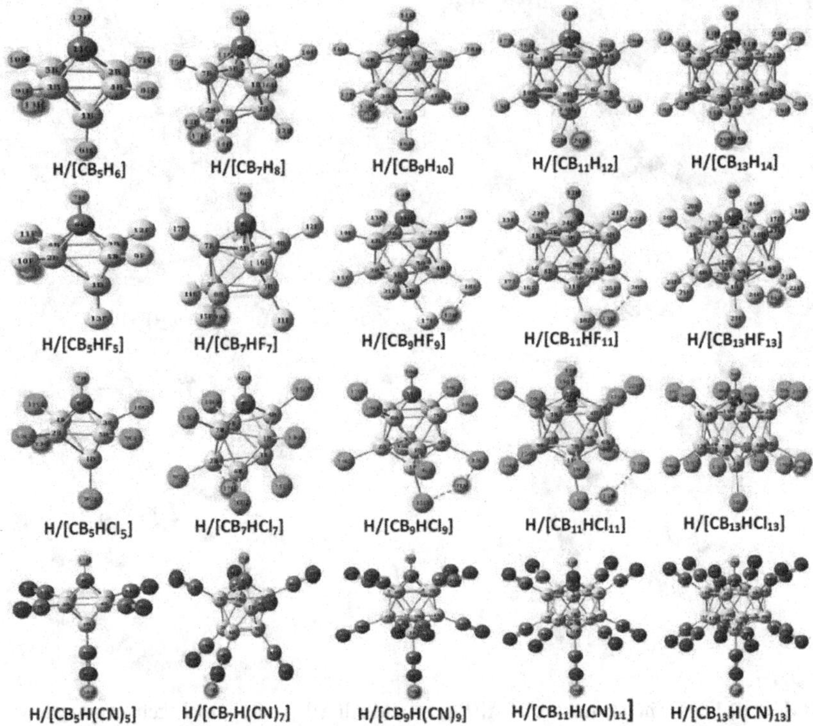

FIGURE 11.6 The structures of protonated carborane-based anions from Ref. [[46]].

two, were found to be superacids. One can note that the complexes based on silaboranes exhibit relatively more acidic than the complexes based on carboranes. For instance, the ΔG_{acid} for H/[$SiB_{12}HF_{12}$] is measured to be 205 kcalmol^{-1}, which is approximately 10 kcalmol^{-1} lower than H/[$CB_{11}F_{12}$] (as shown in Table 11.2). This is not only true for the gas phase but also in the solution. The authors suggest that both gas phase ΔG_{acid} and pK_a values should be reported for full description of acidity. Note that the pK_a value measures the acidity in solution and any compound is considered to be superacid if it has a lower pK_a value in a solvent than that of $HClO_4$ in the same solvent. They also noticed that the complexes with CN substitutions are more stabilized in the solvent, which results in their higher ΔG_{acid} values than those with F or Cl substitutions. According to them, the VDE of anions appears crucial in the acidity of their protonated complexes as suggested by Srivastava and Misra [[35,42]].

Superhalogens can also be designed by exploiting the Hückel rule of aromaticity. According to this rule, a cyclic conjugated molecule is considered to be aromatic if it possesses $(4n + 2)$ π-electrons. Zhao et al. [[48]] made the substitutions of F, CN and NO_2 in some conjugated anions such as $CB_2H_3^-$, $C_5H_5^-$, $BC_5H_6^-$, $BC_9H_8^-$, *etc.* While all of these anions conform to the Huckel rule and exhibit aromaticity,

TABLE 11.2
The ΔG_{acid} values of H/[CB_nHX_n] (n = 5-13; X= H, F, Cl, CN) from Ref. [46] and H/[$SiB_nH_mX_{n-m+1}$] composites (m = 1 and 0; n = 5–13; X = H, F, Cl, CN) from Ref. [47].

Protonated species	ΔG_{acid} (kcalmol^{-1})								
	n = 5	n = 6	n = 7	n = 8	n = 9	n = 10	n = 11	n = 12	n = 13
H/[CB_nH_{n+1}]	307.4	300.7	302.1	286.1	280.6	290.0	267.2	281.9	261.5
H/[CB_nHF_n]	287.6	282.5	267.8	261.7	227.6	261.3	219.2	229.0	218.6
H/[CB_nF_{n+1}]	280.2	277.1	261.3	253.2	223.9	257.1	214.7	225.2	215.5
H/[CB_nHCl_n]	269.0	264.0	251.7	247.3	235.7	249.4	230.1	238.4	232.7
H/[$CB_nH(CN)_n$]	264.4	259.5	247.7	243.7	233.7	247.9	229.1	236.9	224.9
H/[$CB_n(CN)_{n+1}$]	250.9	244.3	244.3	237.1	237.4	232.5	229.1	229.7	222.0
H/[CB_nCl_{n+1}]	243.1	238.4	238.9	234.6	232.0	228.5	225.6	226.4	219.1
H/[SiB_nH_{n+1}]	304.7	296.4	302.1	292.6	284.3	295.3	268.7	282.3	261.7
H/[SiB_nHF_n]	281.8	270.9	281.1	269.7	231.3	275.3	221.7	204.9	218.5
H/[SiB_nHCl_n]	267.9	253.5	254.1	261.5	239.6	262.9	233.6	238.4	227.3
H/[$SiB_nH(CN)_n$]	256.9	246.9	255.3	243.2	241.1	223.5	232.5	232.2	224.3
H/[SiB_nF_{n+1}]	288.1	266.8	253.2	261.7	228.5	255.5	219.4	228.0	216.0
H/[SiB_nCl_{n+1}]	269.8	251.3	239.0	252.5	238.4	231.8	233.5	259.9	227.3
H/[$SiB_n(CN)_{n+1}$]	250.3	243.1	236.4	235.3	237.5	232.5	230.6	229.5	221.6

only half of them are superhalogens. To be specific, only the substitutions of CN and NO_2 can increase the VDE of anions to make them superhalogens as listed in Table 11.1. The authors studied the protonated complexes of these superhalogen anions and noticed that all have lower ΔG_{acid} than 300 kcalmol^{-1} also listed in Table 11.1. In particular, H/[$BC_5(NO_2)_6$], H/a-[$BC_9(NO_2)_8$], H/a-[$BC_9(CN)_8$], H/b-[$BC_9(NO_2)_8$] and H/[$BC_5(CN)_6$] are found to be strongest superacids having ΔG_{acid} values in the range 243.7–248.1 kcalmol^{-1}. The authors also emphasized that the VDE of the anions involved is the most decisive factor for the acidities of the complexes even in the solution phase. They underlined the fact that the design of superacids through the combining a proton (H^+) and superhalogen anion is not coincidence but a well-thought-out approach, as hypothesized by Srivastava and Misra [35].

11.4 CONCLUSION

This chapter explored the significance of superhalogens in the generation of superacids. The idea was promoted by observing that the most potent superacid, $HSbF_6$ consists of SbF_6^-, a superhalogen anion. This led to the hypothesis that the superacids could be considered as protonated complexes of superhalogen anions. To validate this hypothesis, protonation of various types of superhalogens was considered. In the case of polynuclear superhalogen anions, it was observed that the acidity increases consistently with the rise in the vertical detachment energy (VDE) of the anions. This trend was demonstrated by investigating $B_nH_{3n+1}^-$ superhalogens and the corresponding acidity of their protonated complexes, B_nH_{3n+2}. While these findings are based on gas phase calculations, they can be reasonably extended to liquid phases. Moreover, the findings strongly support the idea that the protonation of superhalogen anion offers an effective approach to develop new and even stronger superacids. The strongest superacid, in this way, is yet to come.

REFERENCES

1. Hall, N. F.; Conant, J. B.; (1927). A study of superacid solutions. I: The use of the chloranil electrode in glacial acetic acid and the strength of certain weak bases. *Journal of the American Chemical Society*, 49(12), 3047–3061.
2. Olah, G. A.; Prakash, G. K. S.; Sommer, J.; (1985). *Superacids*. Wiley, New York.
3. Olah, G. A.; Prakash, G. S.; Sommer, J.; (1979). Acids up to billions of times stronger than sulfuric acid have opened up fascinating new areas of chemistry. *Science*, 206, 13–20.
4. Koppel, I. A.; Taft, R. W.; Anvia, F.; Zhu, S. Z.; Hu, L. Q.; Sung, K. S.; Kondratenko, N. V.; Volkonskii, V.M.; Vlasov, R.; Notario, P. C.; Maria. (1994). The gas-phase acidities of very strong neutral Brønsted acids, *Journal of the American Chemical Society*, 116, 3047.
5. Koppel, I. A.; Burk, P.; Koppel, I.; Leito, I.; Sonoda, T.; Mishima, M.; (2000). Gas-phase acidities of some neutral Brønsted superacids: A DFT and ab initio study. *Journal of the American Chemical Society*, 122(21), 5114–5124.
6. Bour, C.; Guillot, R.; Gandon, V.; (2015). First evidence for the existence of hexafluoroantimonic (V) acid. *Chemistry – A European Journal*, 21(16), 6066–6069.
7. Olah, G. A.; Lukas, J.; (1967). Stable carbonium ions. XXXIX. 1 Formation of alkylcarbonium ions via hydride ion abstraction from alkanes in fluorosulfonic acid-antimony pentafluoride solution: Isolation of some crystalline alkylcarbonium ion salts. *Journal of the American Chemical Society*, 89(9), 2227–2228.
8. Martin, A.; Arda, A.; Désiré, J.; Martin-Mingot, A.; Probst, N.; Sinaÿ, P.; Blériot, Y.; (2016). Catching elusive glycosyl cations in a condensed phase with HF/SbF5 superacid. *Nature Chemistry*, 8(2), 186–191.
9. Prakash, G. S.; Rasul, G.; Burrichter, A.; Laali, K. K.; Olah, G. A.; (1996). Ab Initio/IGLO/GIAO-MP2 studies of fluorocarbocations: Experimental and theoretical investigation of the cleavage reaction of trifluoroacetic acid in superacids. *The Journal of Organic Chemistry*, 61(26), 9253–9258.
10. Saito, S.; Ohwada, T.; Shudo, K.; (1996). Superacid-catalyzed reaction of substituted benzaldehydes with benzene. *The Journal of Organic Chemistry*, 61(23), 8089–8093.
11. Olah, G. A.; Török, B.; Shamma, T.; Török, M.; Prakash, G. S.; (1996). Solid acid (superacid) catalyzed regioselective adamantylation of substituted benzenes. *Catalysis Letters*, 42, 5–13.
12. Razaq, M.; Razaq, A.; Yeager, E.; DesMarteau, D. D.; Singh, S.; (1989). Perfluorosulfonimide as an additive in phosphoric acid fuel cell. *Journal of the Electrochemical Society*, 136(2), 385.
13. Park, H. W.; Kwon, S. S.; Lim, K. J.; Suh, K. S.; (1996). Characteristics of fast charge-discharge and performances of activated carbon electrode for electric double layer capacitor containing sulfuric acid aqueous solution. *Journal of KIEE*, 9(4), 205–210.

14. Gutsev, G. L.; Boldyrev, A. I.; (1981). DVM-Xα calculations on the ionization potentials of MX_{k+1}^- complex anions and the electron affinities of MX_{k+1} "superhalogens". *Chemical Physics*, 56(3), 277–283.
15. Srivastava, A. K.; (2023) *Superhalogens: Properties and applications, Springer Briefs in Molecular Science*, Springer, Cham, Switzerland.
16. Srivastava, A. K.; Misra, N.; (2016). Superhalogens as building blocks of complex hydrides for hydrogen storage. *Electrochemistry Communications*, 68, 99–103.
17. Giri, S.; Behera, S.; Jena, P.; (2014). Superhalogens as building blocks of halogen-free electrolytes in lithium-ion batteries (*Angew. Chem. Int. Ed.* 50/2014). *Angewandte Chemie International Edition*, 53(50), 13942–13942.
18. Srivastava, A. K.; Misra, N.; (2016). Designing new electrolytic salts for lithium ion batteries using superhalogen anions. *Polyhedron*, 117, 422–426.
19. Sun, Y. Y.; Li, J. F.; Zhou, F. Q.; Li, J. L.; Yin, B.; (2016). Probing the potential of halogen-free superhalogen anions as effective electrolytes of Li-ion batteries: A theoretical prospect from combined ab initio and DFT studies. *Physical Chemistry Chemical Physics*, 18(41), 28576–28584.
20. Srivastava, A. K.; Kumar, A.; Tiwari, S. N.; Misra, N.; (2017). Application of superhalogens in the design of organic superconductors. *New Journal of Chemistry*, 41(24), 14847–14850.
21. Jena, P.; Sun, Q.; (2018). Super atomic clusters: Design rules and potential for building blocks of materials. *Chemical Reviews*, 118(11), 5755–5870.
22. Gutsev, G. L.; Bartlett, R. J.; Boldyrev, A. I.; Simons, J.; (1997). Adiabatic electron affinities of small superhalogens: LiF_2, $LiCl_2$, NaF_2, and $NaCl_2$. *The Journal of Chemical Physics*, 107(10), 3867–3875.
23. Smuczynska, S.; Skurski, P.; (2009). Halogenoids as ligands in superhalogen anions. *Inorganic Chemistry*, 48(21), 10231–10238.
24. Pathak, B.; Samanta, D.; Ahuja, R.; Jena, P.; (2011). Borane derivatives: A new class of super-and hyperhalogens. *ChemPhysChem*, 12(13), 2423–2428.
25. Child, B. Z.; Giri, S.; Gronert, S.; Jena, P.; (2014). Aromatic superhalogens. *Chemistry–A European Journal*, 20(16), 4736–4745.
26. Tripathi, J. K.; Srivastava, A. K.; (2022). $CF_{4-n}(SO_3)_n$; (n= 1–4): A new series of organic superhalogens. *Molecular Physics*, 120(21), e2123748.
27. Srivastava, A. K.; (2023). Recent progress on the design and application of superhalogens. *Chemical Communications*, 59, 5961.
28. Frisch, M. J.; Trucks, G. W.; Schlegel, H. B.; (2009). *Gaussian 09, Revision D. 01*. Gaussian, Inc., Wallingford, CT.
29. Topol, I. A.; Tawa, G. J.; Burt, S. K.; Rashin, A. A.; (1997). Calculation of absolute and relative acidities of substituted imidazoles in aqueous solvent. *The Journal of Physical Chemistry A*, 101(51), 10075–10081.
30. Sikorska, C.; Skurski, P.; (2012). The saturation of the excess electron binding energy in Al_nF_{3n+1}- (n = 1–5) anions. *Chemical Physics Letters*, 536, 34–38.
31. Czapla, M.; Skurski, P.; (2015). The existence and gas phase acidity of the HAl_nF_{3n+1} superacids (n = 1–4). *Chemical Physics Letters*, 630, 1–5.
32. Møller, C.; Plesset, M. S.; (1934). Note on an approximation treatment for many-electron systems. *Physical Review*, 46(7), 618.
33. Czapla, M.; Skurski, P.; (2015). Strength of the Lewis–Brønsted superacids containing In, Sn, and Sb and the electron binding energies of their corresponding superhalogen anions. *The Journal of Physical Chemistry A*, 119(51), 12868–12875.
34. Lu, L.; Hu, H.; Hou, H.; Wang, B.; (2013). An improved B3LYP method in the calculation of organic thermochemistry and reactivity. *Computational and Theoretical Chemistry*, 1015, 64–71.

35. Srivastava, A. K.; Misra, N.; (2015). Hydrogenated superhalogens behave as superacids. *Polyhedron*, 102, 711–714.
36. Wang, X. B.; Ding, C. F.; Wang, L. S.; Boldyrev, A. I.; Simons, J.; (1999). First experimental photoelectron spectra of superhalogens and their theoretical interpretations. *The Journal of Chemical Physics*, 110(10), 4763–4771.
37. Anusiewicz, I.; Skurski, P.; (2002). An ab initio study on BeX_3^- superhalogen anions (X= F, Cl, Br). *Chemical Physics Letters*, 358(5–6), 426–434.
38. Anusiewicz, I.; Sobczyk, M.; Dąbkowska, I.; Skurski, P.; (2003). An ab initio study on MgX_3^- and CaX_3^- superhalogen anions (X= F, Cl, Br). *Chemical Physics*, 291(2), 171–180.
39. Sikorska, C.; Smuczynska, S.; Skurski, P.; Anusiewicz, I.; (2008). BX_4^- and AlX_4^- superhalogen anions (X= F, Cl, Br): An ab initio study. *Inorganic Chemistry*, 47(16), 7348–7354.
40. Czapla, M.; Anusiewicz, I.; Skurski, P. (2016). Does the protonation of superhalogen anions always lead to superacids? *Chemical Physics*, 465, 46–51.
41. Srivastava, A. K.; (2023). A simple strategy to design polycyclic superhalogens. *The Journal of Physical Chemistry*, 127, 22, 4867–4872.
42. Srivastava, A. K.; Kumar, A.; Misra, N.; (2017). Superhalogens as building blocks of a new series of superacids. *New Journal of Chemistry*, 41(13), 5445–5449.
43. Willis, M.; Götz, M.; Kandalam, A. K.; Ganteför, G. F.; Jena, P.; (2010). Hyperhalogens: Discovery of a new class of highly electronegative species. *Angewandte Chemie International Edition*, 49(47), 8966–8970.
44. Srivastava, A. K.; Kumar, A.; Misra, N.; (2017). A path to design stronger superacids by using superhalogens. *Journal of Fluorine Chemistry*, 197, 59–62.
45. Chai, J. D.; Head-Gordon, M.; (2008). Systematic optimization of long-range corrected hybrid density functionals. *The Journal of Chemical Physics*, 128(8).
46. Zhao, R. F.; Zhou, F. Q.; Xu, W. H.; Li, J. F.; Li, C. C.; Li, J. L.; Yin, B.; (2018). Superhalogen-based composite with strong acidity-a crossing point between two topics. *Inorganic Chemistry Frontiers*, 5(11), 2934–2947.
47. Luo, L.; Zhou, F. Q.; Zhao, R. F.; Li, J. F.; Wu, L. Y.; Li, J. L.; Yin, B.; (2019). Combining proton and silaborane-based superhalogen anions – an effective route to new superacids as verified via systematic DFT calculations. *Dalton Transactions*, 48(43), 16184–16198.
48. Zhou, F. Q.; Zhao, R. F.; Li, J. F.; Xu, W. H.; Li, C. C.; Luo, L.; Yin, B.; (2019). Constructing organic superacids from superhalogens is a rational route as verified by DFT calculations. *Physical Chemistry Chemical Physics*, 21(5), 2804–2815.

Index

Page locators in **bold** indicate a table
Page locators in *italics* indicate a figure

O

P

Q

R

Printed in the United States
by Baker & Taylor Publisher Services

Printed in the United States
by Baker & Taylor Publisher Services